D1826875

Emerging Technology-Based Services and Systems in Libraries, Educational Institutions, and Non-Profit Organizations

Dickson K. W. Chiu
The University of Hong Kong, Hong Kong

Kevin K. W. Ho
University of Tsukuba, Japan

A volume in the Advances in Library and Information Science (ALIS) Book Series

Published in the United States of America by
IGI Global
Information Science Reference (an imprint of IGI Global)
701 E. Chocolate Avenue
Hershey PA, USA 17033
Tel: 717-533-8845
Fax: 717-533-8661
E-mail: cust@igi-global.com
Web site: http://www.igi-global.com

Library of Congress Cataloging-in-Publication Data

Names: Chiu, Dickson K. W., 1966- editor. | Ho, Kevin K. W., 1969- editor.
Title: Emerging technology-based services and systems in libraries,
  educational institutions, and non-profit organizations / edited by
  Dickson K.W. Chiu and Kevin K.W. Ho.
Description: Hershey, PA : Engineering Science Reference, [2023] | Includes
  bibliographical references and index. | Summary: "This book covers
  IT-enabled creation, curation, representation, communication, storage,
  retrieval, analysis, and use of records, documents, files, data,
  learning objects, and other contents. It also acts as a forum for
  interdisciplinary and emerging topics such as socio-information studies,
  educational technologies, knowledge management, big data, artificial
  intelligence, personal information protection, digital literacy, other
  media, and technology innovation topics in their applications to
  libraries, as well as other education, information, government, and
  NGOs. This book welcomes research that applies a broad array of
  approaches and epistemologies, including any mix of qualitative,
  quantitative, mixed methods, action, participatory, evaluation, design,
  development, or other established methodologies"-- Provided by
  publisher.
Identifiers: LCCN 2023004866 (print) | LCCN 2023004867 (ebook) | ISBN
  9781668486719 (hardcover) | ISBN 9781668486726 (paperback) | ISBN
  9781668486733 (ebook)
Subjects: LCSH: Libraries--Information technology. | Information
  science--Technological innovations. | Educational technology. |
  Educational innovations. | Nonprofit organizations--Information
  technology. | Knowledge management.
Classification: LCC Z678.9 .E44 2023 (print) | LCC Z678.9 (ebook) | DDC
  025.00285--dc23/eng/20230411
LC record available at https://lccn.loc.gov/2023004866
LC ebook record available at https://lccn.loc.gov/2023004867

This book is published in the IGI Global book series Advances in Library and Information Science (ALIS) (ISSN: 2326-4136; eISSN: 2326-4144)

British Cataloguing in Publication Data
A Cataloguing in Publication record for this book is available from the British Library.

All work contributed to this book is new, previously-unpublished material.
The views expressed in this book are those of the authors, but not necessarily of the publisher.

For electronic access to this publication, please contact: eresources@igi-global.com.

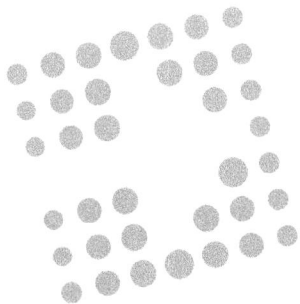

Advances in Library and Information Science (ALIS) Book Series

ISSN:2326-4136
EISSN:2326-4144

Editor-in-Chief: Alfonso Ippolito, Sapienza University-Rome, Italy; Carlo Inglese, Sapienza University-Rome, Italy

MISSION

The **Advances in Library and Information Science (ALIS) Book Series** is comprised of high quality, research-oriented publications on the continuing developments and trends affecting the public, school, and academic fields, as well as specialized libraries and librarians globally. These discussions on professional and organizational considerations in library and information resource development and management assist in showcasing the latest methodologies and tools in the field.

The **ALIS Book Series** aims to expand the body of library science literature by covering a wide range of topics affecting the profession and field at large. The series also seeks to provide readers with an essential resource for uncovering the latest research in library and information science management, development, and technologies.

COVERAGE

- Human Resources Management
- Licensing Issues
- Cataloging and Classification
- Libraries as Community Centers
- Journal Collections
- Visual Literacy
- Succession Planning in Libraries
- Children and Youth Services
- User Experience (UX)
- Remote Access Technologies

IGI Global is currently accepting manuscripts for publication within this series. To submit a proposal for a volume in this series, please contact our Acquisition Editors at Acquisitions@igi-global.com or visit: http://www.igi-global.com/publish/.

Titles in this Series

For a list of additional titles in this series, please visit:
http://www.igi-global.com/book-series/advances-library-information-science/73002

Applying Positivist and Interpretivist Philosophies to Social Research Practices
C.C. Jayasundara (University of Kelaniya, Sri Lanka)
Information Science Reference • copyright 2023 • 300pp • H/C (ISBN: 9781668454312)
• US $215.00 (our price)

Handbook of Research on Advancements of Contactless Technology and Service Innovation in Library and Information Science
Barbara Holland (Brooklyn Public Library, USA (retired))
Information Science Reference • copyright 2023 • 388pp • H/C (ISBN: 9781668476932)
• US $270.00 (our price)

Perspectives on Justice, Equity, Diversity, and Inclusion in Libraries
Nandita S. Mani (University of Massachusetts, Amherst, USA) Michelle A. Cawley (University of North Carolina at Chapel Hill, USA) and Emily P. Jones (University of North Carolina at Chapel Hill, USA)
Information Science Reference • copyright 2023 • 320pp • H/C (ISBN: 9781668472552)
• US $215.00 (our price)

Handbook of Research on Technological Advances of Library and Information Science in Industry 5.0
Barbara Jane Holland (Independent Researcher, USA)
Information Science Reference • copyright 2023 • 549pp • H/C (ISBN: 9781668447550)
• US $270.00 (our price)

Global Perspectives on Sustainable Library Practices
Victoria Okojie (University of Abuja, Nigeria) and Magnus Osahon Igbinovia (Ambrose Alli University, Nigeria)
Information Science Reference • copyright 2023 • 376pp • H/C (ISBN: 9781668459645)
• US $240.00 (our price)

For an entire list of titles in this series, please visit:
http://www.igi-global.com/book-series/advances-library-information-science/73002

701 East Chocolate Avenue, Hershey, PA 17033, USA
Tel: 717-533-8845 x100 • Fax: 717-533-8661
E-Mail: cust@igi-global.com • www.igi-global.com

List of Reviewers

Daniel Akwasi Afrane, *University of Media, Arts, and Communication, Ghana*
Asefeh Asemi, *Institute of Data Analytics and Information Systems, Hungary*
Helen Chan, *University of Hong Kong, Hong Kong*
Chao-chen Chen, *National Taiwan Normal University, China*
Antonia Bernadette Donkor, *University of Ghana, Ghana*
Patrick Hung, *Ontario Tech University, Canada*
Apple H.C. Lam, *University of Hong Kong, Hong Kong*
Elisha Mupaikwa, *National University of Science and Technology, Zimbabwe*
Javaid Ahmad Wani, *University of Kashmir, India*
Xu Wang, *Yanshan University, China*
Jenny Wong, *University of Hong Kong, Hong Kong*
Zerong Xie, *University of Hong Kong, Hong Kong*

Table of Contents

Detailed Table of Contents

Chapter 1
> *Antonia Bernadette Donkor, University of Ghana, Ghana*
> *Daniel Akwasi Afrane, University of Media, Arts, and Communication,*
>   *Ghana*

Academic libraries are mandated to support the information needs of their students, staff, and faculty. In the current technological environment and the influx of artificial intelligence technology into services provision and delivery, this study assessed the knowledge and perception of librarians on the use of AI in library services provision in Ghana. The population for the study consisted of librarians from all 15 public universities in Ghana. Findings from the study revealed that the librarians were knowledgeable of the emergence of AI and sources their information on AI mainly from research articles. It was also revealed that educational level had a significant positive ($r= 0.3$, $p<0.01$) relationship with the sources of knowledge on AI tools and applications. There was a positive significant relationship ($r= 0.533$, $p<0.01$) between educational level and the frequency of knowledge acquisition on the application of AI in academic library services provision. Academic libraries are encouraged to invest in AI tools and applications to leverage their advantages.

Chapter 2
> *Elisha Mupaikwa, National University of Science and Technology,*
>   *Zimbabwe*

Over the past few decades, there has been an increase in ongoing research into the use of artificial intelligence in education. The usage of artificial intelligence in educational institutions has grown during the past few years. This chapter examines how artificial intelligence is used in education. The establishment of inclusive learning environments, collaborative learning and research, social robots, student assessment, automated grading, recommender systems, and student performance prediction were identified to be the main applications of artificial intelligence in education. Lack of funding, lack of faith in the system, lack of portability of systems, infrastructure issues, ethical issues regarding data privacy and security, and a lack of experience were some of the difficulties noted. After that, the chapter included recommendations for infrastructure improvements, ICT policy changes at the national and institutional levels, curriculum revisions for teacher- and school-level preparation, and additional research.

Chapter 3

Research on Artificial Intelligence Risk Prevention and Control System for
Xu Wang, Yanshan University, China
Binbin Liu, Yanshan University, China

With the rapid development of artificial intelligence (AI) science and technology, people's demand for library information services and risk prevention and control construction has increased in both directions, affecting profound changes in the construction of libraries in various countries. Artificial intelligence is reshaping the management model and service model of smart libraries, and libraries must pay attention to risk prevention and control while welcoming the model change. This chapter is based on the background of artificial intelligence era. To achieve the deep synergy between artificial intelligence and smart libraries, it combines risk management and multiple synergy theory, and it proposes that the types of AI risks in smart libraries are divided into their own risks and application risks. This chapter constructs the AI risk prevention and control system for smart libraries and proposes feasible AI risk prevention and control countermeasures for smart libraries.

Chapter 4

A Review on Intelligent Library Services and Systems: Utilizing
Asefeh Asemi, Institute of Data Analytics and Information Systems,
Hungary

The chapter includes a comprehensive review of relevant published literature from the past five years. Through this review, the chapter explores the use of emerging technology-based services and systems such as intelligent libraries, recommender systems, fuzzy logic, and recommender system chatbots in the field of library and information science. The aim of this review is to provide up-to-date insights and information for professionals and researchers interested in utilizing these technologies to improve library services. The chapter highlights how these technologies can enhance the user experience and resource discovery. The chapter begins by introducing the concept of intelligent libraries and the role of emerging technology-based services and systems. It then explores the use of recommender systems and fuzzy logic in intelligent libraries, citing relevant published literature and discussing the benefits they bring to users. In the end, it discusses the potential challenges and limitations of these technologies, as well as best practices for their implementation and maintenance.

In libraries, the internet of things (IoT) has enormous promise. It is not the sensor on the object, but it does have the capacity for electronic tracking and data exchange. This has created a plethora of new opportunities to improve the efficiency of libraries and, as a result, the user experience of various services. IoT has played a critical role in transforming libraries into Smart places by improving services such as "collection management," "instruction," "data security," and so on. It can also allow real-time global connection of a large library system. In this context, the chapter looks to explain IoT and its numerous technologies. The study additionally indicates possible library areas for implementation and how they affect library effectiveness in terms of patrons, operations, and technological innovation. This chapter will serve as a road map for scholars, practitioners, and readers interested in IoT, technosphere, and tech habitat.

One of the many promising developments in data innovation is chatbots. They were created as a brand-new user interface that would let users communicate with services merely through chat, replacing or enhancing the need for applications or website visits. Since digitalisation is pacing up around the globe, usage of chatbots is prevalent in many domains for the interface between customers and entities. However, there is still another industry where chatbots have a considerable amount of prospect: education. In the educational sector, chatbots are equipped but to a certain extent only. The aim of the study is to create awareness about potential of chatbots in learning processes and its various applications in the academic sector. Chatbots various verticals promote digitalisation in the future of education support systems enhancing the productivity and boosting growth as the education sector of any country is the foundation of its growing future.

The most popular uses of blockchain technology in the art market are related to non-fungible tokens (NFTs). This chapter explores the adoption of NFT in the digital art market and its future development. The authors explore NFT adoption in the digital art market with the five pillars of the digital entrepreneurship model, including knowledge base, business environment, finance, technology, and culture. Their scarcity and utilities determine the value of NFTs. Collectors and artists should also be aware of the benefits and drawbacks of blockchain technology and take appropriate steps to guarantee their rights are protected. This chapter provides a fundamental review of the current development and outlook of NFT and opens new opportunities for future study. Scant research focuses on the present condition and adoption of NFTs within the Hong Kong digital art market. This review offers a much-needed exploration and understanding, particularly beneficial for potential investors and participants seeking a comprehensive insight into NFT adoption.

This study reviews the current parents' perspectives and habits of using graphic novels, including electronic equipment, in Hong Kong to support children's reading and learning. Parents were invited to participate in semi-structured group interviews based on our five research questions. The authors employ a thematic analysis approach to analyze the data and determine the similarities and differences between the literature review and the current mobile digital environment. The findings indicated that respondents were not keen on using graphic novels to support their children's reading and learning. The two main reasons were inadequate understanding or bias on graphic novels and the wide availability of online digital materials for reading and learning support. Scant studies focus on graphic novels in parent-child reading from parents' perspectives, especially in the Asian context, though graphic novels are becoming a worldwide trend. Parents' opinions about this issue are invaluable for educators and librarians in curriculum design and collections development.

Chapter 9

 Cho Yiu Cheung, The University of Hong Kong, Hong Kong
 Apple Hiu Ching Lam, The University of Hong Kong, Hong Kong
 Dickson K. W. Chiu, The University of Hong Kong, Hong Kong

Makerspaces have developed from a trend to a core service in higher education libraries. Many academic libraries have been actively expanding the "makerspace" within the physical library and revitalizing the library as a center of learning and innovation. This case study investigates the application of makerspace technologies in a major comprehensive library in Hong Kong, which has designed a specific makerspace to encourage innovation and creativity. Few studies have focused on in-depth studies of makerspaces in East Asian academic libraries and how patrons perceive makerspace services and innovative spaces. A survey instrument was developed using the 5E instructional model (engage, explore, explain, elaborate, and evaluate) to evaluate makers' experiences systematically. The finding revealed that respondents applauded the importance of innovative spaces and the perceived outcomes from makerspaces, including nurturing creativity and critical thinking. However, they did not have sufficient skills to use emerging technologies, resulting in low usage.

Chapter 10

 Helen Chui Ling Kee, The University of Hong Kong, Hong Kong
 Mimi Mei Wa Chan, The University of Hong Kong, Hong Kong
 Dickson K. W. Chiu, The University of Hong Kong, Hong Kong

As the US is a highly ethnically diverse country and the origin of public libraries, this chapter explores how US public libraries construct social capital for the public via various services and activities. This chapter selected nine cases of US public libraries with interviews of their management for analysis of library services and activities related to empowering library users and building their social capital. The findings indicate that libraries are community meeting places connecting community members through library programs, activities, and information services as community educational institutions that empower underprivileged people and new immigrants by satisfying their information needs. The process contributes to the social development of library users and their communities, building social capital. Scant studies summarize the good practice of renowned public libraries in building social capital. This chapter contributes to understanding the good practice of US public libraries as a creator of social capital, serving as a reference for public libraries worldwide.

Preface

As editors of this edited reference book, we are delighted to present a comprehensive collection of research contributions that delve into the exciting intersection of information and communication technologies, service-oriented economies, and non-commercial applications. The global landscape has been witnessing a paradigm shift, where technology plays a pivotal role in shaping the way organizations function, transcending beyond traditional commercial work. In this context, this book aims to shed light on emerging technological advancements and their transformative impact on libraries, educational institutions, and non-profit organizations.

The mission of this book is to provide a platform for empirical, conceptual, and methodological studies that explore contemporary topics within the broad disciplines of information and communication technologies. Our focus extends to diverse areas, including libraries and cultural memory, education and academia, government and public sectors, as well as non-governmental organizations (NGOs). We seek to unravel the complexities of IT-enabled creation, curation, representation, communication, storage, retrieval, analysis, and use of various forms of content, such as records, documents, files, data, learning objects, and more.

CHAPTER OVERVIEW

Chapter 1 delves into the adoption of Artificial Intelligence (AI) in academic libraries in Ghana. The study focuses on assessing the knowledge and perception of librarians regarding AI usage. Findings reveal that librarians are knowledgeable about AI, with research articles being their primary source of information. Moreover, the study establishes a positive relationship between educational level and knowledge acquisition on AI tools and applications. The chapter emphasizes the potential benefits of investing in AI tools to enhance academic library services.

Chapter 2 explores the growing usage of Artificial Intelligence in educational institutions. It highlights several applications of AI, such as inclusive learning environments, collaborative learning, social robots, student assessment, and automated grading. The chapter also acknowledges the challenges faced in implementing AI, such as funding, data privacy, and ethical concerns. To address these challenges, the chapter proposes recommendations for infrastructure improvements, policy changes, and further research.

Chapter 3 examines how the rapid development of AI impacts smart libraries' management and service models. It introduces the concept of AI risks, dividing them into own risks and application risks. Drawing on multiple synergy theory, the chapter constructs an AI risk prevention and control system for smart libraries. Furthermore, it offers practical AI risk prevention measures to ensure a smooth integration of AI in library services.

Chapter 4 presents a comprehensive review of recent literature on emerging technology-based services and systems in library and information science. It delves into intelligent libraries, recommender systems, fuzzy logic, and recommender system chatbots, discussing their applications and benefits for users. The chapter also addresses potential challenges and provides best practices for implementation and maintenance.

Focusing on the Internet of Things (IoT) in libraries, Chapter 5 explains various IoT technologies and their potential impact on library effectiveness, operations, and technological innovation. It highlights IoT's capacity to improve services like collection management, instruction, and data security, transforming libraries into Smart places. The chapter serves as a roadmap for readers interested in understanding and utilizing IoT in the library context.

Chapter 6 explores the potential of chatbots in the educational sector, as a new user interface facilitating communication between users and services. It highlights various applications of chatbots in the academic environment and their role in promoting digitalization in the future of education support systems. The chapter aims to raise awareness about chatbots' efficacy and impact on learning processes.

In Chapter 7, the adoption of Non-Fungible Tokens (NFTs) in the digital art market is examined. The chapter employs the Digital Entrepreneurship Model to analyze NFT adoption, focusing on aspects like scarcity and utility in determining the value of NFTs. It also explores the benefits and drawbacks of blockchain technology and offers insights into the current development and future prospects of NFT adoption.

Chapter 8 investigates parents' perspectives and habits regarding the use of graphic novels and electronic devices in supporting children's reading and learning in Hong Kong. The study employs thematic analysis to reveal similarities and differences between literature reviews and the current digital environment. The chapter provides valuable insights for educators and librarians in curriculum design and collection development.

Focusing on makerspaces in a major comprehensive library in Hong Kong, Chapter 9 investigates their application in encouraging innovation and creativity. The study evaluates makers' experiences using the 5E Instructional Model and highlights the perceived outcomes, including nurturing creativity and critical thinking. The chapter emphasizes the importance of innovative spaces and user skills in using emerging technologies.

Chapter 10 explores how US public libraries contribute to building social capital through various services and activities. The study analyzes nine case studies of US public libraries, highlighting their role as community meeting places and information providers for underprivileged people and new immigrants. The chapter serves as a valuable reference for public libraries worldwide aiming to empower their communities and foster social development.

Furthermore, this book serves as a forum for interdisciplinary discussions, welcoming research on emerging topics like socio-information studies, educational technologies, knowledge management, big data, artificial intelligence, personal information protection, digital literacy, other media, and technology innovations as they find applications in libraries, education, information management, government, and NGOs.

The intended audience for this book comprises researchers and practitioners who are already active in fields such as computer science, information systems, system/industrial engineering, and related disciplines. We also encourage industry specialists and consultants operating in similar areas to explore the knowledge presented in these pages. Beyond the conventional boundaries, we extend an invitation to professionals from various service provision domains, including marketing, operations research, logistics, economics, law, and other service industries. We hope that they will draw from their cross-disciplinary expertise to enrich the non-commercial settings of libraries, educational institutions, and non-profit organizations.

Given the scarcity of case studies in such specific settings, this book also serves as valuable teaching material for university programs centered around these themes. By featuring cutting-edge research, we aim to inspire and educate the next generation of professionals, equipping them with the knowledge and insights needed to navigate the dynamic landscape of technology-based services and systems in non-commercial domains.

The book encompasses a diverse range of themes, including but not limited to:

- Contemporary information system innovation, planning, development, adoption, diffusion, and maintenance for libraries, education, and non-profit organizations.
- Emerging human-computer interactions, such as virtual and augmented reality, and their applications in libraries, education, and non-profit organizations.
- Technology-based service design and innovation tailored to the needs of libraries, education, and non-profit organizations.

- Understanding user information behaviors on cloud, social, and mobile platforms relevant to libraries and non-profit organizations.
- Data analytics, curation, and visualization for non-commercial applications, empowering organizations to make data-driven decisions.
- Privacy, security, reliability, and technology management and governance in contemporary information systems, safeguarding sensitive information.
- Exploring the impact of digital literacy, reading, digitalization, social informatics, culture, and digital humanities in non-commercial contexts.
- Knowledge management, preservation, and linked data strategies for non-commercial applications, fostering efficient information sharing and collaboration.
- Investigating the role of artificial intelligence and intelligent systems in enhancing non-commercial services.
- Exploring the potential of emerging technologies, such as the Internet of Things (IoT) and Blockchains, in non-profit applications.

We express our gratitude to the contributing authors for their dedication and commitment in producing insightful chapters that collectively advance our understanding of technology's transformative influence in non-commercial settings. We hope that this volume will serve as a valuable resource for academics, practitioners, and policymakers alike, stimulating further research and innovation in the realm of emerging technology-based services and systems.

Dickson K. W. Chiu
The University of Hong Kong, Hong Kong

Kevin K. W. Ho
University of Tsukuba, Japan

Chapter 1

Application of AI in Academic Library Services:
Prospects and Implications for Quality Service Delivery

Antonia Bernadette Donkor
https://orcid.org/0000-0002-2372-6125
University of Ghana, Ghana

Daniel Akwasi Afrane
https://orcid.org/0000-0001-5259-4507
University of Media, Arts, and Communication, Ghana

ABSTRACT

Academic libraries are mandated to support the information needs of their students, staff, and faculty. In the current technological environment and the influx of artificial intelligence technology into services provision and delivery, this study assessed the knowledge and perception of librarians on the use of AI in library services provision in Ghana. The population for the study consisted of librarians from all 15 public universities in Ghana. Findings from the study revealed that the librarians were knowledgeable of the emergence of AI and sources their information on AI mainly from research articles. It was also revealed that educational level had a significant positive (r= 0.3, p<0.01) relationship with the sources of knowledge on AI tools and applications. There was a positive significant relationship (r= 0.533, p<0.01) between educational level and the frequency of knowledge acquisition on the application of AI in academic library services provision. Academic libraries are encouraged to invest in AI tools and applications to leverage their advantages.

DOI: 10.4018/978-1-6684-8671-9.ch001

INTRODUCTION

Information and its role in development is a necessity in every country. Information is a significant factor of production, playing a major role in the industrial revolution with technological developments in all spheres of human endeavours. Growth is a priority for universities as they pioneer the development and advancement of knowledge. Educational institutions invest in technologies and information resources through their libraries to support their teaching, learning, and research functions. These libraries play a significant role in the collection, organization, dissemination, preservation, and retrieval of information in a timely and organized manner.

Academic libraries are mandated to meet the information needs of their parent institutions by providing access to information resources, information, and reference services and implementing and investing in emerging technologies to optimize their service provision for user satisfaction. Technological advancements and the proliferation of artificial intelligence have transformed the academic library fundamentally by providing invaluable resources with increased access to information.

Artificial Intelligence is an essential technology whose machinery is enhanced with the support of state-of-the-art technological innovations and improvements. Artificial intelligence has emerged as a pervasive form of technology in our everyday lives and now forms a part of various activities in the academic world. The term artificial intelligence (AI), according to Kaliraj and Devi (2022), was coined by John McCarthy and is defined as "the science and engineering of making intelligent machines and brilliant software programs." It is also "the application of computers and utilization of computer-based products and services in the performance of different library operations and functions or the provision of various services and production of output products" (Tsabedze et al., 2022).

The emergence of AI and its implications on academic library activities, services, and products is critical in the present information-dependent society. The degree to which AI is adopted in the library space is likely to be driven by widespread technical developments and so cannot be controlled by librarians. Its advancement becomes a prevailing threat to the need for professional librarians. The intensive burden on librarians to provide high-quality services to library users due to the information explosion in our present society has led to the integration of modern technologies.

This chapter assesses Artificial Intelligence's prospective use and implications in academic libraries and analyses librarians' competencies, perceptions, and skills to use AI supported technologies. The objectives of the chapter are to:

1.   Assess the knowledge and perception of librarians in using AI in Academic libraries.
2.   Assess the use of AI in service delivery in Academic libraries.
3.   Identify the prospects and implications of AI in Academic libraries.

These hypotheses were also tested:

H1: There is a relationship between one's level of education and their source of knowledge of AI.

H2: There is a relationship between the level of one's education and the frequency of their knowledge of AI acquisition.

REVIEW OF LITERATURE

To serve as a standard against which the finding from the study was measured, the literature was reviewed under the following topics:

Introduction to AI

AI (Artificial Intelligence) is an aspect of computer science committed to developing machines and software programs to perform human-centred tasks dependent on human intelligence. AI systems use algorithms and data to learn from experience and make decisions based on that learning. Several attempts have been made to define AI and what it encompasses. One such definition is that UK Research and Innovation (UKRI) defines AI as a "suite of technologies and tools that aim to reproduce or surpass abilities in computational systems that would require intelligence if humans were to perform them, including learning and adapting, sensing, understanding, interacting, reasoning and planning, acting autonomously, or even creating. It enables us to use and make sense of data" (UKRI, 2021: 4).

There are two broad categories of AI. These are Narrow AI and General AI. Narrow AI, also known as weak AI, is designed to perform specific tasks, such as voice recognition, image recognition, or natural language processing. On the other hand, General AI, also known as strong AI or AGI, is designed to perform any intellectual task that humans can perform. The different forms of AI include machine learning, natural language processing, computer vision, robotics, and expert systems. Machine learning allows computers to learn and improve based on their experience without being explicitly programmed. The underlying concept is the use of machines and intelligence, the human attribute distinguishing them from other creations. Machines are programmed or taught to use similar human intelligence to process work and achieve a reasonable outcome (Kaliraj & Devi, 2022)

Through the different types of AI, a broad range of applications have been designed and implemented in different spheres of society to augment human capabilities. In healthcare, AI is used to augment disease diagnosis and drug development. It is used for fraud detection and risk analysis in finance. It is also used in developing self-driven cars and traffic management (Gasparini & Kautonen, 2022) in the transportation sector and for personalized learning and student performance assessment in education. It is also integrated into knowledge discovery and chatbot applications to offer virtual and remote reference services in academic libraries (Cox, 2022; Gasparini & Kautonen, 2022). Although the use of AI innovations is raising ethical, legal, and social implications, including concerns around job displacement, bias, and privacy in recent times, AI applications are rapidly evolving, offering society, institutions, and organizations the opportunities for innovation, growth, and efficiency (Lund et al., 2020).

Knowledge and Use of AI in the Delivery of Library Services

AI (Artificial Intelligence) is becoming increasingly relevant in academic libraries due to its potential to improve library services and user experiences, creating a keen interest in acquiring knowledge among librarians (Lund et al., 2020). Information on AI is accessed from print and electronic sources of information. Studies by Majanja and Kiplang'at 2003, Aesaert and Van Braak 2015, and Yamson et al. 2018 reveal the preference for electronic resources among young adults and studies. The different ways in which AI technology has been implemented in academic libraries, evident in the literature, are discussed in this section.

- *Chatbots and Virtual Assistants:* Through chatbots and virtual assistants, libraries can assist their users in finding information, answering reference questions, and always provide support remotely and virtually. Virtual assistance provided through chatbots was beneficial to libraries during the outbreak of COVID-19 when lockdowns were enforced in most countries around the world to curb the spread of the coronavirus (Johnson, 2020; Donkor, 2021). While a study by Aklabi (2021) described how the King Saud University Libraries used virtual assistance to provide vital responses and information virtually and remotely to the satisfaction and admiration of their graduate students who were conducting research, Afrane et al. (2022) also report on how the integration of AI applications and tools has transformed the academic library spaces in some selected university libraries in Ghana. Both scholars mentioned how AI implementations in their information and learning commons facilitate research and learning collaborations and improve teaching and learning in the universities they studied.

- *Recommendation Systems*: AI algorithms can be used to analyze user behavior and recommend relevant resources, such as books, articles, and databases. Admissibly, human behavior analysis is a huge and daunting task especially when one needs that analysis to make informed decisions and provide adequate resources to users. In recent years, when most young and older adults turn to social media and the internet, Tariq et al. (2021) proposed an architecture to gather and analyse massive social media data using Big Data analytics mechanisms. Their proposed architecture is useful in monitoring and analyzing students' behavior and search preferences to predict their academic information needs.

- *Metadata Management:* AI can be used to enhance metadata records, making it easier for users to find the resources they need. AI-powered metadata uses artificial intelligence (AI) techniques to automatically generate, analyse and annotate metadata. Metadata is information that describes data, such as title, author, and keywords, and it is used to help users find, use, and understand data. AI-powered metadata can be generated in several ways. Natural language processing (NLP) algorithms are used to extract keywords, concepts, and other relevant information from the text (Tariq et al., 2021). This can be done by analyzing the content of documents, web pages, or other data sources. Another approach is to use computer vision algorithms to analyse images and videos and extract metadata from them, such as identifying objects, people, and locations in the visual content. AI-powered metadata can also be used to enrich existing metadata by adding new information or correcting errors. For example, AI can be used to identify missing or incorrect metadata in a database of images and videos and automatically update the metadata accordingly (Tsabedze et al., 2022). Overall, AI-powered metadata can help libraries manage large amounts of data more efficiently by automating the metadata generation and maintenance process. It can also improve the accuracy and completeness of metadata, which can help users find and use data more effectively.

- *Digital Preservation*: AI can help libraries preserve digital materials by identifying, classifying, and analyzing content to ensure long-term access and usability. AI technology can be used in digital archives to automate and improve various aspects of the archival process. Some of how AI can integrate into digital archiving, as outlined by Cushing and Osti (2022), are metadata generation; AI algorithms can be used to generate metadata for digital archives automatically. This can help improve the accuracy and completeness of metadata, as well as save time and resources that would be required for manual metadata creation. Text recognition using AI algorithms can be used to recognize and transcribe text in scanned documents, making it searchable and easier to access. This can be particularly useful for historical archives that contain handwritten documents or other hard-to-read text. Image recognition with AI algorithms can be used to recognize and tag images in digital archives, making them easier to search and browse (Tariq et al., 2021). Content analysis with AI algorithms can be used to analyze the content of digital archives, identifying patterns, trends, and other insights that may not be apparent through manual analysis and can help to uncover hidden connections between different pieces of data or identify gaps in the archive that may need to be filled. Preservation algorithms can be used to monitor digital archives and detect potential risks to data integrity, such as hardware failures or software bugs to ensure that digital archives remain accessible and usable over time. Through AI technology, digital archives can be more efficient, accessible, and useful to researchers, scholars, and other users. The ethical and responsible way AI algorithms are developed and used is key, considering privacy, bias, and transparency issues (Lund et al., 2020).

- *Text Mining and Analysis*: Text mining and analysis use computational techniques to extract insights and information from unstructured or semi-structured textual data. This can include analyzing large collections of documents, social media posts, emails, customer feedback, and other types of textual data (Arshad et al., 2020). Through AI, large amounts of text and data can be analyzed, allowing libraries to identify patterns and trends that can inform collection development, research, and other technical areas. Tariq et al. (2021) argue that text mining helps organizations to gain valuable insights from textual data that would be difficult or impossible to obtain through manual analysis. They, however, caution on the importance of ensuring that the analysis is conducted ethically and responsibly, considering privacy, bias, and accuracy.

- *Intelligent Search*: Intelligent search uses advanced algorithms and artificial intelligence (AI) technologies to improve the accuracy, relevance, and efficiency of search results, unlike Traditional search engines, which rely on keyword-based searches, which can often return irrelevant or low-quality results. Intelligent search uses machine learning, natural language processing, and other AI techniques to understand the context and intent behind search queries and provide more personalized and accurate results (Fu et al., 2011). One of the key benefits of intelligent search is that it can adapt to the user's behaviors and preferences over time, making it easier to find the information they are looking for. For example, a search engine could learn which types of results a user clicks on most frequently and prioritize those types of results in future searches. This can save users time and effort and help them find the information they need more quickly. Intelligent search is used widely in e-commerce, enterprise search, and online advertising. AI algorithms can improve search accuracy and relevance, making it easier for users to find the needed information (Fu et al., 2011). AI can transform academic libraries by improving the quality and efficiency of library services, enhancing user experiences, and enabling librarians to focus on high-value tasks such as research support and instruction.

Skills and Competencies Needed for the Implementation of AI in Libraries

The implementation of AI in libraries requires a range of skills and competencies. Afrane et al. (2022) outlined some basic IT skills librarians require to improve their service delivery and learning. They mentioned that web design, management, and programming language were essential skills librarians needed in the current technological environment. Among the skills and competencies needed for the effective implementation of AI in libraries include:

- *Data Management and Analysis*: Libraries need staff with expertise in data management, including data cleaning, transformation, and analysis, to support the implementation of AI algorithms (Perrier et al., 2017). Data management involves various activities: collection, storage, retrieval, and sharing. Organizations should ensure their data is accurate, complete, up-to-date, and stored securely to protect it from unauthorized access or loss (Yoon & Schultz, 2017). Librarians building their skills in data management also need to comply with regulatory requirements related to data privacy and security (Frederick et al., 2019).
- *Programming and Technology*: Librarians and information professionals need skills in programming languages, such as Python or R, and familiarity with AI and machine learning tools and platforms (Afrane et al., 2022). Programming and technology are also closely linked to emerging fields of artificial intelligence (AI), machine learning, and data science. These fields require specialized skills and knowledge in programming and technology to develop algorithms, analyze data, and build intelligent systems. Programming and technology are critical in the development of modern software and applications. By staying current with the latest programming languages and technologies, librarians can create innovative and effective solutions for complex problems in the research and academic domain (Tariq et al., 2021).

- *Information Architecture*: Librarians need knowledge of information architecture, including metadata standards, taxonomy, and ontology development, to ensure that data and resources are structured and organized for AI algorithms (Huvila et al., 2016). Information architecture aims to create a clear and intuitive structure for information that allows users to conveniently find what they are looking for, understand the relationships between different pieces of information, and navigate through the content with ease. This involves identifying user needs and goals, creating a taxonomy or classification scheme, and designing the overall structure and navigation system (Robins, 2002).

- *User Experience Design*: Librarians need to understand user needs and have expertise in user experience design to ensure that AI applications are user-friendly and intuitive. With the user in mind, this involves designing digital products, such as websites, mobile apps, and software. It involves understanding the needs and goals of users and designing an interface that is intuitive, easy to use, and engaging, meets users' needs, and helps them accomplish their tasks efficiently and effectively to create a positive user experience. To accomplish this requires a deep understanding of user behavior, psychology, and technology (Yoon & Schultz, 2017).

- *Ethical and Legal Issues*: Librarians should be familiar with ethical and legal issues surrounding AI, including data privacy, bias, and accountability, to ensure that AI applications are developed and implemented ethically and responsibly. Ethical and legal issues are important considerations in librarianship, particularly when applying AI to their work. Librarians must be aware of the potential impacts of AI on their work, users, and society at large. Incorporating ethical and legal principles into their work, librarians can create products and services that are safe, inclusive, and beneficial to users (Bradley, 2022).

- *Communication and Collaboration*: Librarians must be skilled in communicating and collaborating to work effectively with IT staff, vendors, and other stakeholders in implementing AI applications in the library. Effective communication and collaboration are essential for librarians to provide high-quality services and resources to their users. By working together and communicating effectively, librarians can better understand and meet their users' needs and enhance the library's overall impact and user community (Bottorff et al., 2008). Implementing AI in libraries requires a multidisciplinary approach involving librarians and information professionals with diverse skills and competencies.

Prospects of AI on the Delivery of Service Quality and Customer Satisfaction

AI's impact on delivering quality service and customer satisfaction can be significant if implemented in academic libraries. These include:

- *Increased Efficiency and Speed*: A study by Bradley (2022) shows that AI-powered systems can help automate routine tasks, such as book recommendations or answering frequently asked questions in libraries, freeing up librarians' time to focus on more complex tasks. This can result in faster and more efficient services, increasing customer satisfaction. AI can automate the process of cataloging books and other materials, reducing the time and effort required for manual cataloging. This can be achieved through natural language processing and machine learning techniqueswhich can extract key information from texts and classify them accordingly (Bradley, 2022). Also, AI-powered chatbots can handle routine patron queries such as checking library hours and renewing books, freeing up library staff to focus on more complex tasks. AI can also analyze library usage data to predict which books and resources are likely in demand, enabling libraries to allocate their resources better and plan for future needs (Yoon & Schultz, 2017).
- *Improved Personalization*: AI algorithms can analyze user behaviors and preferences to provide selective dissemination of information, personalized recommendations, and services, such as personalized reading lists or customized research support. This can enhance the user experience and increase customer satisfaction. AI can help libraries provide personalized book recommendations to patrons based on their reading history and preferences. This can improve the patron's experience and increase the likelihood of repeat visits (Robins, 2002).
- *Better Resource Management*: AI can assist with managing library resources, such as recommending the most popular resources, identifying the most relevant materials, and managing inventory. This can improve the quality and availability of resources, leading to higher customer satisfaction. AI-powered natural language processing can help libraries improve search functionality, making it easier for patrons to find the necessary books and resources. AI can automate sorting returned books, reducing the time and effort required for manual sorting (Fu et al., 2011).
- *Enhanced User Experience*: AI can improve the user's experience by providing support, answering questions promptly and accurately, and providing customized services based on user preferences. This can result in higher customer satisfaction and loyalty. AI has also helped libraries to

enhance the user experience by providing personalized and efficient services, making it easier for users to find and access the resources they need (Cushing & Osti, 2022).

- *Reduced Errors and Bias*: AI can reduce errors and bias in the delivery of library services by automating tasks and providing consistent and objective responses. This can improve service quality and lead to higher customer satisfaction. AI algorithms can make decisions without human intervention, reducing the potential for human error or bias. AI also has the potential to reduce errors and biases in various fields by providing more accurate and consistent results and detecting and eliminating errors and biases in data and processes (Pannucci et al., 2010).

- *Greater Collaboration*: AI can facilitate collaboration among librarians, researchers, and students, by providing real-time insights and data analysis. AI-powered recommender systems can suggest resources to users based on their interests and needs. This can encourage collaboration between users with similar interests and lead to new resources discovery. AI can analyze data on library usage and resources to identify patterns and trends. Such information can be used to improve library services and resources, encouraging collaboration between users. AI has the potential to enhance collaboration in libraries by providing personalized services, freeing up staff time, and improving access to resources. This can encourage librarians and users to work together and collaborate on projects, leading to increased knowledge sharing and innovation (Pham & Tanner, 2015).

- *Advancement in Research*: Artificial intelligence (AI) is transforming libraries by enhancing their ability to collect, organize, and analyze vast amounts of data, making research more efficient and effective. AI can enable new research and discoveries by analyzing data sets that would be too large or complex for humans to analyze manually. With AI, libraries can extract useful insights from vast amounts of data quickly and accurately. AI algorithms can analyze a researcher's preferences and recommend relevant articles, books, and other resources. This helps researchers find information they may have missed or overlooked, leading to more comprehensive research. This also helps researchers access relevant information quickly, making the research process more efficient. AI tools can analyze large volumes of text, including books, journals, and research papers, to identify patterns, trends, and relationships that may be difficult to spot manually. This can help researchers identify new research areas or find connections between seemingly unrelated topics (Bottorff et al., 2008).

Notwithstanding, there are also potential drawbacks to using AI in library services, such as concerns about privacy, data security, and ethical considerations. It is essential to consider the potential risks and benefits of AI in library services and to implement AI systems ethically and responsibly to ensure they enhance the quality of library services and customer satisfaction.

Implications of AI in Libraries

In as much as the use of AI offers libraries diverse prospects, there are several implications and challenges to its use in libraries, including:

- *Ethical and Legal Issues*: AI raises ethical and legal concerns related to data privacy, bias, and accountability. Libraries are to ensure the ethical and responsible development and implementation of AI, not to perpetuate and even amplify existing societal prejudices and discrimination. For instance, facial recognition technology has been found to have higher error rates for people of color and women, bringing up concerns about racial and gender biases. Again, AI systems can collect and analyze large amounts of personal data, raising privacy and data protection concerns (Hussain, 2023). Organizations should observe transparency about collecting and using data and protecting and safeguarding the rights and privacy of the people they serve and interact with. Many AI systems are considered "black boxes" because their decision-making processes are not transparent or explainable. This lack of transparency can make it difficult to hold organizations accountable for their decisions and erode public trust in AI. AI systems can generate and manipulate intellectual property, raising questions about ownership and control over these assets (Okunlaya et al., 2022).
- *Workforce Displacement*: The use of AI in libraries may lead to the displacement of some jobs that can be automated, such as data entry or book recommendations. It is important to consider the potential impact on the workforce and plan appropriately to overcome resistance and apathy to change. AI can automate repetitive and routine tasks, such as data entry, customer service, and basic analysis, which humans currently perform. This can lead to job displacement, particularly in industries where these tasks are prevalent. The effect of AI in the workplace is complex and multifaceted. While its implementation will likely displace workers, it can create new job opportunities and competitive advantages for organizations. Therefore, policymakers, employers, and workers must anticipate and adapt to these changes to ensure a smooth transition to the new technological landscape (Țundrea et al., 2020).

- *Technical Challenges*: Implementing AI requires significant technical expertise, and there may be challenges related to integrating AI with existing library systems. AI systems can be vulnerable to cyber-attacks, and organizations should implement the necessary safeguards to protect their systems from hacking and other malicious activities. AI algorithms require large amounts of high-quality data to train effectively. However, obtaining such data can be difficult and time-consuming. Moreover, the data quality can affect the accuracy and reliability of the AI model. AI algorithms can be biased and unfair if the data used to train them is not objective. This can result in discriminatory outcomes, which can affect people in society. Addressing bias and fairness is a significant technical challenge for AI (Farag et al., 2021).

Notwithstanding the challenges of implementing AI in organizations, academic libraries can leverage its advantages of improved efficiency, enhanced user experience, and capabilities to advance its user community's teaching, learning and research pursuits.

METHODOLOGY

The quantitative research method guided this study. Using the survey approach, the population and sample were determined. The population for the study consisted of all fifteen (15) public universities in Ghana since they produce a chunk of the nation's human resources. The researchers designed the questionnaire with insight from the literature and distributed it among librarians of the sampled university libraries. The questionnaire consisted of five sections consisting of open-ended and close-ended questions. Section A collected Demographic information on respondents, Section B: Awareness of the emergent use of AI in libraries, Section C: Awareness of AI Applications in Academic Libraries, Section D: Perception of the application of AI in academic libraries, Section E: Use of AI in service delivery in Academic libraries and Section F: Skills and competencies of librarians in the implementation of AI in services and Section G: Challenges in adopting AI in academic library services delivery. The link to the Google form was distributed electronically to librarians from January 2023 to the middle of February 2023.

DATA ANALYSIS AND FINDINGS

One hundred fourteen (114) questionnaires were completed, returned, and used in data analysis. Data collected using the questionnaire was cleaned, coded, and analyzed using the Statistical Package for Social Sciences (SPSS) Version 26 to obtain descriptive and inferential analyses.

Demographics

Table 1 shows that 79 (69.3%) of the study respondents were females, and 35(30.7%) were males.

Table 1. Gender of respondents

Gender	n	%
Male	35	30.7
Female	79	69.3
Total	**114**	**100.0**
Educational Level	**n**	**%**
BA in Information Studies/Equivalent	42	36.8
MA in Information Studies/Equivalent	26	22.8
MPhil in Information Studies/Equivalent	21	18.4
PhD in Information Studies/Equivalent	11	9.6
Diploma	14	12.3
Total	**114**	**100.0**

Also, enquiring about the educational level of respondents, it is evident from Table 1 that 42 (36.8%) of the respondents had a Bachelor's Degree in Information Studies and its equivalent. Twenty-six (22.8%) had a Master's Degree in Information Studies and its equivalent, 14 (12.3%) had a Diploma in Information Studies and its equivalent, and 21 (18.4%) had a Master of Philosophy Degree, and 11 (9.6%) had a Doctor of Philosophy Degree in Information Studies and its equivalent.

Knowledge and Perception of Librarians' Use of AI in Academic Libraries

The study's first objective was to assess the knowledge and perception of librarians' use of AI in Academic libraries. To do so, the following sub-objectives were assessed.

i. Knowledge of the emergence of AI in libraries
ii. Source of knowledge on AI tools and applications
iii. Perception of librarians of the use of AI in Academic libraries

H1: to determine the relationship between one's level of education and their source of knowledge of AI in Academic Libraries

H2: to determine the relationship between the level of education and the frequency of one's knowledge acquisition.

KNOWLEDGE OF THE EMERGENCE OF AI

Librarians' knowledge of the emergence of AI was assessed. Table 2 revealed that all 114(100.0%) respondents indicated they knew about AI in academic libraries.

Table 2. Awareness of artificial intelligence tools and applications

	n	%
Yes	114	100.0

Source of Knowledge on AI Tools and Applications

Probing further the source of librarian's knowledge of AI tools and applications was assessed to reveal that the majority, 55 (48.2%), obtained their knowledge from Research Articles, 33(28.9%) from Workshops/symposiums/webinars and 26(22.8%) from Formal education as indicated in Table 3.

Table 3. Source of knowledge on AI tools and applications

	n	%
Formal education	26	22.8
Workshops/symposiums/webinars	33	28.9
Research articles	55	48.2
Total	**114**	**100.0**

H1: To Determine the Relationship Between One's Level of Education and Their Source of Knowledge of AI in Academic Libraries

To ascertain if there exists a relationship between education and one's source of knowledge on AI, a collection analysis between respondents' educational level and the source of their knowledge was assessed. It was revealed that educational level has a significant positive (r= 0.3, p<0.01) relationship with the sources of knowledge on AI tools and Applications. This means the sources of knowledge on AI tools increase as librarians acquire higher education and vice versa.

Table 4. Correlations between education and source of knowledge

	Educational Level	Source of Knowledge on AI Tools and Applications
Educational level	1	0.396**
Source of knowledge on AI tools and applications	0.396**	1

**. Correlation is significant at the 0.01 level (2-tailed).

Frequency of Knowledge Acquisition on the Application of AI in Academic Library Services Provision

The frequency at which respondents acquired knowledge on the application of AI in academic libraries was assessed. Table 5 revealed that a majority, 76 (66.7%), often read about the application of AI in academic libraries, while 2 (1.8%) always read about AI in academic libraries. Twenty-eight (24.6%) respondents sometimes acquired knowledge of the application of AI, while 4 (3.5%) respondents each never and rarely acquired knowledge of AI.

Table 5. Frequency of knowledge acquisition on the application of AI in academic library services provision

	Frequency	Percent
Never	4	3.5
Rarely	4	3.5
Sometimes	28	24.6
Always	2	1.8
Often	76	66.7
Total		114

H2: To Determine the Relationship Between Educational Level and the Frequency of One's Knowledge Acquisition

Again, a correlation analysis was conducted to ascertain if a relationship exists between educational level and the frequency of one's knowledge acquisition. A correlation analysis was conducted to ascertain the relationship between educational level and the frequency of knowledge acquisition on the application of AI in academic library services provision was assessed. Results indicated a positive significant relationship ($r= 0.533$, $p<0.01$) between educational level and the frequency of knowledge acquisition on the application of AI in academic library services provision. This means the more one gets educated, the more he acquires knowledge on applying AI tools in academic libraries.

Table 6. Correlation analysis between educational level and frequency of knowledge acquisition

	Educational Level	Frequency of Knowledge Acquisition on the Application of AI in Academic Library Services Provision
Educational level	1	.533**
Frequency of knowledge acquisition on the application of AI in academic library services provision	.533**	1

**. Correlation is significant at the 0.01 level (2-tailed).

Perception of Librarians of the Use of AI in Academic Libraries

Assessing the perception of librarians of the use of AI in Academic libraries, the Mean and Standard deviation results from Table 7 showed that a majority (Mean=4.3684, SD=0.66868) agreed that AI Tools would improve service provision to users in academic libraries. Again, a large majority (Mean=4.3333, SD=0.73673) agreed that applying AI tools will improve their work efficiency in academic libraries. A majority (Mean=4.3158, SD=0.75628) also agreed that AI tools would enable quick and efficient retrieval of information for users.

PERCEPTION OF LIBRARIANS ON THE IMPORTANCE OF AI IN ACADEMIC LIBRARIES

Additionally, while a majority (Mean=4.2193, SD=0.83887) agreed that the implementation of AI would enhance the work of librarians, another majority (Mean=3.9561, SD=0.93497) agreed that AI would increase the patronage of the library, however (Mean=2.9123, SD=1.31406) were neutral of the view that the application of AI will take over their jobs and make them redundant.

Table 7. Perception of librarians of the use of AI in Academic libraries

	Mean	Std. Deviation
AI Tools will improve service provision to users in academic libraries	4.3684	0.66868
AI tools will improve the efficiency of my work in academic libraries	4.3333	0.73673
AI tools will enable quick and efficient retrieval of information for users	4.3158	0.75628
AI will increase the patronage of the library	3.9561	0.93497
AI will take over jobs and make people redundant	2.9123	1.31406
AI will enhance the work of librarians	4.2193	0.83887

Use of AI in Academic Libraries

The study's second objective was to assess the use of AI in service delivery in Academic libraries. The study revealed that a majority did not know the use of Pattern Recognition for the acquisition, processing of books, and circulation of library materials used in your library (Mean=1.8421, SD=0.36625). Another majority (Mean=1.7895, SD=0.40948) also disagreed that they knew about using Chatbot for reference services (Alexa, Siri, and IBM Watson) in your library. Another majority (Mean=1.7281, SD=0.44692) were not knowledgeable about Big Data Analytics to manage library digital data processing, database usage, and the sharing of data used in your library.

Again, Table 8 shows that the majority (Mean=1.7193, SD=0.45133) disagreed with the use of Natural Language Processing in Voice Search and Search Assistance of library and online materials used in your library. Another majority (Mean=1.6491, SD=0.47935) disagreed with Robotics for shelving and retrieving books used in your library. The majority (Mean=1.5965, SD=0.49277) each disagreed that Text Data Mining for citation counts, Altimetric scores, library trends, social media Tagging, etc. used in your library and Expert Systems for the effective management of the system library used in your library.

Table 8. Knowledge of the uses of AI in academic libraries

	Mean	Std. Deviation
Expert Systems for the effective management of the system library used in your library	1.5965	0.49277
Natural Language Processing in Voice Search and Search Assistance of library and online materials used in your library	1.7193	0.45133
Chatbot for reference services (Alexa, Siri, and IBM Watson) used in your library	1.7895	0.40948
Pattern Recognition for acquisition, processing of books, and circulation of library materials used in your library	1.8421	0.36625
Big Data Analytics to manage library digital data processing, database usage, and the sharing of data used in your library	1.7281	0.44692
Text Data Mining for citation counts, Altimetric scores, library trends, social media Tagging etc., used in your library	1.5965	0.49277
Robotics for shelving and retrieving books used in your library	1.6491	0.47935

Prospects and Implications of AI in Academic Libraries

The third objective of the study was to assess the prospects and implications of AI in academic libraries. In doing so, it was revealed that a majority (Mean=4.3860, SD=0.54055) agree that Pattern Recognition will enhance the acquisition, processing of books, and circulation of library materials. Again, a majority (Mean=4.3158, SD=0.62847) agreed that Big Data Analytics enhances the management of library digital data processing, database usage, and data sharing. Results also revealed that Robotics would enhance shelving and the retrieval of books (Mean=4.2632, SD=0.85241).

Again, a majority (Mean=4.2105, SD=0.75813) agreed that Text Data Mining would enhance information collection on citation counts, Altimetric scores, library trends, social media Tagging, etc. Another majority (Mean=4.2456, SD=0.73547) agreed that Chatbot would enhance reference services (Alexa, Siri, and IBM Watson). Another majority (Mean=4.2368, SD=0.75613) agreed that Expert Systems would assist in making effective organizational management decisions for the system library. Another majority (Mean=4.1053, SD=0.86596) agreed that Natural Language Processing in Voice Search and Search Assistance would enhance the searching and retrieval of library and online materials. Yet, another majority (Mean=4.0088, SD=1.02617) agreed that Robotics would guide weeding decision-making, as shown in Table 7.

Table 9. Prospects and implications of AI in academic libraries

	Mean	Std. Deviation
Expert Systems will assist in making effective organizational management decisions for the system library	4.2368	0.75613
Natural Language Processing in Voice Search and Search Assistance will enhance the search and retrieval of library and online materials	4.1053	0.86596
Chatbot will enhance reference services (Alexa, Siri, and IBM Watson)	4.2456	0.73547
Pattern Recognition will enhance the acquisition, processing of books, and circulation of library materials	4.3860	0.54055
Big Data Analytics enhance the management of library digital data processing, database usage, and the sharing of data	4.3158	0.62847
Text Data Mining will enhance information collection on citation counts, Altimetric scores, library trends, social media Tagging, etc.	4.2105	0.75813
Robotics will enhance shelving and the retrieval of books	4.2632	0.85241
Robotics will guide weeding decision making	4.0088	1.02617

DISCUSSION OF FINDINGS

This study was conducted to assess librarians' knowledge of the emergence of AI, their perception of the use of AI in academic libraries in libraries, and their source of knowledge of AI tools and applications in academic libraries. The study involved 114 respondents, with a majority of 79 (69.3%) being females and 35 (30.7%) being males. This can be attributed to the fact that the information studies and librarianship profession is female-dominated (Majanja and Kiplang'at, 2003). The educational level of the respondents varied, with the majority having a bachelor's degree in information studies (36.8%), followed by a Master's degree (22.8%) and a Diploma in Information Studies (12.3%).

Results indicated that all 114 respondents knew about the emergence of AI in academic libraries. The majority of librarians obtained their knowledge from research articles (48.2%), followed by workshops/symposiums/webinars (28.9%), and formal education (22.8%). Research articles being the chief source of information on AI among librarians can be attributed to them knowing that such articles contain up-to-date information on research and studies carried out in the field. Again, these research articles may be in their subscribed journal and databases and will have ready and easy access. The proliferation of technology has pushed librarians and individuals to prefer electronic information sources compared to their print counterparts. This finding corroborates with studies by Batool and Ameen (2010), Aesaert and Van Braak (2015), and Yamson et al. (2018).

Regarding perception, the librarians generally agreed that AI tools would improve service provision to users (Mean=4.3684, SD=0.66868), improve work efficiency (Mean=4.3333, SD=0.73673), and enable quick and efficient retrieval of information for users (Mean=4.3158, SD=0.75628). This finding confirms studies by Bradley (2022), Yoon & Schultz (2017), and Robins (2002), who believe AI implementation will enable librarians to provide effective and satisfactory services to their user community.

Again, while most agreed that the implementation of AI would enhance the work of librarians (Mean=4.2193, SD=0.83887), some agreed that AI would increase library patronage (Mean=3.9561, SD=0.93497). A study by Robins (2002) suggested that the implementation of AI in academic libraries will make it possible for librarians to provide personalized services such as selective dissemination of information and personalized research support, among other services customized to meet the specific needs of their users which will eventually draw more users to the library.

The study also showed that some librarians were, however, neutral about the view that the application of AI would take over their jobs and make them redundant (Mean=2.9123, SD=1.31406) and could be attributed to the novelty of the AI technologies and applications and the uncertainties that surround its implementation.

Results further indicated that librarians generally had little or no knowledge of the use of Pattern Recognition for the acquisition, processing of books, and circulation of library materials (Mean=1.8421, SD=0.36625), Chatbot for reference services (Mean=1.7895, SD=0.40948), Big Data Analytics to manage library digital data processing, database usage, and the sharing of data (Mean=1.7281, SD=0.44692), Natural Language Processing in Voice Search and Search Assistance of library and online materials (Mean=1.7193, SD=0.45133), Robotics for shelving and retrieving books (Mean=1.6491, SD=0.47935), Text Data Mining for citation counts, Altimetric scores, library trends, social media tagging, etc. (Mean=1.5965, SD=0.49277), and Expert Systems for the effective management of the system library (Mean=1.5965, SD=0.49277). This generally low knowledge of the users of AI applications in academic libraries can be attributed to the fact that most libraries have not implemented AI. However, it confirms the initial revelation from the study that AI knowledge was acquired primarily through research articles limiting their practical knowledge of AI applications and tools. This confirms a study by Afrane et al. (2022) which suggested that librarians and information professionals needed to build their skills and capacity in programming languages, such as Python or R, and their familiarity with AI and machine learning tools and platforms.

CONCLUSION AND RECOMMENDATIONS

The study sought to assess the knowledge and perception of librarians who worked in academic libraries in Ghana's use of AI in academic library services provision. The study revealed that librarians in Ghana generally had a positive perception of using AI in academic libraries, but their knowledge of AI tools and applications was limited. The study also revealed a significant positive relationship between educational level and the sources from which information on AI was sought, with research articles being the most sought-after information source. The study recommends that librarians be exposed to AI tools and applications through workshops, training, and further education to fully utilize AI's potential benefits in academic libraries. Academic libraries are encouraged to invest in AI tools and applications to leverage their advantages.

LIMITATIONS AND FUTURE RESEARCH

Artificial Intelligence is a broad and emerging technology. This study looked at librarians' knowledge and perception of the use of AI in the delivery of academic library services, thereby limiting the extent to which the researchers could delve into the use of AI applications in academic libraries. Further research into implementing AI tools and applications in delivering the different library services is appropriate and recommended.

REFERENCES

Afrane, D. A., Donkor, A. B., & Yamson, G. C. (2022). Libraries for Tomorrow: The Use of ICT and Space Transformation in Some Academic Libraries in Ghana. *Mousaion: South African Journal of Information Science*, *40*(2). Advance online publication. doi:10.25159/2663-659X/9896

Afrane, D. A., Van Der Walt, T., & Donkor, A. B. (2022). Student's assessment of Balme Library's use of information technology in providing quick and efficient library services. *South African Journal of Library and Information Science*, *88*(1). Advance online publication. doi:10.7553/88-1-1975

Al Aklabi, A. T. (2021). An Artificial Intelligence (AI) and Digital Services at King Saud University Libraries in COVID-19 Pandemic. *Journal of Engineering and Applied Sciences Technology*, 1–8. doi:10.47363/JEAST/2021(3)131

Arshad, M., Khan, A., Ahmed, P., Abbas Shah, N., & Ahmad, P. (2020). Next Generation Data Analytics: Text Mining in Library Practice and Research. *Library Philosophy and Practice (e-Journal)*.

Bottorff, T., Glaser, R., Todd, A., & Alderman, B. (2008). Branching out: Communication and collaboration among librarians at multi-campus institutions. *Journal of Library Administration*, *48*(4), 329–363. doi:10.1080/01930820802289391

Bradley, F. (2022). *Representation of Libraries in Artificial Intelligence Regulations and Implications for Ethics and Practice*. doi:10.1080/24750158.2022.2101911

Cox, A. (2022). How artificial intelligence might change academic library work: Applying the competencies literature and the theory of the professions. *Journal of the Association for Information Science and Technology*. Advance online publication. doi:10.1002/asi.24635

Cox, A. M., & Mazumdar, S. (2022). Defining artificial intelligence for librarians. *Journal of Librarianship and Information Science*. Advance online publication. doi:10.1177/09610006221142029

Cushing, A. L., & Osti, G. (2022). "So how do we balance all of these needs?": How the concept of AI technology impacts digital archival expertise. *The Journal of Documentation*, *79*(7), 12–29. Advance online publication. doi:10.1108/JD-08-2022-0170

Farag, H. A., Mahfouz, S. N., & Alhajri, S. (2021). *Artificial Intelligence Investing in Academic Libraries: Reality and Artificial Intelligence Investing in Academic Libraries: Reality and Challenges Challenges*. https://digitalcommons.unl.edu/libphilprac

Frederick, A., Run, Y., Frederick, A., & Run, Y. (2019). The Role of Academic Libraries in Research Data Management: A Case in Ghanaian University Libraries. *OAlib*, *6*(3), 1–16. doi:10.4236/oalib.1105286

Fu, J., Lv, J., & Li, X. (2011). The Research about Intelligent Search Engine and in Digital Library Personalization Services. *Advanced Materials Research*, *143–144*, 333–337. doi:10.4028/WWW.SCIENTIFIC.NET/AMR.143-144.333

Gasparini, A., & Kautonen, H. (2022). Understanding Artificial Intelligence in Research Libraries: An Extensive Literature Review. *LIBER Quarterly*, *32*(1), 1–36. doi:10.53377/lq.10934

Hussain, A. (2023). Use of artificial intelligence in the library services: Prospects and challenges. *Library Hi Tech News*, *40*(2), 15–17. doi:10.1108/LHTN-11-2022-0125

Huvila, I., Lloyd, A., Budd, J. M., Palmer, C., & Toms, E. (2016). Information work in information science research and practice. *Proceedings of the Association for Information Science and Technology, 53*(1), 1–5. doi:10.1002/pra2.2016.14505301004

Kaliraj, P., & Devi, T. (2022). *Artificial Intelligence Theory, Models, And Applications* (P. Kaliraj & T. Devi, Eds.; 1st ed.). CRC Press.

Lund, B. D., Omame, I., Tijani, S., & Agbaji, D. (2020). Perceptions toward Artificial Intelligence among Academic Library Employees and Alignment with the Diffusion of Innovations' Adopter Categories. In *College & Research Libraries* (Vol. 81, Issue 5). https://crl.acrl.org/index.php/crl/rt/printerFriendly/24516/32350

Okunlaya, R. O., Syed Abdullah, N., & Alias, R. A. (2022). Artificial intelligence (AI) library services innovative conceptual framework for the digital transformation of university education. *Library Hi Tech, 40*(6), 1869–1892. doi:10.1108/LHT-07-2021-0242

Pannucci, C. J., Wilkins, E. G., & Pannucci, C. (2010). Identifying and Avoiding Bias in Research. *Plastic and Reconstructive Surgery, 126*(2), 619–625. doi:10.1097/PRS.0b013e3181de24bc PMID:20679844

Perrier, L., Blondal, E., Ayala, A. P., Dearborn, D., Kenny, T., Lightfoot, D., Reka, R., Thuna, M., Trimble, L., & MacDonald, H. (2017). Research data management in academic institutions: A scoping review. *PLoS One, 12*(5), e0178261. Advance online publication. doi:10.1371/journal.pone.0178261 PMID:28542450

Pham, H. T., & Tanner, K. (2015). Collaboration Between Academics and Library Staff: A Structurationist Perspective. *Australian Academic and Research Libraries, 46*(1), 2–18. doi:10.1080/00048623.2014.989661

Robins, D. (2002). Information Architecture in Library and Information Science Curricula. *Bulletin of the American Society for Information Science, 28*(2), 20–22. doi:10.1002/bult.231

Tariq, M. U., Babar, M., Poulin, M., Khattak, A. S., Alshehri, M. D., & Kaleem, S. (2021). Human Behavior Analysis Using Intelligent Big Data Analytics. *Frontiers in Psychology, 12*, 686610. Advance online publication. doi:10.3389/fpsyg.2021.686610 PMID:34295289

Tsabedze, V. W., Mathabela, N. N., & Ademola, S. S. (2022). A framework for integrating artificial intelligence into library and information science curricula. In *Innovative Technologies for Enhancing Knowledge Access in Academic Libraries* (pp. 233–246). IGI Global. doi:10.4018/978-1-6684-3364-5.ch014

Ţundrea, E., Turcuţ, F., & Fotea, S. L. (2020). Challenges and Opportunities When Integrating Artificial Intelligence in the Development of Library Management Systems. *Springer Proceedings in Business and Economics*, 369–382. doi:10.1007/978-3-030-43449-6_22

UKRI. (2021). *Transforming our world with AI*. Available at: https://www.ukri.org/wp-content/uploads/2021/02/UKRI-120221-TransformingOurWorldWithAI.pdf

Yoon, A., & Schultz, T. (2017). Research data management services in academic libraries in the US: A content analysis of libraries' websites. *College & Research Libraries*, *78*(7), 920–933. doi:10.5860/crl.78.7.920

KEY TERMS AND DEFINITIONS

Academic Libraries: Academic libraries are libraries that are affiliated with academic institutions such as colleges and universities. These libraries support the research and academic activities of their institution by providing access to a wide range of resources, including books, journals, databases, and other materials.

Artificial Intelligence: Refers to the development of computer systems that can perform tasks that would typically require human intelligence to accomplish. These tasks may include visual perception, speech recognition, decision-making, and language translation. AI systems can be programmed to learn from experience and adapt to new situations, which allows them to continually improve their performance.

Chatbots: Chatbots are computer programs designed to simulate conversation with human users through text or voice-based interfaces. Chatbots use natural language processing (NLP) to understand and interpret user input and generate appropriate responses.

Library Assistant: A library assistant is a support staff member who works in a library to help with a variety of tasks related to library operations.

Library Services: Library services refer to the various services and resources provided by libraries to support the needs of their users.

Metadata: Metadata is information that describes data, providing additional context and meaning to help users understand and manage the data.

Natural Language Processing: A subfield of artificial intelligence that focuses on enabling computers to understand, interpret, and generate human language. It involves a range of techniques and tools used to process and analyse human language data, including text, speech, and even images or videos with textual content.

Text Mining: Text mining is the process of analysing and extracting useful information from unstructured text data. It involves using machine learning and natural language processing techniques to automatically identify and extract patterns, relationships, and insights from large volumes of text data.

Chapter 2

The Use of Artificial Intelligence in Education:
Applications, Challenges, and the Way Forward

Elisha Mupaikwa
https://orcid.org/0000-0002-0313-7139
National University of Science and Technology, Zimbabwe

ABSTRACT

Over the past few decades, there has been an increase in ongoing research into the use of artificial intelligence in education. The usage of artificial intelligence in educational institutions has grown during the past few years. This chapter examines how artificial intelligence is used in education. The establishment of inclusive learning environments, collaborative learning and research, social robots, student assessment, automated grading, recommender systems, and student performance prediction were identified to be the main applications of artificial intelligence in education. Lack of funding, lack of faith in the system, lack of portability of systems, infrastructure issues, ethical issues regarding data privacy and security, and a lack of experience were some of the difficulties noted. After that, the chapter included recommendations for infrastructure improvements, ICT policy changes at the national and institutional levels, curriculum revisions for teacher- and school-level preparation, and additional research.

DOI: 10.4018/978-1-6684-8671-9.ch002

INTRODUCTION

The goal of artificial intelligence (AI) is to create computer systems that can reason and solve issues the same way humans would. Different subfields have formed from the discipline of AI as it has developed. These advancements resulted from the necessity to create technology that facilitates problem-solving in a variety of economic areas, including manufacturing, engineering, healthcare, medicine, agriculture, finance, and education. Learning institutions today include technology like learning management systems (LMS), intelligent torture systems (ITS), and online learning to make it easier for students to obtain knowledge and develop their abilities (Khan et al., 2021). Applications of these technologies in education are essential for improving teaching and learning at all educational levels, including tertiary education, where AI as an emerging technology has aided in institutional and student research to produce new knowledge and advance practice in a variety of fields of study and practice. The use of these technologies in education has had a transformational impact, and educators who have used these technologies have empowered themselves and their learners, creating a flexible teaching and learning environment that is user-centred. While some authors have characterized emerging technologies as disruptive because they forced the user to abandon their traditional means of solving problems, the application of these technologies in education has played a transformational role.

Recent years have seen a rise in the popularity of research on technologies that mimic human behaviour and reasoning to address both organized and unstructured situations in education. In addition, prediction models have been created as part of this research to track student performance in virtual learning environments (VLEs), and these models make use of machine learning (ML) techniques to forecast students' performance and outcomes. These AI technologies offer prospects for efficient task execution and problem solutions in the educational environment. These technologies easily fall under the heading of "emerging technologies" since they provide fresh approaches to tackling issues as well as several potential applications in the future. While there have been some experimental projects in fields including e-business, information science, manufacturing, health care, and the military, there appears to be a gap in the application of these technologies in education. With extensive studies on artificial intelligence being done, it is now clear how AI could change teaching approaches at the junior, secondary, and postsecondary levels of education. When they have been employed, they supported learning for people with physical disabilities by using tools like speech recognition, text translation, and voice recognition, transforming the traditional roles of teachers in primary and secondary schools, as well as lecturer instructors at university institutions.

While studies have looked into how AI might be used to revolutionize education, little is known about the fundamental building blocks of AI. Therefore, the purpose of this chapter is to investigate how various AI components might alter the overall educational environments, including everything from the delivery of teaching to the administration of the educational environment. In higher education, AI has been used for profiling and prediction, assessment and evaluation, adaptive systems and personalization, and intelligent tutoring systems, according to a review of the literature by Zawacki-Richter et al., (2019).

BACKGROUND

In several sectors, AI usage and development have increased during the last few decades. A new area of study known as "AI in Education (AIED)" has emerged as a result of the usage and research of AI in the fields of education and training. According to Tan et al., (2022), AIED refers to the use of AI in educational environments to enhance teaching and learning by emulating human intelligence to infer, judge, and make decisions, among other activities of education and training as well as research. Three dimensions have developed in the field of artificial intelligence in education (AIED). These are AI-directed, learner-as-recipient; AI-supported, learner-as-collaborator; and AI-empowered, learner-as-leader Ouyang & Jiao (2021). For the first paradigm, the learner is reduced to a passive observer while the AI system assumes a leading and directing role in the educational process, imparting knowledge and setting the learning agenda. A collaborative relationship exists between the AI system and the user in the second paradigm. The third paradigm aims to provide the learner with more control over their education by personalizing the learning experience and assigning each student a specially designed learning assignment.

According to Zawacki-Richter et al., (2019), there was a considerable increase in the use of AI in education between 2018 and 2022. They projected this trend would continue significantly, stating that this is the way higher education will develop in the future. Given the previously predominant applications of AI in the domains of business and engineering, researchers' rising interest in the application of AI in education may have contributed to this recognition of the potential role AI may play in education. Pence (2019) reported that in American contexts, AI had grown in popularity even in homes, while in colleges and universities, it had been classified as applications dealing with the management of the institution in areas of recruitment, marketing, admission, determining financial aid, and answering queries from accepted students. The other category takes into account how AI is used to organize lessons and analyze data gathered from learning management systems.

THE BASICS OF AI

Khemani (2013) claims that AI is a synthesis of several fields and that research in this area appears to have taken two different directions, one following cognitive sciences and concentrating on the study of intelligent behaviour, and the other concentrating on engineering and the creation of intelligent machines. Researchers and philosophers have debated whether it is possible to program robots to think and function like humans for a very long time. This is because from the beginning of the concept of AI, the idea was to use computers' processing capability to make them think and act like people. This has resulted in several attempts, inseveral fields abd disciplines, to better understand human thought, decision-making, and behaviour to create computational systems that simulate those traits. The use of AI, thus, has become more common in a variety of fields, including gaming, speech recognition, natural language processing, manufacturing and engineering, health care, vision systems, and agricultural systems. Recently, the application of AI in training and education has increasingly gained popularity.

While several authors have proposed multiple definitions of AI, there is general agreement that any system must demonstrate its ability in the following areas to be characterized as intelligent: i) reasoning; ii) Learning; iii) Understanding; iv) grasping truths; v) seeing relationships; vi)considering meanings; and, vii)separating fact from beliefs (Mueller and Massaron, 2018). These characteristics are genereally associated with how people observe, comprehend, foresee, and control the world and their surroundings to address daily challenges. According to Negnevitsky (2005), intelligence is characterized by the capacity to comprehend and learn new information as well as the capacity to reason and reason instead of acting instinctively or reflexively. How people perceive their environment and their experiences within it, as well as how they reason using inferences taken from visible conditions and the outcomes of their actions, form the basis for the development of AI technology. AI, programs therefore, are said to be intelligent if they display behaviours that would be considered intelligent if they were displayed by humans (Khemani, 2013). These technologies are machine-based systems that have variable degrees of autonomy and are capable of making predictions, suggestions, or judgments that affect real or virtual environments for a specified set of human-defined objectives (OECED, 2020). Some authors contend that the primary objectives of studying AI go beyond simply simulating intelligence and instead aim to create sophisticated imitations of intelligence (Khameni, 2013). Since the term "artificial" implies that it is not real, some people prefer the terms "synthetic intelligence" or "machine intelligence". To summarise the characteristics of AI, several texts describe AI as being capable of "acting humanely", "thinking humanely", "thinking rationally" and "acting rationally". Using this description as a foundation, other authors have defined AI as:

Figure 1. Fields of artificial intelligence

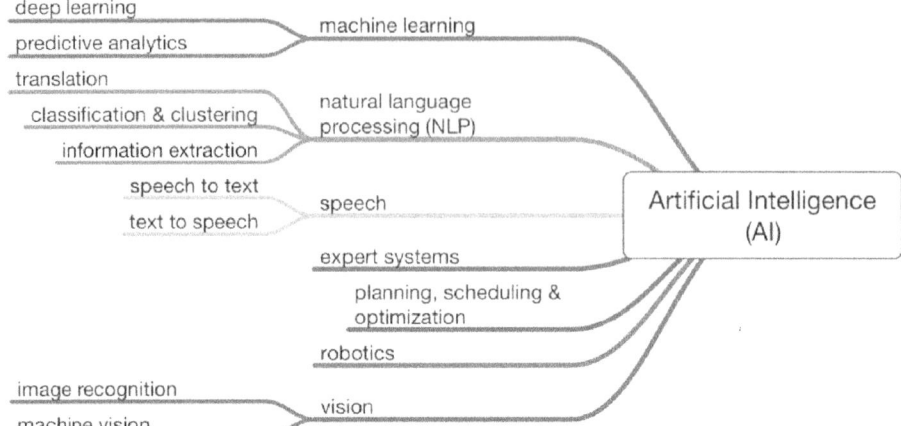

a)   Thinking humanly: Artificial intelligence is an effort to make computers think (Haugeland, 1985).

b)   Thinking rationally: Artificial intelligence is the study of computations that make it possible to perceive, reason and act (Winston, 1992)

c)   Acting humanly: The art of creating machines that perform functions that require intelligence when performed by people (Kurzweil, 1990). It may also be described as the study of how to make computers do things at which at the moment people are better (Rich and Knight, 1991).

d)   Acting rationally: AI is concerned with intelligent behaviour and artefacts (Nilsson, 1995). Poole et al., (1998) also define computational intelligence as the study of the design of intelligent agents.

Beqiri (2016) identified the major fields and application areas of AI as; Machine learning, Natural language processing, Speech processing, Expert systems, Robotics, Vision processing and Planning, scheduling and optimisation. These fields are further subdivided into several disciplines as shown in Figure 1.

From Figure 1, these various disciplines of AI have slowly been gaining recognition across various sectors such as engineering, manufacturing and production, education and agriculture among many others.

AI IN EDUCATION

The variety of applications of AI in education is demonstrated by the wealth of literature on the subject. However, although progressively becoming a part of the learning process and transforming research, teaching, and learning, these technologies seem to have yet to revolutionize education and research (Baker, 2021). Intelligent tutoring systems, NLP for language learning, educational and social robots, profiling, personalization, predicting performance, retention and dropout, collaborative learning and research, student performance assessment, personalization of education and enhanced training, curricular planning; evaluation and grading, quality management, accreditation, sentiment analysis, recommendation systems, classroom monitoring and visual analysis and educational publication statistics are the dominant topics related to the applications of AI in education (Zawacki-Richter et al., 2019). These applications are then classified as learner-facing, teacher-facing, and system-facing. Without the help of the teacher, learner-facing systems are software solutions that offer pupils adaptive and personalized learning while teacher-facing systems aim to lessen the workload of the teacher by automating the teacher's duties. Administrators and managers at the institutional level can obtain information through system-facing tools by, for instance, tracking patterns among faculties and colleges.

Today, the majority of conversations about technology and education centre on computers and digitization, with some talks also including more traditional teaching methods like radio and television that gained popularity during the Covid-19 epidemic. However, Chen et al., (2020) in their review of the application of AI in education provide scenarios for AI applications in education and related technologies and these are shown in Table 1 below

Table 1. AI Scenarios and related technologies

Scenarios of AI Education	AI-Related Techniques
Assessment of Students and Schools	Adaptive learning method and personalized learning approach, academic analytics
Grading and evaluation of papers and examinations	Image recognition, computer vision, prediction systems
Personalised intelligent teaching	Data mining or Bayesian knowledge interface, intelligent teaching systems, learning analytics
Smart school	Face recognition, speech recognition, Virtual labs, hearing and sensing technologies
Online and mobile remote education	Edge computing, virtual personalized assistants, real-time analytics

Figure 2. Architectural framework of AI in education

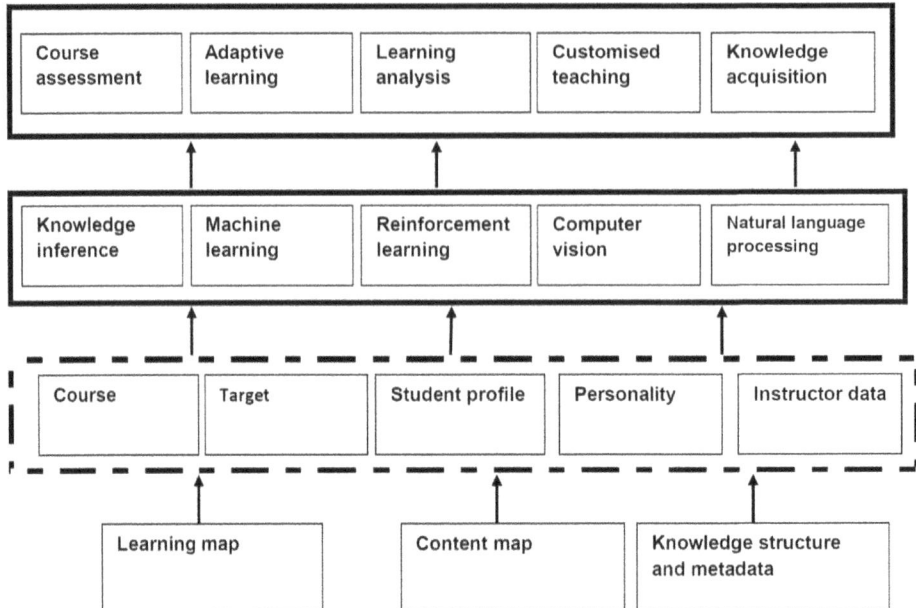

From the scenarios and related AI technologies shown in Table 1, Chen, Chen and Lin (2020) drew an architectural framework representing the technological structure of AI in education as shown in Figure 2 below.

INTELLIGENT TUTORING SYSTEMS

By their very nature, intelligent tutoring systems (ITS) are adaptive tutoring systems that offer users personalized (personalized) and dynamic services, information resources, direction, and feedback. By giving students tailored instruction, automating common work for teachers, and facilitating adaptive evaluations, AI systems have significantly aided online teaching and learning (Seo et al., 2021). These systems provide learners with individualized help based on computerized evaluations of their performance and general student information gleaned from a dataset. To improve instant individual feedback and increase access to tutoring and other educational resources, learning is adapted to each student's preferences or unique needs. Additionally, it is feasible to tailor feedback and content to each user based on their knowledge level and to save time for the teacher and other students by avoiding discussing issues with those who do not need them. Through the usage of NLP tools like chatbots, they have become more well-known in the fields of science, technology, engineering, mathematics (STEM) and language education (Chen et al., 2022). Computer-aided learning (CAI) has developed into sophisticated tutoring programs.

Ahmad et al., (2021) claim that ITS, which interacts with students and delivers information and assessments to learners based on their knowledge and progress, is an example of AI applied to education. Learning materials are offered to students by intelligent tutoring systems following what the systems have discovered from the students' learning behaviours as obtained during the students' learning experiences (Crompton and Song, 2021). Students are given the material on which they will be evaluated, and whether or not they go on to the following learning units depends on how well they perform. These programs choose activities for each student and lead them through a range of curriculum-related tasks. When doing so, AI systems can assess a student's degree of comprehension thanks to the system's contact with the pupils. Additionally, content and feedback are presented in a tailored way. The Cognitive Teacher is cited by Koedinger et al., (2013) as an illustration of an intelligent tutoring system that has demonstrated effectiveness in teaching mathematics and has been useful in assisting students with their homework, educational games, and online resource learning. By combining this technology with web resources, Datashop, a big data platform with millions of student records that can be studied to understand student learning, was created. The Data Science / Machine Learning platform (DSMLP), which uses machine learning techniques to provide access to resources and student projects, the Canvas LMS from UC San Diego, and Boost, a mobile-based application for supporting intelligent online learning, are among the other intelligent tutoring systems mentioned by González-Calatayud et al., (2021). English writing, listening, and speaking abilities are among the languages in that artificial intelligence has been shown to facilitate learning (Dewi et al., 2021). Netflix, Joox Music, Duolingo, Google Translate, and Grammarly are just a few examples of intelligent tutoring systems that make learning languages easier. Additionally, chatbots have been employed to assist with language learning in a virtual learning environment. AI is a helpful tool in computer-assisted language learning because it enables text analysis using NLP to determine meanings and produce explanations for text strings.

FACILITATING COLLABORATION

Collaborative learning refers to a learning environment in which students and researchers work together to solve problems. When older students collaborate on learning in higher education settings, the focus is on the students, whereas when younger students collaborate on learning, the focus is frequently on the teacher, who guides the learning process activities. Both categories benefit from the deployment of AI technologies, which increase cognitive resources. For example, computer-supported collaborative learning systems encourage interaction among students via a network or shared computers (Tan et al., 2022). AI is used by teachers in the classroom to set up student collaborative groups. Crompton and Song (2021) claim that AI makes use of student knowledge to provide matches for various groups based on what is known about the students and their prior experiences in educational settings. According to research by Tan et al. (2022), there has been a growth in the use of AI for collaborative learning at all educational levels, especially from 2007 to 2022. Universities dominated this trend, using online platforms.

STUDENT ASSESSMENT

Due to the requirement to enhance the assessment of effectiveness and validity in a setting of big datasets in a digital learning environment, developments in AI-related assessment have gained popularity (Gardner, O'Leary and Yuan, 2020). Assessment in pedagogy can be either formative or summative. Although AI has been included in both types of assessment, González-Calatayud, Prendes-Esipinola, and Roig-Vila (2021) contend that AI-based assessments are better suited for formative assessment at all stages of learning. However, automated assessments have primarily been used with short responses, which have been found to offer more information about the student's understanding than multiple-choice questions. In scenarios with huge datasets or groups, these automated evaluations have enhanced assessment efficacy and validation (Gardner, O'Leary and Yuan, 2021). Traditional exams have been automated with AI-based tests. In addition to evaluating students' performance through exams or assessments, these systems can also create assessment tasks and identify qualified peers to evaluate the work (Swiecki et al., 2022). These assessment methods are mostly used in online learning, and AI capabilities that go along with them have increased assessment effectiveness and decreased challenges associated with open-ended question assessment.

One of the breakthroughs in the use of AI in student assessment is the Al-based students' assessment and recommendation (AISAR), which includes suggestion, performance prediction, grouping, and score estimation. Using clustering approaches, instructors can identify students who face comparable learning issues and offer group support by grouping students based on their performance levels. Tutors can use this information to recognize exceptional students and forecast each student's performance. The AISAR divides pupils into two main groups: average and poor, and it uses these classifications to suggest new content or other changes as well as corrective measures. Automated student assessments can also be used to forecast and track student performance to develop preventative strategies for students with learning challenges. When the teacher notices unsatisfactory development, the teacher intervenes to help the weaker children (Khan et al., 2021).

AUTOMATED GRADING

Assessments in educational settings frequently lead to the performance-based grading of students. Artificial intelligence is frequently used to automate tutoring work grading. Open-ended questions are frequently assessed by teachers in higher education, and these evaluations have historically been difficult, time-consuming, and subjective. By automating the procedure, the application of AI has made the grading of pupils simpler. Through Automated Essay Scoring Systems (AES), such as WriteToLearn and Research Writing Tutor, this can be applied to a variety of assessment types, including essays and multiple-choice questions. These services also give students comments so they can edit their works before final submission (Crompton and Song, 2021). AES is criticized by Gardner, O'Leary, and Yuan (2020) for being less suited to evaluating the creative and high-order aspects of writing. These technologies include feedback components that are personalized to match the needs of certain users in addition to automating grading.

SENTIMENT ANALYSIS

The ability to identify student attitudes and feelings has become crucial in education departments when evaluating the effectiveness of a teaching strategy. Ngoc and Thi (2021) go on to define sentiment analysis as a description of attitudes, thoughts, or judgments on a certain view or notion that are motivated by feelings. Although this has been utilized in commerce and politics, its usage in education offers insights into the efficacy of the instruction and feedback, giving teachers a chance to reflect on their performance. Since tutors may more easily direct and offer corrective action to support students' learning through observations of students' emotions displayed through facial expressions, the interaction between people in face-to-face learning environments has frequently been credited with outstanding results. However, the lack of these facilities in VLE learning environments compromises the usefulness of the VLE.

The goal of sentiment analysis among students, according to Devi et al., (2022), is to investigate the mechanisms through which language and text elicit negative emotions like wrath and positive ones like joy, grief, and happiness. Barron-

Estrada, Zatarain-Cabada, and Oramas-Bustillos (2018) acknowledge that while the majority of online teaching and learning systems are capable of detecting basic emotions like "happy," "sad," or "angry," these systems are constrained in their ability to detect secondary emotions like "boring," or "frustrated." The emotions can be identified through the use of convolutional neural networks characterised by adorable architectures. However, according to Altrabsheh (2016), there hasn't been much research done on how to analyze embedded students' sentiments from student feedback for online instruction. Amused, bored, and enthusiastic are three emotions that can be quickly identified using an AI model for polarity detection, according to Altrabsheh (2016). Analysis of student feedback reveals a variety of concerns that students may have with the lecturer, such as students who may not fully comprehend a lesson. Communication between the student and the teacher is enhanced as a result, enabling teachers to evaluate the impact of online feedback and education on their students by using opinion mining (sentiment analysis). According to Devi et al., (2022), popular AI methods that have been applied for sentiment analysis include dynamic convolutional neural networks (DCNN). The literature on sentiment analysis emphasizes the use of NLP and deep learning techniques in this field.

Social Robots

These are robots that communicate with people while learning, supporting human teachers. However, Ahmad et al., (2021) note that there is still disagreement over the usefulness of social robots in the classroom, with other academics stating that human instructors are preferable. Despite the disagreements surrounding their usage in education, social robots have gained popularity and have been given a variety of tasks. These resources have been employed as peer tutors, and it has been demonstrated that using them in teaching and learning improves cognitive and affective outcomes on par with human tutors (Belpaeme et al., 2018).

AI IN INCLUSIVE LEARNING

For a very long time, education researchers have bemoaned the shortcomings of conventional teaching approaches. These teacher-centred approaches are only available to physically abled students, leaving students who are physically underprivileged to continue to experience unfair access to educational resources. By providing accessible tools to instructors and students who are physically challenged, the use of AI in education has the potential to remove these obstacles and promote fairness and equality for all pupils. According to Salas-Pilco, Xiao, and Oshima (2022), inclusive education is relevant to a variety of learner populations, including those with special needs and disabilities, females and women, pupils at risk, and members of ethnic minorities. The use of AI in inclusive education fosters student engagement across the senses, generates low-risk environments, scaffolds learning objectives, includes students with disabilities in authentic contexts, encourages collaborative learning, and reinforces good behaviour (Salas-Pilco et al., 2022). The use of AI in a variety of teaching and learning contexts may be advantageous. Providing differentiated instruction and feedback on a mixed-ability class, giving prompt feedback on students' assessments, and identifying struggling students in their learning process are the three core areas of teaching that Murphy (2019) focused on in his assessment of AI applications in supporting K–12 teachers. These technologies can support computer-assisted instruction, adapting the content to the learner's level of understanding based on their responses and the feedback they receive. Such expert systems are rule-based, knowledge-based, and inferential, using heuristics. Among the popular applications are ALEKS, MATHia, Dreambox Learning, STMath and Achieve3000 (Murphy, 2019). These systems evaluate student performance based on their responses and categorize them by their performance, supporting teachers in identifying difficult kids. Systems that grade students' essay writing skills and early warning detection systems that spot students having difficulty with their essays are two other well-known AI applications built on machine learning.

PREDICTING STUDENT PERFORMANCE

In educational institutions, a student's performance is frequently evaluated using their academic standing in tests, reading assignments, and class activities, among other chores that a tutor could provide. El-Keiey, Elmenshawy, and Hassanein (2022) counter that other factors, such as personality traits, IQ, and basic personal information, also have an impact on students' success. Datasets for learning analytics include student information including age, sex, department of study, parents' educational background, occupation, and place of living. When analyzed, these characteristics from the literature offer perceptions of the potential performance of students (Barik, Barukab and Ahmed, 2020). According to Rashid and Aziz (2016), a variety of demographic and environmental characteristics, including social background, gender, economic status, residential area, and the medium of instruction, have an impact on students' performance in learning environments. Rashid and Aziz (2016) used prior exam results, students' parents' socioeconomic position, tutor categories, prior high school scores, high school type, and teaching languages to predict students' performance using AI. Additionally, IQ levels and personality qualities have both been used to forecast student achievement. Students can choose courses that fit their level of knowledge, intelligence, and competencies according to AI-based performance prediction. With the help of this, teachers may keep track of their student's learning progress and suggest relevant modules and corrective measures for those who need them. Higher education institutions have created early interventions for students at risk using this historical data from AI (Cheng, 2021).

AI IN HIGHER EDUCATION

Higher education learning environments are complicated, with students from a variety of social and economic origins and a range of issues. Researchers claim that AI plays a transformational role, altering how teaching and research are conducted, improving researcher collaboration, and increasing the effectiveness of policy development. There have been various reasons for incorporating AI in educational instruction in higher education. For instance, according to the literature, AI in higher education aims to:

a)   Increase outcomes in teaching, learning and research;
b)   Increase access to educational resources;
c)   Increase retention of learners;
d)   Lowering educational and operational costs; and,
e)   Decrease time to completion of tasks (Akinwalere and Ivanov, 2022).

Others have concurred that AI offers a chance to manage admissions, counselling, library, and research services for both undergraduate and postgraduate levels of education. Personalized support services are now essential at higher education institutions for different types of students to ensure effective program completion. Because of this, using AI in such situations enables the profiling of students as well as the prediction of their performance and issues in the classroom. According to Zawacki-Richter et al. (2019), AI techniques may be utilized for student performance profiling and prediction, as well as the possibility that a student will enrol in a program or drop out of it. This directs curriculum content and curriculum activities and offers chances for quick, supportive interventions. Classification and regression are fundamental components of data mining and are important machine learning algorithms for achieving this. AI has aided teaching and learning in disciplines at higher education institutions that are connected to industries that have historically used AI; examples of these industries include engineering and manufacturing. The teaching of engineering and scientific programs at universities and colleges now includes instruction in artificial intelligence. This has given engineering students access to educational tools that are sustainable, effective, efficient, and accessible (Nu'nez and Lantada, 2020). These technologies are helpful in higher education institutions like colleges where admissions departments frequently get a significant volume of applications due to AI's capacity for managing large datasets.

The learning in these schools has become modularized, allowing students to take customized collections of modules in addition to the high volume of applications. This creates difficulties for teachers in conventional learning settings that typically do not offer spaces for individualized tutoring, and it raises the possibility that the use of AI could be the answer to problems with course scheduling and student acceptance to learning programs. According to other studies cited by Zawacki et al. (2019), support vector machines (SVM) have been utilized in some studies to accurately categorize students and predict admission decisions. Based on spatial trends among applications from particular geographic locations, SVM was also helpful for forecasting student placements. To choose retention and support initiatives, prediction models have also been employed to identify students who are likely to discontinue their studies. For instance, Zawacki et al. (2019) reported that using classification techniques like decision trees and logic regression based on students' demographic, academic, and financial characteristics, AI had also been used to predict at-risk students in their first years at colleges and had also been used to predict attrition of undergraduate students. The University of New South Wales (UNSW) in Sydney uses a Microsoft Teams-based Question Bot to respond to questions in computer science and engineering, Pearson Group uses "LongWe," and MIP Politecnico di Milano Graduate School of Business uses FLEXA to provide personalized content to provide skills related to student's career goals are just a few of the institutions of higher learning that have incorporated AI into their instruction according to Akinwalere and Ivanov (2022).

CHATGPT IN ESSAY WRITING

Recent developments in AI tools of education have seen the emergence of ChatGPT, a natural language processing tool allowing people to interact with a computer naturally and conversationally. This technology learns from internet data and is capable of producing original results through generative AI, thus the name "Generative Pre-trained Transformer" (Sabzalieva and Valentini, 2023). A study by Zhai (2023) on ChatGPT user experience in higher education has shown that this technology carries great potential in transforming learning goals, learning activities, and assessment and evaluation techniques in teaching and learning. This is because the technology is capable of helping researchers write papers, that are coherent, accurate, informative and systematic and this is done in a shorter time compared to human effort. From these capabilities ChatGPT has been credited with offering intrinsic motivation, automating and rapidly designing lessons, redesigning assessment strategies and providing easy access to information and providing innovative opportunities for students to reproduce existing knowledge and to enact higher-order thinking, providing new information literacies for academics and researchers, repositioning academics and researchers in the knowledge creation process. From research on the efficacy of ChatGPT, Halaweh (2023) reported that research on whether plagiarism detection tools would detect high levels of similarity scores, the results showed that most essays generated using ChatGPT had a similarity score which was less than 205 suggesting there was a high degree of originality of essays generated by AI.

RECOMMENDER SYSTEMS

Recommender systems have emerged as a result of developments in AI-based web technologies. Recommender systems are well known for giving regular users of the web meaningful and effective web experiences by recommending services and goods including restaurants, music, and tourism attractions. In the academic world, recommender algorithms have directed users to tailored reading material and online journal and textbook materials. According to the supervisor's research interests and the students' proposals, recommender systems have also been used in higher education institutions to help students choose supervisors. For teaching reasons, collaborative research has facilitated course allocation among faculty members in national institutions based on agreed course outlines, while it has also employed recommender systems to pair researchers with comparable research interests and topics (Samin and Azim, 2019).

Classroom Monitoring and Visual Analysis

Classroom observation and visual analysis, according to Ahmad et al., (2021), help teachers identify students' actions, difficulties, and levels of learning in the classroom. AI-based solutions accomplish this by using data from a variety of sources, including student pictures and GPS data. Teachers employ AI-based methods to anticipate kids who may harass others in real-world settings or online communities by observing students' behavioural traits. Additionally, ITS has been used to keep tabs on students' behaviour, track their academic progress, evaluate their performance, and offer individualized incentives and support to them. The predictive analysis enables computer-based solutions to monitor the student body, give proactive interventions, and lower fatigue for administrators (Diebold & Han, 2022). These applications have proven to be crucial in special education classes (Chen et al., 2022). Expert systems are used in educational diagnosis to find learners who are struggling and to pinpoint the reasons behind their mistakes so that suitable training can be created for them. This is accomplished by reviewing the dataset's previously learned information (Jones, 1985).

CHALLENGES IN USING AI IN EDUCATION

Pence (2019) suggested that attempts to integrate AI into the syllabus and fill roles were unlikely to be successful and a significant squandering of possibilities to resuscitate higher education, despite noting the pervasive use of AI in American cultures and even in education. Even though it is widely acknowledged that computer-aided instruction and intelligent computer-aided instruction have revolutionized instruction and improved and diversified service delivery at different levels of education and across disciplines and knowledge domains, educational institutions still face difficulties in the use and adoption of these technologies. Depending on national policies and economies, these difficulties have changed from institution to institution and from country to country. The socioeconomic factors affecting educators, students, and institutions, the economic factors, the technical factors related to infrastructure and expertise, the user attitudes and perceptions, the demographic characteristics, the institutional policies, and the environmental factors, such as the ethical concerns surrounding data collection and processing, are crucial among these challenges.

Using AI offers difficulties due to the uneven distribution of digital resources, although AI promises to address issues facing the education sector generally. AI use in education calls for sophisticated infrastructure. These difficulties among potential AI users have raised issues with access, equity, and the employment of technology in educational institutions. These difficulties are made worse by the insufficient preparation of university lecturers and school teachers. As a result, the underprivileged sections of society are kept out of AI advancements, which exacerbates already-existing disparities (UNESCO, 2019). This necessitates the creation of policies by policymakers to guarantee justice and equality in the use of digital resources in education. Such issues are prevalent in developing countries, where there are also difficulties with access to hardware and electricity, unstable internet connectivity, expensive data rates, and inappropriate material.

The way AI functions in a classroom setting pose problems that are particular to most black-box technologies. These technologies abstract operations, which makes them opaque. According to Kumar (2018), one of the difficulties preventing the successful use of AI in educational instruction is the inability to explain how specific judgments would have been made. These technologies typically don't give the problem's context, and they often offer the same ideas and solutions for issues that appear to be related. Therefore, it follows that human specialists who can provide relevant justifications for decisions and recommendations cannot be replaced by machines. The employment of AI in education has thus been impeded by problems with openness, explainability, and accountability. Human instructors are renowned for being able to meet the demands of different learner types, from those with specific needs to those who have attained higher levels of conceptual mastery.

The main difficulty facing AI developers is to create systems that simulate human instructors in managing varied students and corresponding with them in a human-like manner (Woolf et al., 2013). The learners at today's educational institutions require a variety of abilities, as opposed to the conventional students who demanded a standardized method of instruction, according to Woolf et al. (2013). The non-routine tasks that today's students frequently take on call for critical thinking, systems thinking, interpersonal skills, presentation skills, communication skills, and conflict management. When compared to human tutors, AI educational technologies are frequently inadequate to meet these demands from students.

Large datasets are a component of AI infrastructure, however many institutions in developing countries struggle to create inclusive and high-quality databases. Artificial intelligence-based solutions have limited functionality and frequently produce undesirable results in the absence of enough data. Therefore, training systems using big data sets is necessary for the development of AI and its utilization. The lack of ethics among these technologies' users and creators is cited by Kumar (2018). Important moral questions involving data protection and privacy. Learning analytics (LA) and educational data mining (EDM) offer chances for a full understanding of learners in a method that gives data subjects no control over how their personal information is collected, processed, stored, and shared.

The degree of support provided by national ICT policy for education determines how effectively A-based innovations in education and training are diffused. There aren't enough policies in place in many developing nations to encourage the development of AI ecosystems (UNESCO, 2019). Government initiatives are quite small and still in their infancy concerning AI breakthroughs in education, with the majority coming from the commercial sector.

Other difficulties in using AI as reported by Akinwalere and Ivanov (2022) include unfavourable consequences, a lack of comprehensiveness, inaccurate conclusions, and implementation issues. When used in diverse contexts, AI frequently lacks portability because it was built for a specific purpose. Additionally, during the knowledge engineering process, it might not be possible to gather all the facts related to a problem, which frequently leads to inadequate analysis of the problem and undesirable solutions. Furthermore, if systems are unable to successfully construct data patterns, a lack of precision may make it impossible to establish correlations between variables during system development.

The replacement of the humanistic value chain by a technology-driven utilitarian or pragmatic approach was listed as one of the issues faced by users in the usage of

AI in education by Gocen and Aydemir (2020) in their research on AI in education and schools. The establishment of social connections in a learning environment as a result of the humanistic value chain led to the development of solid communication channels that are useful for knowledge sharing in teaching and learning. Eliminating the humanistic value chain frequently leads to uncontrolled intelligence that puts users and policymakers in difficult ethical situations.

While Chat GPT has been acknowledged to improve the productivity and quality of students in writing essays and examinations as well as improving productivity for editors through assisting in correcting grammatical errors and reducing editors' biases, Lo (2023) and Lund et al.,(2023) reports that among AI researchers, there are concerns about AI-assisted cheating and this has resulted in some schools and universities blocking or banning students' access to ChatGPT to prevent them from outsourcing written assignments and examinations. The use of ChatGPT in education has threatened the profession by compromising academic integrity, unequal access to AI technologies, bias and fairness, emotional manipulation, unintended consequences and misuses, ignoring the context in generating the content, perpetuating gender, racial and tribal discrimination and declining high-order cognitive skills (Farrokhnia et al, 2023; Ray, 2023). Additionally, some universities have started to use software tools that detect AI-generated essays. Although it is generally agreed that AI has become a facilitator in education and training and that it will forever be vital for the future workforce, in most African countries and other developing communities across the globe, AI initiatives are often foreign and therefore lack contextualized relevance because according to Ayanwale et al.,(2022), these initiatives tend to ignore cultural and infrastructural challenges of communities or countries of intended users and beneficiaries.

IMPACT OF AI ON EDUCATION

Global adoption of AI-based instruction and management of the learning environment has created limitless prospects and had a significant impact on curriculum and general education. These effects can be explained historically. Modern learners are more adaptable than traditional learners from decades before. According to Kengam (2020), modern learners frequently spend a lot of time online and on their smartphones, which creates a favourable environment for the use of AI. In contrast to the inflexible traditional atmosphere, a more flexible learning environment is developed where students can choose when and what to learn. Automating processes that formerly required human labour is how AI systems work. The transfer of labour is the effect, as shown by historical statistics (Ilkka, 2018). In addition to organizations reengineering their business processes and integrating AI into their core business operations, this transformation necessitates that employees develop new skills to fit the new working conditions. Additionally, inclusive AI technologies have assisted those special needs children who previously frequently found themselves out of school due to a lack of educational support services in the school. For the benefit of the physically disabled, these devices provide voice recognition and speech recognition capabilities.

In the past, students have relied on the teacher to instruct them on the material and activities found in the curriculum, making them passive students. The teacher, who in some circumstances was unable to recognize the needs of individual students in big groups of students, frequently determined their learning rate and individual needs. However, Gocen and Aydemir (2020) assert that the usage of AI has made it possible for students to learn at their own pace and for teachers to recognize the needs of specific students because instruction is personalized.

Like any organization that processes data in some way, schools deal with massive data, which presents difficulties for institutions that have long-standing manual filing and processing methods. Making decisions has also been made more difficult, and information is frequently inaccurate. However, according to Gocen and Aydemir (2020), the deployment of AI-based solutions minimizes paperwork in schools, hence facilitating labour through automation. These systems frequently have powerful computing capabilities to handle large, complex data sets, assisting in the planning of instruction that is tailored to the needs of each learner, choosing instructional strategies and subject matter using learning analytics, and producing efficient instruction and learning processes that are advantageous to policymakers as well. The computer power of people is frequently insufficient for handling massive data collections.AI added to these capabilities has enhanced student evaluation and grading.

The traditional functions of a teacher have undergone various changes thanks to AI. According to Pence's 2019 evaluation, social robots, intelligent tutors, AI helpers, and open online learning systems all have the potential to take the position of human teachers. But the new function poses fresh difficulties for the educational system. According to Celik et al., (2022), instructors have had a variety of roles in the creation and application of AI-based learning solutions in the educational setting. Some of these roles include:

a) Being role models for AI education in which instructors serve as data sources for systems;
b) Providing data about professional development to AI systems ;
c) Supplying student data and behaviour to AI algorithms;
d) Providing data on student behaviours and knowledge to AI algorithms;
e) Establishing the evaluation standards for learners;

f)   presenting pedagogical steerage for the choice of gaining knowledge of substances; and.

g)   Providing feedback on technical issues.

CONCLUSION

The field of artificial intelligence and its applications in education has developed to encompass the teaching, learning, research, and general administration fields across all levels of education and training. Teachers, students, and school administrators have benefited from using these technologies in the classroom, which has led to wider adoption and use of these tools through collaborative learning and research, social robots, sentiment analysis, automated grading, student assessment, and performance prediction. The marketing of educational programs, the creation of curricula, the enrolment of students, the monitoring of their progress, the delivery of instruction, and the performance of both teachers and students have all improved as a result. Despite these advantages, the deployment of AI has encountered difficulties including a lack of financing, a lack of trust, a lack of experience, a lack of infrastructure, and general reluctance from potential users due to concerns about the technology's dependability and job security.

WAY FORWARD

In general, literature acknowledges the significant contribution AI has made to education while also acknowledging the difficulties that institutions and people encounter when attempting to adopt or apply AI in education. Nations and educational institutions must create ICT-supportive policies for AI-backed instruction to overcome these issues and fully reap the benefits of these technologies. The implementation of these models of instruction is still sporadic in certain countries, thus this needs to be backed up by considerable studies on AI in education. Previous and current research, on the other hand, appears to have fallen short in solving the technologically-driven instruction challenges. The dependability and portability of AI technologies must be improved, as this would boost user confidence and minimize their apprehension about using these tools. This is the other difficulty that AI research must tackle. Furthermore, while teacher-training institutions continue to use conventional teaching methods, it would not be beneficial to develop AI educational systems. Through the application of AI-based instruction and the integration of AI into the training curriculum, change must start at institutions that prepare teachers. This will also be extended to schools so that instructors can use A-instruction. This has to be supplemented with instruction in AI for students in high schools and higher education settings like colleges and universities.

Ethical issues related to the use of AI technologies call for a paradigm shift in student assessment and evaluation in higher education institutions. While higher education institutions have lamented declining levels of integrity in students' essays, the use of assessment methods such as short answer questions and multiple choice questions provides an opportunity for assessors to address the challenges of AI-generated essays. In addition, the use of multiple choice questions may go beyond the testing of rote facts but also provides instructors' capabilities to measure higher-order cognitive abilities.

REFERENCES

Ahmad, K., Iqbal, W., El-Hassan, A., Qadir, J., Benhaddou, D., Ayyash, M ., & Al-Fuqaha, A. (2021). *Artificial Intelligence in Education: A Panoramic Review.* Academic Press.

Akinwalere, S. N., & Ivanov, V. (2022). Artificial intelligence in higher education and opportunities. *Boarder Crosssing*, *12*(1), 1–15. doi:10.33182/bc.v12i1.2015

Altrabsheh, N. (2016). *Sentiment analysis and students' real-time feedback* [Doctor of Philosophy Thesis]. The University of Portsmouth.

Ayanwale, M. A., Sanusi, I. T., Adelana, O. P., Aruleba, K. D., & Oyelere, S. S. (2022). Teachers' Readiness and Intention to teach artificial intelligence in Schools. *Computers and Education: Artificial Intelligence*, *3*, 100099. doi:10.1016/j.caeai.2022.100099

Baker, R. S. (2021). *Artificial intelligence and Bringing it all together. OECD Digital Education Outlook: Pushing the frontier with Artificial intelligence, Blockchain and robots.* IECD.

Barik, L., Barukab, O., & Ahmed, A. A. (2020). Employing artificial intelligence techniques for student Performance evaluation and teaching strategy enrichment: An innovative approach. *International Journal of Advanced and Applied Sciences*, *7*(11), 10–24. doi:10.21833/ijaas.2020.11.002

Barron-Estrada, M., Zatarain-Cabada, & Oramas-Bustillos, R. (2019). Emotion recognition for education using sentiment analysis. *Research in Computing Science, 148*(5).

Belpaeme, T., Kennedy, J., Ramachandran, A., Scassellati, B., & Tanaka, F. (2018). Social robots for education: *A review. Science Robotics*, *3*(21), eaat5954. Advance online publication. doi:10.1126cirobotics.aat5954 PMID:33141719

Beqiri, R. (2016). *Artificial intelligence Architecture*. Longread.

Celik, I., Dindar, M., Muukkonen, H., & Järvelä, S. (2022). The Promises and Challenges of Artificial Intelligence for Teachers: A Systematic Review of Research. *TechTrends*, *66*(4), 616–630. doi:10.100711528-022-00715-y

Chen, X., Zou, D., Xie, H., Cheng, G., & Liu, C. (2022). Two Decades of Artificial Intelligence in Education: Contributors, Collaborations, Research Topics, Challenges, and Future Directions. *Journal of Educational Technology & Society*, *25*(1), 28–47.

Cheng, Y. (2021). *Improving students' academic performance with artificial intelligence and Semantic Technologies*. The Australian National University.

Crompton, H., & Song, D. (2021). The potential of artificial intelligence in higher education. *Revista Virtual Universidad Católica Del Norte*, *62*(62), 1–4. doi:10.35575/rvucn.n62a1

Devi, J. V., Yamini, U. M., Sanjuri, K., & Virinchi, G. (2022). Traditional sentiment and emotion identification system to improve teaching and learning. *Journal of Engineering Sciences*, *13*(5).

Dewi, H. K., Rahim, N. A., Putri, R. F., Wardani, T. I., & Pandin, M. G. R. (2021). *The use of artificial intelligence in English learning among university students: A case study in English Department*. Univerrsitas Airlagga.

Diebold, G., & Han, C. (2022). *How Artificial intelligence can improve K-12 Education in the United States*. Center for Data Innovations.

El-Keiey, S., ElMenshawy, D., & Hassanein, E. (2022). Student's Performance Prediction based on Personality Traits and Intelligence Quotient using Machine Learning. *International Journal of Advanced Computer Science and Applications*, *13*(9). Advance online publication. doi:10.14569/IJACSA.2022.0130934

Farrokhnia, M., Banihashem, S. K., Noroozi, O., & Wals, A. (2023). A SWOT analysis of ChatGPT: Implications for educational practice and research. *Innovations in Education and Teaching International*, 1–15. Advance online publication. doi:10.1080/14703297.2023.2195846

Gardner, J., O'Leary, M., & Yuan, L. (2021). Artificial intelligence in educational assessment: Breakthrough? Or buncombe and ballyhoo? Journal of Computer Assisted Learning. doi:10.1111/jcal.12577

Gocen, A & Aydemir, F. (2020). AI in education and schools. *Research on Education and Media, 12*(1).

González-Calatayud, V., Prendes-Espinosa, P., & Roig-Vila, R. (2021). Artificial Intelligence for Student Assessment: *A Systematic Review. Applied Sciences (Basel, Switzerland), 11*(12), 5467. doi:10.3390/app11125467

Halaweh, M. (2023). ChatGPT in education: Strategies for responsible implementation. *Contemporary Educational Technology, 15*(2), ep421. doi:10.30935/cedtech/13036

Haugeland, J. (Ed.). (1985). *Artificial intelligence.The very idea.* MIT Press.

Ilkka, T. (2018). *The impact of AI on learning, teaching and education: policies for the future. JRS Science for Policy Report.* European Commission.

Jones, M. (1985). Application of artificial intelligence within education. *Computers & Mathematics with Applications (Oxford, England), 11*(5), 517–526. doi:10.1016/0898-1221(85)90054-9

Kengam, J. (2020). *Artificial intelligence in education.* Researchgate.

Khan, I. M., Ahmad, A. R., Jabeur, N & Mahdi, N. (2021) An Artificial intelligence approach to monitoring student performance and devise preventive measures. *Smart Learning Environments, 17.*

Khemani, D. (2013). *A First Course in Artificial Intelligence.* McGraw Hill Education.

Koedinger, K. R., Brunskill, E., Baker, R. S. J. D., McLaughlin, E. A., & Stamper, J. (2013). *New Potentials for Data-Driven Intelligent Tutoring System Development and Optimization.* Association for the Advancement of Artificial Intelligence. doi:10.1609/aimag.v34i3.2484

Kumar, V. (2018). *Role of artificial intelligence in education systems, its challenges and opportunities in India.* Academic Press.

Kurzweil, R. (1990). *The age of intelligent machines.* MIT Press.

Lo, C. K. (2023). *What is the impact of ChatGPT on education? A rapid review of the literature.* Education Sciences & MDPI. doi:10.3390/educsci13040410

Lund, B. D., Wang, T., Mannuru, N. R., Nie, B., Shimray, S., & Wang, Z. (2023). ChatGPT and a New Academic Reality: AI-Written Research Papers and the Ethics of the Large Language Models in Scholarly Publishing. *Journal of the Association for Information Science and Technology.* Advance online publication. doi:10.1002/asi.24750

Mueller, J. P., & Massaron, L. (2018). *Artificial intelligence for dummies*. John Wiley & Sons.

Murphy, R. F. (2019). *Artificial intelligence application to suppirtK-12 teachers and teaching, A review of promising applications, opportunities and challenges. Perspectives, Expert insights on tertiary timely*. Rand Corporation.

Negnevitsky, M. (2005). *Artificial intelligence: a guide to intelligent systems* (2nd ed.). Addison-Wesley.

Ngoc, H. V., & Thi, M. N. (2021). Sentiment analysis of students' review on online course. A transfer learning method. *Proceedings of the International Conference on Industrial Engineering and Operations Management*.

Nilsson, N. J. (1998). *Artificial intelligence: A new synthesis*. Morgan Kaufman.

Nunez, J. M., & Lantada, A. D. (2020). Artificial Intelligence Aided Engineering Education: State of the Art, Potentials and Challenges. *International Journal of Engineering Education, 36*(6), 1740–1751.

OECED. (2020). *Trustworthy Ai in Education: Promises and Challenges*. Background paper for the G20 AI Dialogue, Digital Economy Task Force Meeting. Riyadh, Saudi Arabia.

Ouyang, F., & Jiao, P. (2021). *Artificial intelligence in education: The three paradigms. Computers and Education: Artificial intelligence*. Elsevier.

Pence, H. E. (2019). Artificial intelligence in education: New wine in old wineskins. *Journal of Educational Technology Systems, 48*(1), 5–13. doi:10.1177/0047239519865577

Poole, D., Mackworth, A. K., & Goebel, R. (1998). *Computational intelligence. A logical approach*. Oxford University Press.

Rashid, T. A., & Aziz, N. K. (2016). *Student academic performance using artificial Intelligence*. ZANCO Jou.

Rich, E., & Knight, K. (1991). *Artificial intelligence* (2nd ed.). McGraw Hill.

Sabzalieva, E., & Valentini, A. (2023). *ChatGPT and artificial intelligence in higher education. A quick start guide*. UNESCO.

Salas-Pilco, S. Z., Xiao, K., & Oshima, J. (2022). Artificial intelligence and new technologies in inclusive education for minority students: A Systematic Review. *Sustainability (Basel), 14*(13572), 13572. Advance online publication. doi:10.3390u142013572

Samin, H & Azim,T. (2019). Knowledge-based recommender systems for academia using machine learning: a case study on Higher Education landscape of Pakistan. *IEEE. Research Gate.*

Seo, K., Tang, J., Roll, I., Fels, S., & Yoon, D. (2021). The impact of artificial intelligence on learner-instructor interaction in online learning. *International Journal of Educational Technology in Higher Education*, *18*(1), 54. doi:10.118641239-021-00292-9 PMID:34778540

Swiecki, Z., Khosravi, H., Chen, G., Martinez-Maldonado, R., Lodge, J. M., Milligan, S., Selwyn, N., & Gašević, D. (2022). Assessment in the age of artificial intelligence. *Computers and Education: Artificial Intelligence*, *3*, 100075. doi:10.1016/j.caeai.2022.100075

Tan, S. C., Lee, A. V. Y., & Lee, M. (2022). A systematic review of artificial intelligence techniques for collaborative learning over the past two decades. *Computers and Education. Artificial Intelligence*, *3*, 1000097.

UNESCO. (2019). Artificial intelligence in Education: challenges and opportunities for sustainable development. In *Development Working Papers On Education Policy*. United Nations.

Winston, P. H. (1992). *Artificial Intelligence* (3rd ed.). Addison-Wesley Publishing Company.

Woolf, B.P., Lane, C., Chaudhri, V. K., & Kolodner, J. L. (2013). Artificial Intelligent Grand Challenges for Education. *AI Magazine, 10*(13).

Zawacki-Richter, O., Marín, V. I., Bond, M., & Gouverneur, F. (2019). A systematic review of research on artificial intelligence applications in higher education – where are the educators? *International Journal of Educational Technology in Higher Education*, *2019*(16), 39. doi:10.118641239-019-0171-0

Zhai, X. (2023). *ChatGPT user experience: Implications for higher education. National Science Foundation*. NSF.

Chapter 3
Research on Artificial Intelligence Risk Prevention and Control System for Smart Libraries

Xu Wang

https://orcid.org/0000-0002-8789-7951
Yanshan University, China

Binbin Liu
Yanshan University, China

ABSTRACT

With the rapid development of artificial intelligence (AI) science and technology, people's demand for library information services and risk prevention and control construction has increased in both directions, affecting profound changes in the construction of libraries in various countries. Artificial intelligence is reshaping the management model and service model of smart libraries, and libraries must pay attention to risk prevention and control while welcoming the model change. This chapter is based on the background of artificial intelligence era. To achieve the deep synergy between artificial intelligence and smart libraries, it combines risk management and multiple synergy theory, and it proposes that the types of AI risks in smart libraries are divided into their own risks and application risks. This chapter constructs the AI risk prevention and control system for smart libraries and proposes feasible AI risk prevention and control countermeasures for smart libraries.

DOI: 10.4018/978-1-6684-8671-9.ch003

INTRODUCTION

Smart library is another important concept and development model that emerged in the library field after digital library, and the related concept and practice have been around for 20 years (Wang, 2011). A smart library is a concept that is not limited by space but can be perceived realistically (Aittola et al., 2003). Smart libraries refer to a kind of intelligent building formed by applying intelligent technology to library construction (Cox et al., 2019). It is the organic combination and innovation of intelligent building and highly automated management of digital library. The healthy and sustainable development of smart libraries is inseparable from the in-depth application of artificial intelligence. Technology is a double-edged sword, and its risks cannot be ignored. Artificial intelligence technology should be treated dialectically. Artificial intelligence is still in the stage of weak artificial intelligence. Even if it develops to a higher stage, it cannot be arbitrarily evolved from human control (Hussain, 2023). Artificial intelligence technology is changing the way people live, work and learn, and even subverting the original behavior habits and thinking patterns (Sanji et al., 2022). Artificial intelligence not only brings opportunities to libraries and even human development, but also brings more severe challenges. Smart libraries make use of artificial intelligence to create convenience for users, but their risks must be controlled within a reasonable range to achieve library-to-library collaboration, people-to-library collaboration, and everyone collaboration. Smart libraries are the development direction of the library in the new era. Relying on artificial intelligence technology, it continuously strengthens its resource management advantages and service position. AI can foster intelligent decisions for retrieving and sharing information for learning and research. With the help of artificial intelligence technology, the smart library can infinitely perceive the needs of users, make knowledge more accessible, and provide more intelligent and humanized services (Okunlaya et al., 2022). While we are enjoying the benefits that artificial intelligence brings to smart libraries, librarians should maintain an objective and calm attitude to meet the challenges of artificial intelligence.

The scientific development of smart libraries is based on the advantages of artificial intelligence technology and effective control of risks. The study of AI risks will promote the collaborative and sustainable development of multiple subjects of smart libraries from the micro level, and promote the scientific application of AI in the construction of smart cities and smart societies from the macro level to maintain public security and social stability. Artificial intelligence technology liberates the heavy work of "people", improves user service experience, accelerates the dissemination, organization and management of knowledge, promotes the high level of development of smart libraries, and becomes an important indicator for measuring the quality of library development. Before enjoying the benefits created by AI, we should anticipate the possible negative impacts on smart libraries. In the face of AI challenges, i.e., we should not let it go or choke on it, but should actively take effective risk control measures to maximize the advantages of AI. In order to realize the deep synergy between AI and smart libraries, this chapter analyzes AI risks, combines risk management and multiple synergy theories, builds a risk prevention and control system adapted to the construction and development of smart libraries, and proposes AI risk prevention and control countermeasures for smart libraries from the technical and management levels to provide effective guidance for the high-quality development of AI and smart libraries nowadays.

SMART LIBRARIES AND ARTIFICIAL INTELLIGENCE

Research on Smart Libraries and Artificial Intelligence

Artificial intelligence has become a key technology for building a smart earth, digital China and smart cities, greatly enhancing human's ability to understand and transform the world and promoting the development of various fields towards wisdom, and libraries are no exception. Smart libraries have been and continue to be a research hotspot in the field of graphical intelligence. Wang (2012, 2017a) summarizes the three main characteristics and discusses the five major relationships between connected, convenient, and efficient smart libraries. Zeng and Song (2017), and Zeng and Qin (2018) embed SoLoMo and decentralization ideas into smart libraries' mobile visual search services and management system. To enhance the wisdom of smart libraries, some scholars explore technological innovation such as Internet of Things (Zhou and Yang, 2018), cloud computing (Liu, 2018), RFID (Wang and Wang, 2013), wearable technology (Lu, 2016), blockchain (Chen, 2018), and situational awareness (Yao et al., 2017) in smart library services to accelerate the synergistic development of artificial intelligence and libraries. Stepping into the pan-information society, Dai (2018) analyzes its impact on smart library services and constructs a service system that adapts to the current environment. Chen and Xu (2015) analyze the characteristics of smart libraries, propose a hierarchical model of user-oriented ubiquitous smart services, and discuss the key issues in the construction process. Chu and Duan (2018) argue that technologies such as the Internet of Things and artificial intelligence drive libraries from physical and digital libraries to smart libraries, and that smart libraries are the result of the interaction and integration of smart technology, smart librarians, and library business and management. Wang (2017b) studies the wave of change brought by artificial intelligence and the reinvention of library services, artificial intelligence and the reshaping of library literature resources, human resources, reader-users, service space, and service programs; he believes that AI and library renewal will give a new vitality to the library growth organism, and should uphold the value pursuit of people-oriented, service-oriented, and intelligent for good, and further promote AI and library renewal to achieve a higher level of public cultural service system (Wang, 2019). Su (2018) constructs a framework system of library intelligent services from the perspective of big data and artificial intelligence dual drive, and points out the development dilemmas and strategies of library intelligent services. Fu et al. (2018) use literature analysis method combined with the example method and empirical summary method to review AI technology in the last three decades, and explore its application in library information retrieval, library information classification, and library information management. physical and electronic resources acquisition subscription management and library automation. Wang and Wang (2018) proposed a five-factor model of "human-thing-thing-field-time" for library intelligent space reengineering by combining the application of AI technology in space reengineering and services in various fields of society, and advocated the implementation of various AI service scenarios in intelligent intelligence, intelligent retrieval, human-computer interaction, and knowledge consultation. Fu (2018) studies the current situation of AI application in modern library construction and analyzes the application of AI technology in intelligent library services from resource integration, intelligent retrieval, service model, system analysis and management, and security control of network, etc. Zhu (2019) analyzes only the risks brought by AI application in libraries based on AI theory, and provides an effective response to information leaks, cyber-attacks or the risk of spreading false/malicious information, a Bayesian-based security defense model is proposed. In summary, the existing research on smart libraries mainly focuses on technology innovation and service exploration, and less involves the library AI risk scientific prevention and control.

Research on Collaborative Governance of Smart Libraries and Artificial Intelligence Risk Management

Since the German physicist Hermann (2005) proposed "synergy", synergy theory has been commonly used to study complex systems coordinating multiple subjects from disorderly chaos to orderly synergy. Currently, the theory of multiple synergies has been used in the study of complex systems such as urban public services (Qi, 2013), air pollution control (Yang and Zhang, 2013), urban risk prevention and management (Yao and Tang, 2018), agricultural services (Liu and Gao, 2017), and precise poverty alleviation (Xia, 2018). Synergy theory also drives innovation in the library field. Liu et al. (2017) argue that there is a fit between synergy theory and the complexity, openness, non-linearity and non-equilibrium of library social cooperation, and analyze the dynamics mechanism of library social cooperation;

Wang and Xu (2017) propose library collaborative ecology, build library intelligent service system, and explore its key issues; Wang (2014) explores the possibility and necessity of introducing synergy theory into library information resource services and constructs a library information resource co-creation and sharing service system. Tang and Xie (2013) studied embedded services in university libraries based on synergy theory, analyzed the synergy and supporting technologies of goals, subjects, resources, and processes of embedded services; Ma and Zheng (2015) proposed a clustered general branch library model based on synergy theory and provided guarantees from legislation, system, talents, and technology. Wang et al. (2018) applied the synergy theory to the process control of smart library services. The synergy theory regards both nature and human society as complex systems, and there are two states of order and disorder in the process of development and evolution, and the invisible hand makes each element reach synergistic order and play an overall effect (Hermann, 2005). In the era of artificial intelligence, the technology risks faced by smart libraries and their multiple participating subjects are more complex, with difficult risk management, wide impact and rapid spread, and no single subject is sufficient to cope with such a daunting challenge.

To this end, this paper introduces synergy theory, combined with risk management theory, to govern AI risks from the perspective of the overall system, analyze the main contradictions, avoid the interference of local problems of each subsystem, in order to bring into play the system's holistic effectiveness and guarantee the return of the intelligent library system from the chaotic and disorderly occurrence of risks to a synergistic and orderly equilibrium.

TYPES OF ARTIFICIAL INTELLIGENCE RISK IN SMART LIBRARIES

As early as the 1990s, Ulrich Beck (2004) introduced the concept of "risk society", arguing that technological development, while promoting social progress, poses a threat to the ecological environment and even to people themselves, and that in a risk society, risk has replaced material scarcity as a central social and political concern. Artificial intelligence accelerates social progress and technological innovation, while opportunities and risks coexist in a risky society. Technology is a double-edged sword, and the scientific development of smart libraries requires the support of AI technology and inevitably bears the risks it brings. Risk prevention and control is the response to the uncertainty of artificial intelligence. The risk prevention and control of artificial intelligence in smart libraries must first identify the risks of artificial intelligence. Combined with existing research, this paper divides AI risk in smart libraries into its own risk and application risk.

Self-Risk

From the perspective of intelligent libraries, their own risks include but are not limited to loss of control risk, miscalculation risk, algorithm bias, and attribution of responsibility and power. The risk of loss of control is that AI can lose control of AI due to algorithm vulnerabilities, software and hardware equipment failure, external attacks, and alienation; the risk of system miscalculation caused by design flaws in AI algorithms and incomplete parsing of problems; algorithm bias caused by the algorithm designer itself or by external attacks that lead to algorithm tampering, etc., which makes AI misperceive input events and thus mislead human behavior decisions. The responsibility for problems with artificial intelligence belongs to the manufacturer, designer or artificial intelligence itself, as well as the problem of using artificial intelligence to invent and create, and the sovereignty of its creation belongs to the artificial intelligence or the user or designer, i.e., the potential risk of unclear attribution of responsibility.

Application Risk

From the perspective of smart libraries, application risks include security risks, ethical risks, infringement risks, legal risks, unemployment risks, etc. Security risk refers to the threat of AI applications to data security, network security, public safety, etc. Artificial intelligence is maliciously manipulated by unscrupulous elements causing public safety hazards, and even creating and spreading bad information on its own endangering public order and the physical and mental health of youth; ethical risk refers to the technological alienation generated by artificial intelligence in smart libraries, leading to governance blind spots, such as the emergence of algorithms, gender, racial discrimination and other governance dilemmas of all moral and ethical issues (Liu, 2021). The debate about the ethical impact of AI dates back to the 1960s (Morley et al., 2020). Artificial intelligence provides differentiated services due to gender and race, which can easily raise social inequality issues, such as, ethical risks arising from racial discrimination, sexism, verbal and emotional attacks in intelligent question and answer; infringement risk refers to the possible risks for individuals in terms of contractual license duration, licensee, license method, license territory, license fee and abuse of copyright and patent rights through AI technology mining (Yan, 2022). There are risks of infringement of copyright and patent rights and other infringement risks in the collection and utilization of knowledge resources through AI creation; legal risks are the risks of potential financial losses or other damages occurring to those who are active in the library (librarians, librarians and patrons, etc.) (Zhang, 2022). Artificial intelligence automatically collects user data and sensitive data, recklessly exploits personal data, violates user privacy, etc. The starting point of artificial intelligence applications is to benefit oneself and others, but there are misuse or malicious control to make illegal and criminal activities. Such as robots attacking the public and users, or even spreading public opinion on the network to endanger public safety, etc.; unemployment risk is a risk that the progress of technology makes the work procedures of intelligent libraries gradually mature, and intelligent devices replace manual labor, thus leading to the unemployment of librarians.

Relationships Between Self-Risk and Application Risk

Artificial intelligence replaces a large number of tedious tasks for librarians, reducing the demand for librarians and creating the risk of unemployment. In view of the risks of AI technology itself and application, it needs to be scientifically and reasonably prevented and controlled by smart libraries. As can be seen from Figure 1, the application risk and its own risk are interlinked and transformed, and there is a close connection between the two. The first of the smart library application risk shifts from cyber data security risks to algorithmic ethical risks, i.e., from cyber subject crises to ethical security issues occurring among people and even races within the library. The ethical risk then shifts to the risk of infringement of writings, patents, etc. under the name of the author. When the infringement is conclusively criminalized, it jeopardizes the individual or even the national level depending

Figure 1. Types of artificial intelligence risk in smart libraries

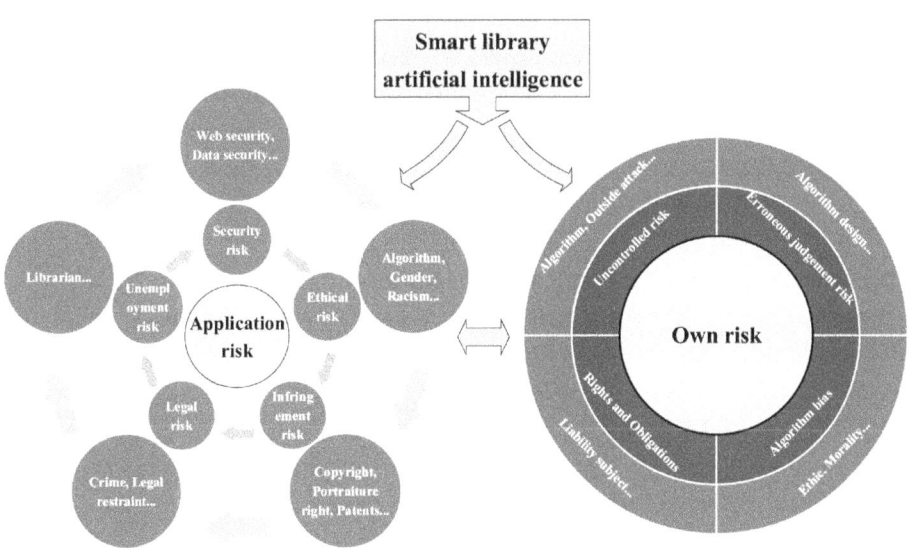

on the severity, which in turn turns to legal risks. AI technology replaces manual prevention, which may lead to the risk of unemployment for librarians.

SYSTEM OF ARTIFICIAL INTELLIGENCE RISK PREVENTION AND CONTROL FOR SMART LIBRARIES

The collaborative and orderly development of multiple subjects of smart libraries is based on the advantages of artificial intelligence, combined with risk identification, risk warning, risk governance, risk evaluation and evaluation of risk control effects to cope with AI itself and application risks according to the characteristics and development needs of smart libraries. AI technology risks in the risk society are symbiotic in nature. That is, the risks that arise in the overall system of smart libraries are equally present in the subsystems (Xie et al., 2019). The study of AI risk is equally applicable to connected libraries, museums, archives, research institutions, etc. of all kinds. We look at smart libraries from a system perspective, coordinate multi-subject subsystems to govern AI risks, take advantage of the system as a whole, and maintain the dynamic balance of AI and smart library development. The multiple subjects of smart libraries are divided into libraries, people (administrators, librarians and readers), museums, archives, scientific research institutions, and regulatory authorities. Artificial intelligence technologies include natural language processing, image recognition, speech recognition, deep learning, reinforcement learning, data mining, etc., and are used in intelligent Q&A, self-checkout, robotics, intelligent search, intelligent recommendation, etc. in smart libraries.

In this section, the AI risk prevention and control system of the smart library is drawn, as shown in Figure 2. Artificial intelligence risk prevention and control in smart libraries relies on visualization and processing means in artificial intelligence applications with algorithmic clustering and analysis techniques within artificial intelligence elements to handle. It not only helps readers to check and retrieve or assists librarians to summarize and organize books, but also can display risk types and risk sizes in the digital platform to achieve the purpose of analyzing and predicting and improving the efficiency of governing risks. First, in the intelligent library artificial intelligence risk prevention and control system, risk identification through artificial intelligence data mining technology, identify their own risk or application risk after the risk warning system alarm prompt. Secondly, multiple subjects of the smart library combine the two platform technologies of AI applications and AI elements to jointly participate in risk management activities. Finally, the smart library database serves as the basis for risk evaluation, while the new prevention and control data after evaluating risk management is imported into the database for storage and updating.

Figure 2. System of artificial intelligence risk prevention and control for smart libraries

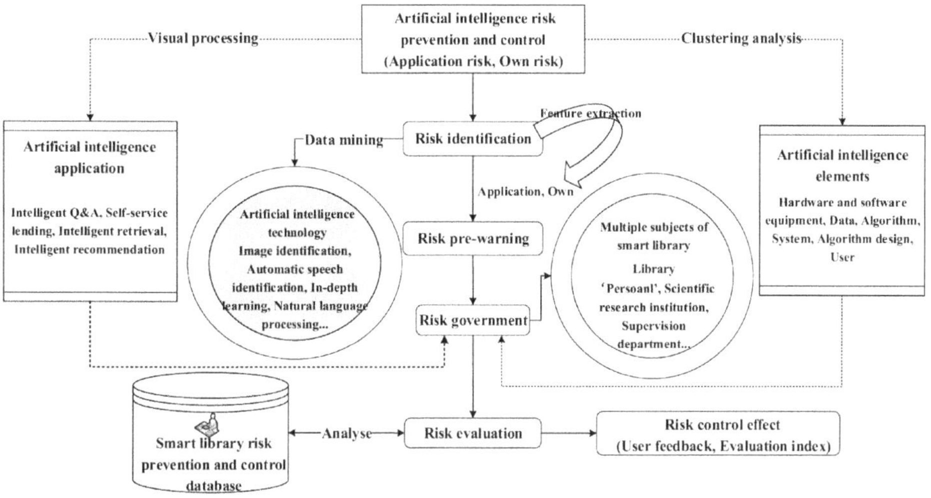

Risk Identification

The so-called risk identification is the process of risk managers conducting serious investigation and scientific research on the process of collecting and organizing, optimizing and integrating, storing and applying, and disseminating and innovating information in smart libraries in the process of providing various services to users, and applying different knowledge and methods to comprehensively identify and systematically categorize the risks and hazards that exist in smart libraries (Shan et al., 2021). Risk identification provides risk identification at the time of risk occurrence, not only the risks of early warning but also the potential associated risks. Risk identification is an important part of AI risk prevention and control. The risk identification process of the smart library is based on the risk signals provided by the risk prevention and control system and equipment by risk management personnel to perform risk inference, relying on AI data mining technology to determine whether the risk comes from AI itself or application risk. Specifically:

First, only by identifying risks can we carry out risk assessment, analyze the source and impact of risks, and then provide scientific and reasonable risk control solutions to reduce risk management costs; second, through risk identification, smart libraries can achieve effective awareness of risks, and ignoring the importance of risk identification can lead to errors in risk management and control programs, and mismatch between the direction of efforts and the source of the problem. In the actual intelligent library service scenario, the risk may not be a single risk, often multiple risks arise simultaneously or sequentially. The principle of risk identification is to analyze AI explicit and implicit risks as comprehensively as possible, and to analyze and mine risk sources and chain reactions.

Risk Warning

Risk warning is a process that analyzes the signs and symptoms of risks and provides emergency alerts. This program sets up warning conditions for factors that may generate risk, and when the warning equipment and system sends a signal to warn that the risk has been or will be generated. In the production of risk warning can be set in advance for the temperature, air pressure, and other thresholds, when the environment reaches the threshold, the system automatically alarm, which can provide guidelines for personnel control or system regulation. In the smart library system, the library should ensure that the risk warning system has strong intelligent analysis, judgment and decision-making capabilities, and can quickly process massive unstructured data and iterative growth data sets to achieve the transformation of traditional data batch processing to stream processing. In addition, the scientific and algorithmic efficiency of the risk warning model should be automatically optimized with the autonomous intelligent learning function of the decision system (Ma, 2016).

For example, the early warning system program sets access rights to the data according to the data security level assigned differently. Artificial intelligence extracts data within the library, and when data such as copyright, copyright, and patent rights are involved, the device automatically warns that the system can block the operation and return access control prompts. When sharing resources with some subjects, the scope of data use and access rights are affirmed and potential risks are disclosed. Risk warning is based on the prediction of risks, and can only provide early warning for predictable and uncertain risks, and cannot provide accurate warning for implied and unknown risks. If a virus invades the system, the type of virus and the way of damage are not known, but malicious attacks and system anomalies can be indicated by early warning devices. Risk warning is benign feedback for risk control and facilitates risk governance. The risk management process should summarize the knowable and unknown factors of risk generation and set early warning signals beforehand, which can prevent the occurrence of risks in time or control the expansion of the degree of risk impact to minimize the cost of loss.

Risk Governance

The risk warning mentioned above serves as a pre-emptive risk alert, and risk management countermeasures should be taken when AI risks are known to occur. However, the choice of risk management countermeasures needs to be judged according to the type of risk, and only when the type of risk is clear can appropriate control methods be adopted. After the risk warning prompt, the personnel of multiple subjects of smart libraries use AI applications and AI element tools for risk management.

Risk governance should provide dual guarantees from technical and management levels, and combine with risk evaluation to reverse the selection of appropriate control strategies. From the technical level, risk governance strengthens the security management of software and hardware devices to guarantee data security, algorithm security and system security; from the management level, it establishes risk management departments, improves risk management systems and cultivates a culture of risk control, etc. Artificial intelligence elements for AI design, production, put into use, destruction and other stages to set up multiple audit mechanisms to ensure the safety of all aspects of AI controllable; while accelerating the development of laws and standards to effectively restrain the design of algorithms, environmental deployment and system use, and clearly define the subject of responsibility, the sovereignty of the creation, etc. For the factors of different risk assessment levels, from the perspective of the overall system of smart libraries, strategies such as risk retention, risk transfer, risk outsourcing and collaborative governance can be adopted. Among these four strategies, risk retention, one is for the indicators of factors with lower risk evaluation levels, which can be handled and completed internally; the other is that the risk is caused by the library itself. Risk transfer means that the risk is caused by a third party, such as the designer or manufacturer, and that risk should be transferred to a third-party organization for control. Risk outsourcing is when the risk is beyond the management capacity of multiple subjects of smart libraries, and the risk is outsourced to an external professional risk management agency without causing the risk to deteriorate. Collaborative governance, i.e., the smart library and other organizations jointly carry out risk control. For complex risks, such as ethical risks, legal risks, etc., collaborative control of multiple subjects is required; for example, legal risks require the judicial department to clearly define the illegal behaviors and standard norms that exist in AI applications in library systems, while libraries need to cooperate with regulation and restraint, and also require AI designers to design AI system devices that meet the norms according to legal requirements and library needs.

Risk Assessment

Effective risk identification needs to be followed by measuring the hazards and impact of the risk, i.e., assessing the risk and providing decision support for risk control. As a huge library system, the time and energy of risk managers are limited. Risk evaluation should be orderly, measured and justified. First, risk evaluation grasps the primary and secondary aspects of risk, and attacks urgent risks in a targeted manner. At the same time, risk evaluation provides effective feedback for risk identification and serves as an effective basis for risk identification. Risk assessment needs to determine how much human and material resources are invested in risk control by multiple subjects of smart libraries. Second, risk evaluation is based on the quantitative assessment of risks, and appropriate risk control strategies can be adopted to guarantee the reasonable allocation and utilization of resources and maintain the synergy and order of the overall system of the smart library. Thirdly, risk evaluation can use the Delphi method to screen the AI risk factor indicators and obtain the risk factor indicators that conform to the known technology risk assessment; then use the hierarchical analysis method to construct the hierarchical structure of risk factors, get the importance of each factor through the judgment matrix, and calculate the weight of each indicator; finally, risk evaluation can be carried out with the help of fuzzy comprehensive evaluation method, and then get the comprehensive of different levels of indicators The risk level of different levels of indicators can be obtained.

In the evaluation stage, the main purpose is to evaluate the risk hazards identified in the preparation stage and assess the risk hazard indicators of the smart library in general accordingly. First, a list of relevant risk hazards is made, and data statistics on their rationality, completeness, practicality, security, etc. are conducted to analyze the more serious risk identification items; second, the evaluation is conducted from the operational application aspect, and the existence and frequency of risk hazards are mined from the operational data; third, some risk hazards that cannot be identified in the first two steps can be identified and evaluated by the risk assessment team members based on their previous Third, some of the risk hazards that cannot be identified in the first two steps can be identified and evaluated by risk assessment team members based on previous practical experience or risk assessment experience (Zhang, 2022).

Risk Control Effect

The risk evaluation is followed by an electronic assessment report, which in turn presents the effect of risk control and summarizes the effect and experience of risk prevention. Taking different risk prevention and control measures inevitably brings different results. Different risk prevention and control effects are generated to measure the reasonableness, timeliness and effectiveness of risk prevention and control measures, and to provide reference for the next risk control. Risk control effect evaluation indicators include user feedback (satisfaction, access, experience, etc.), risk control cost (manpower, time and opportunity, etc.), risk loss (funds, resources, user trust, etc.), etc. The excellent risk control effect eventually directly reflects the problems of risk identification, risk warning, risk evaluation, risk governance and other aspects, which helps to improve the construction of the risk prevention and control system and optimize the risk prevention and control strategy of the smart library. The risk control effect is a demonstration of the effect of the entire risk management process, effectively assisting risk warning to select monitoring indicators for reference and guiding the screening of risk identification, evaluation and other programs. Through the evaluation of risk control effect, it can not only provide a basis for risk managers to make decisions, but also form a virtuous cycle of risk management process and continuously strengthen the perception and prevention and control of artificial intelligence risks in smart libraries.

In addition, the synergistic development and deep integration of smart libraries and artificial intelligence are not achieved overnight, but rather in the prevention and control of risks while ensuring the gradual play of the advantages of artificial intelligence in the smart library system. It is in the process of dynamic balance between the advantages of AI technology and risk prevention and control that the healthy and stable development of the overall system of smart libraries is promoted.

COUNTERMEASURES OF ARTIFICIAL INTELLIGENCE RISKS PREVENTION AND CONTROL FOR SMART LIBRARIES

The IFLA Trend Report lists artificial intelligence as one of the four technological trends (IFLA FAIFE, 2020). AI technology supports library service innovation and opens up a new era of human-computer collaboration, multifaceted integration, sharing, and autonomous control of intelligent services in libraries, empowering libraries to innovate services and challenging their traditional rights, rights and responsibilities, and ethics (Lu et al., 2020). To strengthen the governance of AI technology risks in smart libraries, this paper provides risk prevention and control countermeasures at the technical and management levels to ensure the synergy and orderliness of multiple subjects in smart libraries.

Technical Level

Software and Hardware Equipment Security

The realization of artificial intelligence needs the support of software and hardware devices, and the insecurity of software and hardware devices brings hidden dangers to artificial intelligence. When applying artificial intelligence, smart libraries sense user needs wirelessly through software and hardware devices. If there are security problems with hardware and software devices, it will not only affect user experience, but also lead to leakage of user data and even be attacked by the outside world. Failure of hardware devices such as servers, routers, gateways, storage devices, wearable devices, etc. will directly affect the quality of library services. Equipment failures also allow lawless elements to steal or leak users' private data and destroy important documents and resources. At the same time, the failure of intelligent devices such as robots will threaten the personal safety of users.

The intelligent library is supervised by multiple subjects to ensure that the hardware and software devices are not damaged, and even if they are damaged, they can be recovered in time and effectively, and the risk sources can be diagnosed. Risk diagnosis can be manual diagnosis or system self-diagnosis. Artificial intelligence in the process of operation can also have errors in judgment or loss of code and data damage, which can affect the safe operation of artificial intelligence software and hardware equipment, causing threats to libraries and even users. To establish a safe and reliable device detection system, firstly, multiple firewalls with high-security level should be set up to regularly diagnose the security of devices; secondly, to ensure the security of network communication devices and provide digital encryption to guarantee the security, integrity and consistency of data transmission; and then, it is necessary to set up anomaly detection mechanism for software and hardware devices to detect external attacks and automatically reject abnormal access; finally, to timely update the devices to prevent various uncertain security risks caused by the aging of the equipment. Although the risk of artificial intelligence caused by the device itself and external causes is unavoidable, it is necessary to have recoverability and early warning functions when the software and hardware devices fail and attack, to prevent the security risks brought by artificial intelligence alienation to people.

Data Security

The normal application of artificial intelligence cannot be driven by data and algorithms. The fourth paradigm of data-driven science is changing the way library users conduct research and work. Effective operation of artificial intelligence requires safeguarding data from tampering or malicious attacks, which in turn does not affect the safe use of artificial intelligence. Types of data are classified as user data, open data, metadata, sensitive data, confidential data, etc., and present structured, unstructured, and semi-structured states. To meet the growing needs of users, multimodal data integration is accelerated. As the driving force of artificial intelligence, data security is an important basis for improving the "intelligence" level of artificial intelligence. The data comes from multiple subjects of intelligent libraries, and while ensuring that the data is reasonably shared and utilized, we pay attention to data security. On the one hand, it protects the security of collection resources to prevent users from infringing copyright, copyright, patent rights, etc. and malicious attacks from outside; on the other hand, it strengthens the protection of public security and users' private data to maintain the stability of public order and win users' trust. As a key institution for knowledge organization, management and dissemination, maintaining public security and user privacy is a manifestation of responsibility to society and users, as well as a driving force for the collaborative and orderly development of its multiple subjects. Data security guarantee should be implemented in all stages of data life cycle such as data generation and collection, transmission, storage, application and destruction; accelerate the preparation of data import interface specification, metadata management, open data management, data quality evaluation, sensitive data use and operation specification, clarify rights and obligations of all parties, and ensure manageable and controllable data; provide access control, security audit, security backup, abnormal behavior monitoring early warning, data encryption, data desensitization, and other data security technologies to establish a secure and controllable database (Zhang et al., 2018; Hui et al., 2023).

Algorithm Security

The safe operation of artificial intelligence is inseparable from the support of algorithms. Ensuring the safety of algorithms is the maintenance of copyright and intellectual property rights, but also the scientific management of artificial intelligence risks. Artificial intelligence is jointly driven by algorithms and data, and the risk arises to a large extent from the algorithm black box, which makes it difficult for users to understand the service characteristics of artificial intelligence. In the face of algorithmic black boxes, absolute transparency is unrealistic, and even transparency is relative. However, users have the right to demand algorithmic fairness and get rid of algorithmic bias, discrimination and other unethical and moral ills. Algorithmic transparency pursues a concise description of the algorithm, including its assumptions and limitations, logic, kinds, functions, designers, potential risks, significant changes, etc. (Xu, 2019). Algorithm transparency can have alternative tools such as algorithm filing, algorithm interpretation rights, etc. There should also be algorithm review, evaluation and testing, third-party supervision and other measures to ensure algorithm fairness. Clearly attribute responsibility for problems with AI, disclose design flaws and potential risks of algorithms, etc., so that users can effectively identify and provide early warning feedback. The use of algorithms has a complex interaction between them, and often they are often mixed in the actual application of AI, which makes the algorithms smarter and at the same time can become difficult to understand, especially the intention recognition, fuzzy association, inference judgment and other functions formed by semantic analysis algorithms in natural language processing, and the multi-neuron, distributed parallel computing, multi-layer depth of neural network algorithms in deep learning feedback adjustment, which exacerbates the complexity and opacity of the algorithms (Zhang and Ma, 2021). While artificial intelligence applications are convenient, the design, invocation, and distribution of algorithms should be secure and trustworthy. This is not only responsible for the user, but also for the security and stability of society and even the country.

In the era of artificial intelligence, the design of large-scale algorithms cannot be flawless, and phenomena such as inequality brought by algorithmic discrimination and injustice brought by algorithmic bias all belong to the category of algorithmic defects, which refers to the lack or incompleteness of the algorithm itself, and it is the main manifestation of algorithmic non-robustness (Li, 2019). But errors should be avoided as much as possible; first, set up an algorithm fault tolerance mechanism to improve the fault tolerance of algorithms and reserve time for error correction; second, accelerate the development of standards to form a scientific, standardized, safe and trustworthy AI R&D environment, and strengthen the audit of algorithm design, invocation and update processes; third, conduct multiple testing and verification before deployment and use to avoid as much as possible algorithm bias and the security risks caused by algorithm vulnerabilities Fourthly, after being put into use, it cannot be separated from supervision, because potential risks will be gradually exposed in the process of use, or even be maliciously tampered with. Intelligent search engines, personalized recommendation and intelligent creation are very likely to cause infringement problems. If there is no clear legal provisions and effective preventive measures, it will lead to disorder in libraries and magnify the negative effects of AI technology. At this point, improving the moral education and ethics of AI algorithm designers and developers is fundamental to ensuring the safety of algorithms. Avoid algorithmic bias and eliminate ethical problems in the application process as much as possible.

System Security

System security means resisting external attacks and creating a good experience for users. System design is a complex project and an organic combination of multiple algorithms and platforms. Artificial intelligence construction cannot focus on user experience and ignore system security. Strengthening system security can even sacrifice a certain response time. The interaction between systems should be secure and trustworthy, and a feedback mechanism can be established to ensure timely feedback in case of system failure. On the one hand, the fault tolerance of the system should be strengthened, and the design, deployment and use of the system need to practice industry standards and fulfill legal norms; on the other hand, the system can be warned in time when it fails, allowing time for adjustment and repair and preventing the deterioration of risks. The system needs to go through multiple audits and checks before deployment, and even any minor failures and user feedback in the process of putting it into use must find the root cause of the problem, providing reference for troubleshooting and risk prevention. The system should also be put into use to strengthen security precautions, such as user authentication, program authentication, regular audit of algorithms, authorized access, risk disclosure, etc., and provide instructions for the safe use of users. The development of artificial intelligence should implement the human-centered thinking, with the aim of avoiding harm and benefiting human beings, and improving the overall service level of intelligent libraries.

Management Level

Establish the Risk Management Departments

The vast majority of libraries have not yet established a dedicated security risk management organization or department, and when faced with a crisis event, they only manage it for a short period in response to an unexpected event. And according to the requirements of crisis early warning mechanism, it is necessary to build an organizational structure, establish an early warning management system, and form a kind of unit-wide crisis information atmosphere in the process of dealing with crisis within the organization to ensure the smooth implementation of crisis early warning work (Fei, 2021). Smart libraries are new models of multifaceted collaboration, interconnected efficiency, security and trustworthiness of libraries, and artificial intelligence technology has led to the reconfiguration of their management and service models. Due to the technical limitations and single organizational structure of each subject, it is difficult to cope with the risks brought by AI in a complex environment. The establishment of a risk management department is not only to identify and monitor risks, but also to be able to effectively control and assess them when they occur. The function of the risk management department is to monitor, identify, analyze, warn and manage risks, etc., and collaborate with all participating subjects to improve the efficiency and quality of risk prevention and control; the AI developers should be certified and approved to review the safety information related to AI products, such as source code, training data sets, and test results (Wei and Lu, 2018). The risk management department assumes the responsibility of supervising readers' personal information, social data, and behavioral data acquired by AI to avoid data theft or leakage due to external attacks and product iterations, and to strengthen network equipment supervision and upgrade security testing. The establishment of a risk management department can promote information exchange among multiple subjects, improve the efficiency of risk prevention and control governance of smart libraries, and avoid the deterioration and spread of risks. With the development of technologies such as big data, cloud computing and blockchain, smart libraries are bound to face uncertainty risks brought about by more complex technologies, and the types of risks are more numerous and hidden. Therefore, the risk management department is indispensable in the process of construction and development of smart libraries.

In addition, the establishment of a library risk management department should be integrated into the organizational structure and management model of a smart library. The application of artificial intelligence improves the depth and breadth of smart library services. However, the upgrade of technology also makes risk evolution iterations more complex, which requires professional departmental staff for risk control. Risks are no longer limited to the surface, and sources of risk are more difficult to diagnose. Risk management department personnel are responsible for overseeing the design, manufacture, application, update and upgrade of AI, collaborating with multiple participating subjects to prevent and control the risks brought by AI, maintaining a coordinated and orderly balance, and shouldering the responsibility and mission of maintaining the security and stability of each link.

Improve the Risk Management Rules

The risk management system mainly includes information disclosure, power and responsibility allocation, wind control team, user feedback channel, etc. Information disclosure, i.e., the cautious attitude held by Smart Library towards artificial intelligence, is a responsible performance for users. Disclose the safety information of intelligent products, establish the safety responsibility system, and guarantee the traceability of risk sources and responsible subjects. Under the vision of pluralistic synergy, the participating subjects of smart libraries have equal rights and responsibilities. Ensure timely response in case of risk or AI failure to minimize the impact and loss caused by the risk. The implementation of accountability mechanisms, accountability is only a means, the purpose is to improve the organization members of the AI risk through accountability to the degree of importance.

Libraries should establish a risk control team for risk control, which is an emergency organization for risk prevention and control, and carry out risk identification, early warning, governance, and evaluation based on risk-causing factors, influencing factors, and user feedback (Yu, 2014). The risk control team should be familiar with the organizational structure of smart libraries and be able to coordinate with each other to respond quickly in the face of risk potential. When risks occur, managers are able to control the spread of risks from the management level according to the organizational management model among multiple subjects of the smart library; technical personnel collaborates with managers to effectively carry out risk monitoring and collaborate with third-party artificial intelligence research and development institutions to improve the efficiency of risk governance.

Establish a sound user feedback channel. Artificial intelligence risk prevention requires feedback assistance from users. The construction of smart libraries is all about serving users, and users' feedback is helpful for the optimization of risk management system. Users are not only the objects of smart library services, but also the subjects of multiple participation in smart library governance system. Users' feedback is crucial to improve the ability of smart libraries to govern AI risks. Smart libraries should provide convenient feedback channels for users, so that users are both the recipients of library services and the supervisors and promoters of AI risk prevention and smart library construction.

Improve the Risk Control Information Protection System

The digitalization of information has triggered disruptive changes in traditional libraries, which have gradually transformed from places for lending books and reading newspapers to computer Internet access centers. In this context, the mode of access, storage and utilization of information resources in traditional libraries has changed accordingly, which has greatly enriched the culture of society while also incubating many problems such as information overload, information explosion and information imbalance, which has triggered the collision of information privacy and information security issues on a larger scale (Liu, 2022). At a time when the contradiction between the development of library digital construction and patron privacy is highlighted, the public has reason to question how traceable patron information is in the library data storage system. What will be used for different categories of personal information such as patrons' registration information, borrowing information, and service information? How long will it be kept? What types of security protocols does the library implement on its digital platforms such as mobile apps, websites, etc.? Who has access rights to patron information? The effective protection of library information security depends heavily on librarians, users, and all staff involved in the library (Amini et al., 2020).

When collecting and storing readers' data, handling readers' data, and providing information services, libraries should not only raise librarians' awareness of data security, but also actively guide and raise readers' awareness of data risk prevention. The Public Library Law of the People's Republic of China suggests that public libraries should properly protect readers' personal information, borrowing information, and other information that may involve readers' privacy, and should not sell or otherwise illegally provide it to others. In the specific implementation process, the responsibility should be implemented to the person, for example, the library information security management team or the person in charge can be set up to strictly implement the protection of readers' personal information security through a dedicated special way. Therefore, the library should also guide the readers to pay attention to their own information security and strengthen the awareness of data leakage prevention in the information service multi-process.

Specifically, readers' awareness of data security can be cultivated by having security education specialists, holding regular security seminars, and sending data security attention letters to readers. The risk of reader information leakage may exist in the whole cycle of information processing, such as collection, storage, use and sharing of reader information, and risk regulation at the legal level depends on libraries actively fulfilling their personal information protection obligations. Constructing a trade-off structure for information security and handling with information security assessment as the core can help improve library security efficiency. Libraries can establish reader information protection committees to assess the security of reader information collection methods, content, and procedures, and initiate information security assessment procedures in any library service program. The information committee should have the authority to suspend programs where they believe personal information is not being handled properly until the program's level of information security meets standards or until the committee reasonably believes that the information security issues have been reasonably resolved.

Construct Risk Control Culture and Ethics Education

While improving the risk control system of intelligent libraries, focus on the construction of risk prevention and control culture. Improving the risk prevention and control ability of multiple subjects of smart books, and cultivating the risk prevention awareness of librarians and readers can effectively reduce the infringement of risk. The smart library strengthens the construction of risk prevention and control culture, enhances the sense of synergy of multiple subjects, and builds up its strength whether the risk occurs or not. Prevent and control the risk of artificial intelligence and guarantee the safe and stable development of smart libraries. The aim of providing safe, trustworthy, convenient and humanized services will become the tireless pursuit of library people. Smart libraries should accelerate the deep integration of AI, knowledge resources, devices and people in libraries; the development of AI should form a consensus of openness, sharing and responsibility, and create a fair and just, safe and reliable, collaborative and orderly environment for the development of smart libraries; establish a correct scientific development concept of AI, guide, regulate and restrain the organizations and participating organizations in AI R&D, design, manufacturing and operation subject behaviors (Jiang, 2019; Nugroho et al., 2023), practice the main theme of people-oriented and beneficial to human beings, and increase the depth and breadth of the integration of AI and smart libraries.

Libraries in various countries have historically focused on protecting patron privacy, and a significant portion of this can be traced back to ethical constraints.

The American Library Association (ALA) was the first association in the world to develop a code of ethics for libraries, and the ALA Code of Ethics for Librarians emphasizes professional values, professional ethical responsibilities in a changing information environment, and the mission of librarians to ensure the free flow of information and ideas, of which "Protecting the privacy of patrons" is one of the main provisions. The British Library Association Code of Professional Conduct also emphasizes that "librarians should not disclose patrons' information" (Sha, 2004). At present, there is no coordination between science and technology ethics education and the development of artificial intelligence technology in various countries, so it is necessary to strengthen the level of science and technology ethics education, enrich the content of science and technology ethics education, and focus on the teaching methods of science and technology ethics education (Wang et al., 2022). Secondly, the rational use of Internet computer technology is a key element in the observance of ethics in smart libraries (Onyancha, 2015). Smart libraries should provide ethical education on AI risk management to improve librarians' technical skills and risk prevention and control capabilities (Bradley, 2022). Strengthening the skills training of librarians and complying with relevant laws and regulations and ethical guidelines to ensure that the use of AI does not negatively affect society and humans is in line with the development needs of smart libraries. Only by improving the overall quality and technical level of management and service personnel from an educational perspective can we better adapt to the complex competitive environment of AI. The role of artificial intelligence is not to replace people, but to provide convenience to people so that they can have the time and energy to make more meaningful contributions. Regardless of how technology evolves, it is all about serving people. Smart libraries are user-driven and rely on AI technology to provide human services, but they should not put the cart before the horse and become slaves to technology. In addition, considering that AI-generated content brings information ethics issues, there is a need to promote interdisciplinary research with disciplines such as philosophy, law and sociology in the future to discuss a clear ethical view for AI-generated content and optimize information governance capabilities. In summary, only when the culture of risk control and ethical education is rooted in the diversified subjects of smart libraries can we guarantee that risks are prevented before they occur; when they occur, they are effectively controlled after identification and warning; and after they occur, we can summarize our experience and strengthen risk prevention and control.

CONCLUSION

Based on the analysis of AI risks in smart libraries, this paper constructs a risk prevention and control system and proposes risk control countermeasures from the technical and management levels. Only by controlling AI risks within a reasonable range can the healthy and stable development of smart libraries be guaranteed. Artificial intelligence is reshaping the management mode and service mode of smart libraries, and libraries must pay attention to risk prevention and control while welcoming the model change. Librarians should respond to risks with a positive attitude instead of deliberately avoiding them. Only by facing AI risks squarely and actively taking reasonable measures to control them can we create a safe and comfortable, stable and reliable library service environment for users. Smart libraries will surpass digital libraries as a new stage of library construction and become a new business mode of access to future libraries and a roadmap for library AI. Meanwhile, with smart libraries as a new research fulcrum, librarianship and library community can work together to create learning-to-use smart librarianship (Zhang et al., 2023). To promote high-quality development of smart libraries, we should not be an afterthought but a prognosticator, providing a set of practical risk prevention and control strategies in advance, so that we can deal with the risks regardless of whether they occur or not.

ACKNOWLEDGMENT

This research was supported by the Humanities and Social Science Research Project of Hebei Education Department (No. JCZX2023015), Project of Social Science Development of Hebei Province (No. 20232660) and the National Social Science Found Major Project of China (18ZDA325, 19ZDA348, 21&ZD336).

REFERENCES

Aittola, M., Ryhänen, T., & Ojala, T. (2003, September). SmartLibrary-Location-aware mobile library service. Academic Press.

Amini, M., Vakilimofrad, H., & Saberi, M. K. (2021). Human factors affecting information security in libraries. *The Bottom Line (New York, N.Y.)*, *34*(1), 45–67. doi:10.1108/BL-04-2020-0029

Bradley, F. (2022). Representation of libraries in artificial intelligence regulations and implications for ethics and practice. *Journal of the Australian Library and Information Association*, *71*(3), 189–200. doi:10.1080/24750158.2022.2101911

Chen, X. P. (2018). Application of Blockchain Technology in the Library Smart Services. *Modern Information*, *38*(11), 66–71.

Chen, Y., & Xu, L. (2015). The Construction of Smart Library Aimed at Ubiquitous Consumers' Smart Service. *Library Journal*, *34*(08), 4–9.

Chu, J. L., & Duan, M. Z. (2018). Smart Library and Smart Services. *Library Development*, (04), 85–90.

Cox, A. M., Pinfield, S., & Rutter, S. (2019). The intelligent library: Thought leaders' views on the likely impact of artificial intelligence on academic libraries. *Library Hi Tech*, *37*(3), 418–435. doi:10.1108/LHT-08-2018-0105

Dai, Y. (2018). Research on the Intelligent Service System of Library in the Ubiquitous Information Society. *Journal of Library Science*, *40*(09), 52–55.

Fei, J. (2021). Research on the Management Mode of Library Security Crisis Early-Warning Based on Competitive Intelligence. *New Century Library*, *42*(06), 45–49.

Fu, P., Zou, X. Z., & Wu, D., etal. (2018). Retrospective Analysis and Prediction: Artificial Intelligence and Its Applications in Libraries. *Tushu Qingbao Zhishi*, *35*(02), 50–60.

Fu, Y. X. (2018). Research on Application of Artificial Intelligence in Library Construction. *Library Work and Study*, *40*(09), 47–51.

Haken, H. (2005). *Secrets of Nature's Success Synergetics: the Teaching of Interaction*. Shanghai Translation Publishing House.

Hui, S. C., Kwok, M. Y., Kong, E. W. S., & Chiu, D. K. W. (2023). Information security and technical issues of cloud storage services: A qualitative study on university students in Hong Kong. *Library Hi Tech*. Advance online publication. doi:10.1108/LHT-11-2022-0533

Hussain, A. (2023). Use of artificial intelligence in the library services: prospects and challenges. Library Hi Tech News. doi:10.1108/LHTN-11-2022-0125

IFLA FAIFE. (2020). *IFLA Statement on Libraries and Artificial Intelligence*. https://repository.ifla.org/handle/123456789/1646

Jiang, J. (2019). The main purpose and principles of Artificial Intelligence ethics under the perspective of risk. *Information and Communications Technology and Policy*, *45*(06), 13–16.

Li, Y. K. (2019). *Research on Quality Improvement of Information System Based on Defect Analysis*. Dalian Maritime University.

Liu, G. R., Zhang, L., & Huang, X. Q. (2017). Research on the Dynamic Mechanism of Cooperation between Library and Society—Based on the Synergy Theory. *Library Work and Study*, *39*(07), 26–30.

Liu, H., & Gao, Z. Q. (2017). Construction of Agricultural Science and Technology Service System with Multi-agent Coordination. *Agricultural Engineering*, *7*(02), 146–149.

Liu, Y. (2018). Opinions on the Construction of Smart Libraries Based on Cloud Computing and the Internet of Things. *Journal of Library Science*, *40*(11), 124–127.

Liu, Y. (2021). Research on Data Ethical Risk Identification and Interest Coordination Strategy in Smart Library. *New Century Library*, *42*(12), 11–15.

Liu, Y. (2022). The Risk of Personal Information Breach of Library Readers and Its Scenario-Based Goverance. *Tushuguanxue Yanjiu*, *44*(06), 18–26.

Lu, K., Pan, X. Y., Liu, H., & Ren, B. B. (2020). Research on Ethics Norms of Library Smart Service from the Perspective of AI Governance. *The Library*, *48*(06), 21–28.

Lu, X. H. (2016). Research on Smart Library Management and Service Based on Wearable Technology. *Journal of Library Science*, *38*(10), 114–117.

Ma, X. T. (2016). Design of Library Risk Alarm System Based on Big Data Analysis. *Library Theory and Practice*, *38*(08), 81–84.

Ma, Y., & Zheng, J. M. (2015). Analysis of Main and Branch Library System Mode Based on Synergistic and Cluster Theory. *The Library*, *43*(07), 80–84.

Morley, J., Floridi, L., Kinsey, L., & Elhalal, A. (2020). From what to how: An initial review of publicly available AI ethics tools, methods and research to translate principles into practices. *Science and Engineering Ethics*, *26*(4), 2141–2168. doi:10.100711948-019-00165-5 PMID:31828533

Nugroho, P. A., Anna, N. E. V., & Ismail, N. (2023). The shift in research trends related to artificial intelligence in library repositories during the coronavirus pandemic. *Library Hi Tech*. Advance online publication. doi:10.1108/LHT-07-2022-0326

Okunlaya, R. O., Syed Abdullah, N., & Alias, R. A. (2022). Artificial intelligence (AI) library services innovative conceptual framework for the digital transformation of university education. *Library Hi Tech*, *40*(6), 1869–1892. doi:10.1108/LHT-07-2021-0242

Onyancha, O. B. (2015). An informetrics view of the relationship between internet ethics, computer ethics and cyberethics. *Library Hi Tech*, *33*(3), 387–408. doi:10.1108/LHT-04-2015-0033

Qi, Y. F. (2013). A New Way of Urban Public Service Supply in the Acceleration of Urbanization: Multi-coordinated Network Supply. *Contemporary World and Socialism*, *34*(01), 165–168.

Sanji, M., Behzadi, H., & Gomroki, G. (2022). Chatbot: An intelligent tool for libraries. *Library Hi Tech News*, *39*(3), 17–20. doi:10.1108/LHTN-01-2021-0002

Sha, Y. Z. (2004). A Study of Library Professional Ethics. *Journal of Library Science in China*, (4), 22–25.

Shan, Z., Chen, Y., & Shao, B. (2021). A Case Study of Risk Regulation of Smart Library Transition Abroad. *Tushuguanxue Yanjiu*, *43*(01), 2–8.

Su, Y. (2018). Research on Library Intelligent Service with Drive of Big Data and Artificial Intelligence. *Library and Information*, *38*(05), 103–106.

Tang, Y., & Xie, S. M. (2013). Study on Embedded Services of Academic Libraries Based on the Synergetic Theories. *Library and Information Service*, *57*(08), 78–81.

Ulrich Beck. (2004). Risk Society: On the Road to a New Modernity. Yilin Press.

Wang, J., Song, Y. F., & Du, P. P., etal. (2018). Research on Intelligent Library and Service Process Control Based on Collaborative Theory. *Library Work and Study*, *40*(10), 42–46.

Wang, L., & Xu, X. W. (2017). The Construction of Library Intelligence Service System Based on Collaborative Ecology. *New Century Library*, *39*(4), 28–32.

Wang, P. (2014). Study on Library Information Resources Construction and Sharing Based on Synergetic Theory. *Modern Information*, *34*(4), 33–37.

Wang, S. W. (2011). New Pattern of Future Libraries: The Smart Library. *Library Development*, *34*(12), 1–5.

Wang, S. W. (2012). On Three Main Features of the Smart Library. *Journal of Library Science in China*, *38*(06), 22–28.

Wang, S. W. (2017a). On the Five Relations of the Smart library. *Library Journal*, *36*(04), 4–10.

Wang, S. W. (2017b). Artificial Intelligence and Library Service Reshaping. *Library and Information*, *37*(06), 6–18.

Wang, S. W. (2019). Research on Artificial Intelligence and Library Renewal. *Tushu Qingbao Zhishi*, *36*(04), 35–42.

Wang, S. W., & Wang, T. N. (2018). Library Space Reengineering and Service Based on Artificial Intelligence. *Library and Information*, *38*(03), 50–55.

Wang, X., Feng, X., & Guo, K. (2022). Research hotspots and prospects of ethics education of science and technology in China based on bibliometrics. *Library Hi Tech*. Advance online publication. doi:10.1108/LHT-06-2022-0298

Wang, Y., & Wang, L. (2013). Construction and Application of the Smart Library Based on RFID Technology. *Journal of Library Science*, *35*(12), 98–100.

Wei, Q., & Lu, P. (2018). Innovative Supervision Methods to Prevent AI Risks. *Digital Economy*, *5*(03), 22–27.

Xia, Y. P. (2018). Research on the Cooperative Operation Mechanism of Multiple Subjects in Accurate Poverty Alleviation. *Review of Economic Research*, *40*(37), 72–77.

Xie, Y., Liu, J., Zhu, S., Chong, D., Shi, H., & Chen, Y. (2019). An IoT-based risk warning system for smart libraries. *Library Hi Tech*, *37*(4), 918–912. doi:10.1108/LHT-11-2017-0254

Xu, F. (2019). Regulation on Black Box of AI Algorithms. *Oriental Law*, *12*(06), 78–86.

Yan, Y. C. (2022). Research on the Infringement Risk and Countermeasures in the Text Data Mining of Smart Libraries in China. *Journal of the National Library of China*, *31*(01), 106–113.

Yang, L. H., & Zhang, L. (2016). Comparative Study on Multi-coordinated Treatment of Air Pollution: Cross-Case Analysis of Typical Countries. *Administrative Tribune*, *23*(05), 24–30.

Yao, E. X., & Tang, R. Q. (2018). The Path Optimization of Urban Risk Prevention and Goverance from the Perspective of Multiple Cooperation. *Journal of Hulunbeier Universities*, *26*(01), 49–52.

Yao, N., Xi, C. L., & Huang, X. T. (2017). Research on the Construction of the Situational Awareness Micro-Service Model in Smart Libraries. *Journal of Library Science*, *39*(08), 57–60.

Yu, X. H. (2014). Capability Construction of Risk Management System of Library Knowledge Management. *Library Work and Study*, *36*(09), 60–63.

Zeng, Z. M., & Qin, S. Q. (2018). Decentralized Mobile Visual Search Management System for Smart Library. *Information Science*, *36*(01), 11–15.

Zeng, Z. M., & Song, Y. Y. (2017). Research on Mobile Visual Search Service of Smart Library Based on SoLoMo. *The Library*, *45*(07), 92–98.

Zhang, H., & Ye, Y. (2023). Intelligence, Cognition and Insight: Characterizing Smart Library. *Journal of Library Science in China*, 1-11. http://kns.cnki.net/kcms/detail/11.2746.G2.20230131.1544.001.html

Zhang, T., & Ma, H. Q. (2021). Research on Algorithm Risk and Regulation in Intelligent Intelligence Analysis. *Library and Information Service*, *65*(12), 47–56.

Zhang, Y. J., Wang, J., & Huang, H. Q. (2018). Suggestions on Big Data Risk Management in the Era of AI. *China Economic & Trade Herald*, (06), 63–65.

Zhang, Z. M. (2022). Risk identification and avoidance strategy of intelligent library in the era of artificial intelligence. *Jiangsu Science & Technology Information*, *39*(06), 29–31.

Zhou, Q. X., & Yang, C. J. (2018). Design of smart library monitoring and management system based on internet of things technology. *Zidonghua Yu Yibiao*, *38*(11), 85–88.

Zhu, F. Y. (2019). Security Risk and Solution for Library Applications of Artificial Intelligence. *Tushuguanxue Yanjiu*, *41*(01), 6–11.

Chapter 4

A Review on Intelligent Library Services and Systems:
Utilizing Recommender Systems and Fuzzy Logic Chat Bots

Asefeh Asemi
Institute of Data Analytics and Information Systems, Hungary

ABSTRACT

The chapter includes a comprehensive review of relevant published literature from the past five years. Through this review, the chapter explores the use of emerging technology-based services and systems such as intelligent libraries, recommender systems, fuzzy logic, and recommender system chatbots in the field of library and information science. The aim of this review is to provide up-to-date insights and information for professionals and researchers interested in utilizing these technologies to improve library services. The chapter highlights how these technologies can enhance the user experience and resource discovery. The chapter begins by introducing the concept of intelligent libraries and the role of emerging technology-based services and systems. It then explores the use of recommender systems and fuzzy logic in intelligent libraries, citing relevant published literature and discussing the benefits they bring to users. In the end, it discusses the potential challenges and limitations of these technologies, as well as best practices for their implementation and maintenance.

DOI: 10.4018/978-1-6684-8671-9.ch004

INTRODUCTION

Intelligent libraries are increasingly leveraging emerging technologies to enhance and transform their services and systems. These technologies include artificial intelligence (Ali et al., 2020), the Internet of Things, virtual and augmented reality, robotics, and more. "By integrating these technologies into their operations, libraries can provide new and improved services to their users, such as personalized recommendations, virtual tours, interactive exhibits, and automated processes" (X. Su & Chen, 2022). Additionally, intelligent libraries can collect and analyze data from these technologies to better understand and meet the needs of their users. In the rapidly evolving world of technology, libraries have had to adapt and incorporate new tools and services to remain relevant and useful to their patrons. One area of focus has been the integration of intelligent systems, such as recommender systems and chatbots, to enhance the user experience and provide personalized recommendations (Rodriguez & Mune, 2022). It is necessary to review the emerging technology-based services and systems being implemented in intelligent libraries. We discuss the benefits and challenges of these technologies and their potential impact on the future of libraries. The study provided examples of intelligent libraries that are successfully utilizing these technologies and highlighted best practices for their implementation. Also, this book chapter explored the use of emerging technology-based services and systems in libraries, including the application of fuzzy logic in recommender systems and the development of library chatbots. By examining these trends and technologies, this chapter aims to provide a comprehensive overview of the current state and future potential of intelligent libraries. In this chapter, the various emerging technology-based services and systems discussed that are being implemented in intelligent libraries. The study discussed the benefits and challenges of these technologies and their potential impact on the future of libraries. The study also provided examples of intelligent libraries that are successfully utilizing these technologies and highlighted best practices for their implementation. Table 1 briefly explains the basic concepts in emerging technology-based services and systems in intelligent libraries.

This chapter provides a comprehensive review of relevant literature from the past five years, exploring the use of emerging technology-based services and systems in the field of library and information science, with a particular focus on intelligent libraries. By examining the implementation of intelligent libraries and the integration of technologies such as recommender systems, fuzzy logic, and recommender system chatbots, this chapter offers up-to-date insights and information for professionals and researchers interested in leveraging these technologies to improve library services. Moreover, the chapter delves into the benefits that these technologies bring to users, such as enhancing the user experience and improving resource discovery. Additionally, the chapter acknowledges the potential challenges and limitations of these

Table 1. Basic concepts in emerging technology-based services and systems in intelligent libraries

Concept	Definition
Intelligent Libraries	Libraries that use advanced technologies such as recommender systems, fuzzy logic, and recommender system chatbots to enhance the user experience and improve resource discovery (Asemi et al., 2021)
Emerging Technology-Based Services	Emerging technology-based services refer to new and innovative services that are enabled by advances in technology (Palo & Tähtinen, 2013). These services may leverage emerging technologies such as artificial intelligence, blockchain, virtual reality, and the Internet of Things, among others. They often offer novel solutions to existing problems or create entirely new opportunities for businesses and individuals. Advanced technologies that are being implemented in libraries to improve the user experience and resource discovery
Recommender Systems	Systems that make personalized recommendations to users based on their preferences and behaviors (Asemi & Ko, 2021)
Fuzzy Logic	A type of mathematical logic that deals with uncertainty (Cruz-Cunha et al., 2013) and is used to improve resource recommendations in intelligent libraries
Recommender System Chatbots	Chatbots that use recommender systems (Landbot, 2021) to assist users in their search for resources and materials in intelligent libraries

technologies and provides best practices for their implementation and maintenance. By presenting a comprehensive overview and analysis of these emerging technologies in libraries, this chapter aims to contribute to the advancement of the field and assist practitioners in making informed decisions regarding the adoption and utilization of these technologies. In conclusion, this chapter serves as a vital contribution to the field of library and information science by providing a comprehensive review of the intelligent library services and systems empowered by emerging technologies. By exploring the integration of recommender systems, fuzzy logic, and recommender system chatbots in intelligent libraries, this research offers up-to-date insights and information for professionals and researchers seeking to enhance library services.

BACKGROUND

Over the past decade, there has been a significant increase in the use of technology in libraries to enhance services and improve access to information. Emerging technologies such as virtual reality, artificial intelligence, and the Internet of Things have the potential to revolutionize the way libraries operate and deliver services to users. One example of an emerging technology-based service in libraries is the use of virtual reality (VR) for immersive learning experiences. VR allows users to experience a simulated environment as if they were physically present, providing a unique and interactive learning opportunity. "Virtual Reality (VR) is starting to be used in psychological therapy around the world" (Schuemie et al., 2004). Another emerging technology-based service in libraries is the use of artificial intelligence (AI) to improve information discovery and retrieval (Asemi et al., 2021). AI algorithms can analyze user behavior and search queries to provide personalized recommendations (Shin, 2021) and improve the accuracy of search results. In addition to these services, libraries are also adopting emerging technologies in their systems and infrastructure. The Internet of Things (IoT) allows libraries to connect and manage various devices and systems (Soleimanzade et al., 2019), such as security cameras, heating, and cooling systems, and energy management systems. By integrating these systems through the IoT, libraries can improve efficiency and reduce operational costs. The concept of the library has undergone significant changes in recent years with the emergence of innovative technologies such as recommender systems and fuzzy logic chatbots. These intelligent services and systems have revolutionized the way libraries operate and have significantly impacted the way users' access and interact with library resources. One of the key benefits of these emerging technologies is the ability to personalize the user experience. Recommender systems, for example, use algorithms to suggest resources and materials based on a user's previous interactions and preferences (Huang et al., 2002). This can significantly increase the relevance and usefulness of the resources and materials recommended to users, resulting in increased user satisfaction and engagement. Fuzzy logic chatbots, on the other hand, provide a more interactive and intuitive way for users to access information and assistance (Almansor & Hussain, 2021). These chatbots use natural language processing to understand user queries and provide appropriate responses, making it easier for users to find the information they need. In addition to personalizing the user experience (Trichur Narayanan, 2021), these technologies can improve the efficiency and effectiveness of library operations. Recommender systems can help to reduce the time and effort required for staff to curate and promote resources, while chatbots can help to reduce the workload of staff by providing a first line of assistance for common queries.

There have been several studies conducted on the impact of these technologies on library operations and user experiences (Wilson, 2021). The use of recommender systems in libraries increased usage of electronic resources and reduced unfulfilled requests (Vekaria et al., 2021). Also, the implementation of fuzzy logic chatbots in libraries resulted in increased user satisfaction and a reduction in the workload of staff (Gade & Angal, 2018). Overall, the adoption of emerging technology-based services and systems in libraries has the potential to significantly enhance the user experience and improve the efficiency and effectiveness of library operations. However, it is important for libraries to carefully consider the potential impacts and implications of these technologies, including issues of equity and accessibility. Also, the use of intelligent services and systems utilizing emerging technologies such as recommender systems and fuzzy logic chat bots has the potential to significantly revolutionize the library by improving the user experience and increasing the efficiency and effectiveness of library operations. These study by reviewing the information resources tried to answer these questions:

- What are the most common emerging technology-based services and systems being implemented in intelligent libraries?
- How effective are recommender systems in helping users discover recent resources and materials in intelligent libraries?
- How does the incorporation of fuzzy logic in intelligent libraries enhance the user experience and improve resource recommendation?
- How can recommender system chatbots use to assist users in their search for resources and materials in intelligent libraries?
- How do intelligent libraries compare to traditional libraries in terms of the use of emerging technology-based services and systems?
- What are the potential challenges and limitations of using emerging technology-based services and systems in intelligent libraries?
- How can intelligent libraries effectively integrate and utilize emerging technology-based services and systems to improve user experience and resource discovery?
- What are the best practices for implementing and maintaining emerging technology-based services and systems in intelligent libraries?
- How do user preferences and behaviors impact the effectiveness of emerging technology-based services and systems in intelligent libraries?
- How do intelligent libraries evaluate the success and impact of their implementation of emerging technology-based services and systems?

METHOD

This review focused on the use of emerging technology-based services and systems, specifically intelligent libraries, recommender systems, fuzzy logic, and recommender system chatbots, in library and information science. Then, the study has done in the following steps (Table 2):

- Identify relevant literature: To gather relevant literature, a comprehensive search of academic databases such as WoS and Scopus have conducted by using this search formula: ("intelligent", ("recommender OR "recommendation")

Table 2. Research method steps

Step	Description
1. Identify relevant literature	- Conducted a comprehensive search of academic databases such as Web of Science (WoS) and Scopus. - Utilized the following search formula: ("intelligent", ("recommender" OR "recommendation") AND "systems", "fuzzy logic", "chatbots") and ("library" or "libraries") in the titles of published documents. - Refined the search results by relevance and selected literature that aligns with the review topic. - Ensured inclusion of recent and diverse sources to capture the latest advancements and different perspectives.
2. Evaluate literature quality	- Assessed the relevance of each selected literature to the review topic. Considered the generalizability of the findings, focusing on studies that could apply in various contexts or settings. - Excluded any literature that does not meet the quality criteria established for the review.
3. Discuss findings	- Summarized the main findings derived from the literature. - understanding of emerging technology-based services and systems in libraries. - Provided an outline of the key insights expected to emerge from the research, offering a glimpse into the potential contributions to the field. - Briefly explained how these insights contribute to the existing knowledge base.
4. Conclusions and recommendations	- Summarized the main findings derived from the literature review. Discussed the significance of the findings and their implications for library and information science. - Highlighted gaps or limitations identified in the existing literature. - Provided recommendations for future research directions, suggesting areas that require further investigation or potential improvements in methodologies or approaches. - Considered the practical implications of the research findings and recommend strategies for implementation, if applicable.

AND "systems", "fuzzy logic", "chatbots") and ("library" or "libraries") in the titles of published documents.

- Evaluate the quality of the literature: The quality of the literature evaluated based on the following criteria: relevance to the review topic, methodological rigor, generalizability (the review considered literature that has the potential to apply in different contexts or settings).

- Discuss on findings: The review discussed with a summary of the main findings from the literature and provided an outline the key findings that expect to emerge from the research, and briefly explained how they contribute to the broader understanding of emerging technology-based services and systems in libraries.

- Conclusions and recommendations: The review ended with a summary of the main findings from the literature and recommendations for future research in library and information science.

Table 3. Inclusion and exclusion criteria

Step	Inclusion Criteria	Exclusion Criteria
Literature Search	- Publications of the last five years related to the use of emerging technology-based services and systems in library and information science.	- Publications unrelated to the research topic.
	- Publications discussing intelligent libraries, recommender systems, fuzzy logic, and recommender system chatbots.	- Publications not addressing the specific technologies or systems of interest.
	- Publications with titles containing keywords such as "intelligent," "recommender," "recommendation," "systems," "fuzzy logic," "chatbots," "library," or "libraries."	- Publications lacking relevance to the review topic despite containing the specified keywords. - Non-academic sources, such as blog posts, news articles, and social media content.
Quality Evaluation	- Relevance to the review topic: Publications directly addressing the use of emerging technology-based services and systems in library and information science.	- Irrelevant publications that do not contribute to the research focus.
	- Methodological rigor: Publications with well-designed research methodologies, including clear objectives, appropriate data collection methods, and sound analysis.	- Publications with methodological flaws, such as inadequate sampling, biased data collection, or unreliable analysis techniques.
	- Generalizability: Publications with findings applicable in different contexts or settings, providing insights beyond specific case studies.	- Publications with highly contextual findings limited to specific scenarios or environments.

Table 3 includes a list of inclusion and exclusion criteria used in the literature search and quality evaluation for the Methods section:

FINDINGS AND DISCUSSION

The literature search identified publications focused on the use of emerging technology-based services and systems in library and information science. The search criteria included keywords such as "intelligent," "recommender," "recommendation," "systems," "fuzzy logic," "chatbots," "library," or "libraries." Relevant literature discussing intelligent libraries, recommender systems, fuzzy logic, and recommender system chatbots included in the review. Thus, the following search formula used: ("intelligent", ("recommender OR "recommendation") AND "systems", "fuzzy logic", "chatbots") and ("library" or "libraries") in the titles of documents published in the Web of Science and Scopus databases and scope from 2018 to 2022, retrieved fifty-three documents. These documents stored in Zotero, and an Excel file output was prepared from it. All documents reviewed one by one and based on this review the following findings obtained (Table 4). This initial exploration of literature highlighted the significance of these technologies in the context of libraries. The identified publications shed light on the potential applications and benefits of intelligent libraries and recommender systems, as well as the integration of fuzzy logic and chatbots in library services. These findings provide a foundation for further analysis and exploration of the role of emerging technology-based services and systems in library and information science.

Table 4. Published documents in intelligent library services and systems utilizing recommender systems and fuzzy logic chat bots

Publication Year	Item Type	Author	Title
2022	Journal Article	(Miller & Shortliffe, 2022)	Corrigendum to: The roles of the US National Library of Medicine and Donald A.B. Lindberg in revolutionizing biomedical and health informatics
2022	Journal Article	(L. Sun et al., 2022)	Construction of Cloud Library Intelligent Service Platform Relying on Artificial Neural Network
2022	Journal Article	(Liu, 2022)	Investigating users' willingness of acceptance for background music service in intelligent library
2022	Journal Article	(Xu, 2022)	Intelligent Library Service and Management Based on IoT Assistance and Text Recommendation
2022	Journal Article	(An & Yan, 2022)	Intelligent retrieval method of library document information based on hidden topic mining
2022	Journal Article	(Gu & Tanoue, 2022)	A Research on Library Space Layout and Intelligent Optimization Oriented to Readers' Needs
2022	Journal Article	(Cheng & Hou, 2022)	Intelligent Educational Evaluation of Research Performance between Digital Library and Open Government Data
2022	Conference Paper	(Qian & IEEE, 2022)	The Semantic Framework of Library Intelligent Question Answering System Based on Exploratory Search Behavior
2022	Journal Article	(Pang, 2022)	Intelligent Big Information Retrieval of Smart Library Based on Graph Neural Network (GNN) Algorithm
2022	Journal Article	(H. Wang & Ding, 2022)	Development Strategy of Intelligent Digital Library without Human Service in the Era of "Internet plus "
2022	Journal Article	(Shang et al., 2022)	Intelligent Optimization Method of Resource Recommendation Service of Mobile Library Based on Digital Twin Technology
2022	Journal Article	(X. Su & Chen, 2022)	Intelligent Information Service System of Smart Library Based on Virtual Reality and Eye Movement Technology
2022	Journal Article	(Jiang & Yang, 2022)	Analysis of the Influence of Library Information on the Utilization of Regional Environmental and Ecological Resources: From the Perspective of Intelligent Adaptive Learning
2022	Conference Paper	(Rubei, Di Sipio, et al., 2022)	Endowing third-party libraries recommender systems with explicit user feedback mechanisms
2022	Journal Article	(Rhanoui et al., 2022)	A hybrid recommender system for patron driven library acquisition and weeding
2022	Journal Article	(Coba et al., 2022)	RecoXplainer: A Library for Development and Offline Evaluation of Explainable Recommender Systems
2022	Journal Article	(Rubei, Di Ruscio, et al., 2022)	Providing upgrade plans for third-party libraries: a recommender system using migration graphs
2022	Conference Paper	(Koliarakis et al., 2022)	Modified collaborative filtering for hybrid recommender systems and personalized search: The case of digital library

continued on following page

Table 4. Continued

Publication Year	Item Type	Author	Title
2022	Journal Article	(Rodriguez & Mune, 2022)	Uncoding library chatbots: deploying a new virtual reference tool at the San Jose State University library
2022	Journal Article	(Kaushal & Yadav, 2022)	The Role of Chatbots in Academic Libraries: An Experience-based Perspective
2021	Conference Paper	(Jamil, 2021)	Architecture of an Intelligent Personal Health Library for Improved Health Outcomes
2021	Journal Article	(Asemi et al, 2019)	Intelligent libraries: a review on expert systems, artificial intelligence, and robot
2021	Journal Article	(Chen et al., 2021)	How to improve the university library intelligent knowledge service: A system dynamics model
2021	Journal Article	(Z. Gao & Zou, 2021)	Intelligent Data Mining of Computer-Aided Extension Residential Building Design Based on Algorithm Library
2021	Journal Article	(Su, 2021)	Design of the online platform of intelligent library based on machine learning and image recognition
2021	Conference Paper	(Nguyen et al., 2021)	Adversarial Machine Learning: On the Resilience of Third-party Library Recommender Systems
2021	Journal Article	(Ansari et al., 2021)	Using data mining techniques to predict user's behavior and create recommender systems in the libraries and information centers
2020	Journal Article	(Gao, 2020)	Intelligent library knowledge innovation service system based on multimedia technology
2020	Journal Article	(Sun, 2020)	Research on interest reading recommendation method of intelligent library based on big data technology
2020	Conference Paper	(Paullier et al., 2020)	A Recommender Systems' algorithm evaluation using the Lenskit library and MovieLens databases
2020	Conference Paper	(Anand et al., 2020)	Auto-Surprise: An Automated Recommender-System (AutoRecSys) Library with Tree of Parzens Estimator (TPE) Optimization
2019	Journal Article	(Cox et al., 2019)	The intelligent library Thought leaders' views on the impact of artificial intelligence on academic libraries
2019	Journal Article	(K. Xie et al., 2019)	Internet of Things-based intelligent evacuation protocol in libraries
2019	Conference Paper	(Paullier et al., 2020)	Research on Library functional layout based on Intelligent occupying system
2019	Conference Paper	(Jing & IEEE, 2019)	Logical Construction and Realization of Intelligent Service in AI Library
2019	Conference Paper	(Y. Wang et al., 2019)	Intelligent Library Information Service Terminal Based on BDS and Wireless Communication
2019	Conference Paper	(W. Zhang, 2019)	Research on the Construction of Community Library Intelligent Service System from the Perspective of Smart City
2019	Journal Article	(Chen & Shen, 2019)	The correlation analysis between the service quality of intelligent library and the behavioral intention of users
2019	Journal Article	(Shen, 2019)	Emerging scenarios of data infrastructure and novel concepts of digital libraries in intelligent infrastructure for human-centred communities: A qualitative research
2019	Journal Article	(Y. Zhang & Yan, 2019)	MIL-61 and Eu3+@MIL-61 as Signal Transducers to Construct an Intelligent Boolean Logical Library Based on Visualized Luminescent Metal-Organic Frameworks
2018	Conference Paper	(Nie, 2018)	The Way to Construct the Intelligent Library-Taking Nanyang College Library as an example
2018	Conference Paper	(J. Zhang et al., 2018)	Research of Intelligent Library Based on RFID Technology

continued on following page

Table 4. Continued

Publication Year	Item Type	Author	Title
2018	Conference Paper	(Iantovics et al., 2018)	Intelligent University Library Information Systems to Support Students Efficient Learning
2018	Journal Article	(Y. Gao, 2018)	Implementation of an Intelligent Library System Based on WSN and RFID
2018	Conference Paper	(Gade & Angal, 2018)	Real-Time Intelligent NI myRIO-Based Library Management Robotic System Using LabVIEW
2018	Conference Paper	(Shen & Assoc Comp Machinery, 2018)	Digital Library Systems in Intelligent Infrastructure for Human-Centered Communities: A Qualitative Research
2018	Journal Article	(Bak et al., 2018)	Towards Intelligent Drug Design System: Application of Artificial Dipeptide Receptor Library in QSAR-Oriented Studies
2018	Conference Paper	(Xie, 2018)	Study on the Intelligent Terminal Innovation of Personalized Active Service of Mobile Library-A Case Study of Zhejiang University of Media and Communications
2018	Conference Paper	(Collins et al., 2018)	Position Bias in Recommender Systems for Digital Libraries
2018	Conference Paper	(Jomsri, 2018)	FUCL mining technique for book recommender system in library service
2018	Journal Article	(Simovic, 2018)	A Big Data smart library recommender system for an educational institution
2018	Conference Paper	(Tejeda-Lorente et al., 2018)	Using Bibliometrics and Fuzzy Linguistic Modeling to Deal with Cold Start in Recommender Systems for Digital Libraries
2018	Journal Article	(Asemi & Asemi, 2018)	Artificial Intelligence (AI) application in Library Systems in Iran: A taxonomy study

A.   Most common emerging technology-based services and systems being implemented in intelligent libraries

Intelligent libraries use a variety of technology-based services and systems to improve their operations and enhance the user experience. Some of the most common emerging technologies being implemented in intelligent libraries include (Nguyen et al., 2021; Soleimanzade et al., 2019; Vaishali Kaushal & Rajan Yadav, 2022; Rodriguez & Mune, 2022; Asemi et al., 2021; X. Su & Chen, 2022):

- Artificial intelligence (AI) and machine learning (ML) - AI and ML are being used to analyze library data and improve resource recommendation systems, search and discovery, and digital collections management.
- Internet of Things (IoT) - IoT is being used to monitor and control the physical environment of libraries, such as temperature and humidity, to ensure the preservation of collections.
- Augmented Reality (AR) and Virtual Reality (VR) - AR and VR are being used to create immersive learning experiences, allowing users to interact with digital content in new and innovative ways.

- Chatbots and voice assistants - Chatbots and voice assistants are being used to provide quick and easy access to information, as well as to assist with library services such as borrowing, returning, and reserving books.
- 3D Printing - 3D printing technology is being used in libraries to provide access to makerspaces, where users can design and create physical objects.
- Blockchain technology is being used to secure and manage digital collections and provide users with secure and permanent access to digital content.
- Robotic Process Automation (RPA) - RPA is being used to automate repetitive and time-consuming tasks, freeing library staff to focus on more strategic and creative activities.

These are just a few examples of the many technology-based services and systems being implemented in intelligent libraries. The goal of these technologies is to provide a more personalized and efficient experience for users, and to support the ongoing mission of libraries to preserve and disseminate knowledge and information.

B.   Role of recommender systems in helping users discover recent resources and materials in intelligent libraries

Recommender systems can be highly effective in helping users discover recent resources and materials in intelligent libraries. By analyzing patterns in user behavior and library usage data, recommender systems can provide personalized recommendations to users based on their past preferences and interests. This can help users discover recent resources that they may not have otherwise been aware of or considered. Some of the key benefits of recommender systems in intelligent libraries include (Vekaria et al., 2021, Anand et al., 2020; Ansari et al., 2021; Coba et al., 2022; Collins et al., 2018; Landbot, 2021; Nguyen et al., 2021; Rhanoui et al., 2022; Rubei, Di Ruscio, et al., 2022; Rubei, Di Sipio, et al., 2022; Shang et al., 2022; Simovic, 2018; H. Sun, 2020; Tejeda-Lorente et al., 2018):

- Increased Discovery: Recommender systems can help users discover recent resources and materials that they may not have otherwise been aware of or considered, leading to a more diverse and enriched user experience.
- Improved User Experience: By providing personalized recommendations, recommender systems can make the library experience more engaging and relevant to each individual user.
- Increased Usage: Recommender systems can help drive usage of the library's resources, particularly for less well-known or underutilized items.

- Better Resource Allocation: By providing insight into which resources are most popular, recommender systems can help libraries make more informed decisions about resource allocation and collection development.

The effectiveness of recommender systems can vary depending on factors such as the quality of the data being used, the algorithms being employed, and the user engagement with the system. The best recommender systems use a combination of explicit (e.g., ratings and feedback) and implicit (e.g., usage data) information to generate recommendations, and are continually refined and improved through machine learning and other techniques.

C.   Incorporation of fuzzy logic in intelligent libraries enhances the user experience and improves resource recommendation

The incorporation of fuzzy logic in intelligent libraries can enhance the user experience and improve resource recommendation in several ways (Almansor & Hussain, 2021; Cruz-Cunha et al., 2013; Tejeda-Lorente et al., 2018):

- Handling of uncertainty: Fuzzy logic can help handle uncertainty in the data used for resource recommendation, allowing for more nuanced and accurate recommendations. For example, if a user has only rated a few books, fuzzy logic can make use of related information, such as the user's reading history, to generate a more informed recommendation.
- Personalization: Fuzzy logic can be used to personalize resource recommendations based on a user's preferences, history, and other factors. The system can weigh varied factors differently based on their importance to the user, allowing for more customized recommendations.
- Improved Search Results: Fuzzy logic can be used to improve search results by allowing for the inclusion of synonyms, fuzzy matching, and related terms, resulting in more relevant and accurate search results.
- Better Resource Recommendations: Fuzzy logic can be used to generate resource recommendations based on multiple criteria, such as popularity, relevance, and user preferences. This can result in more diverse and relevant recommendations, helping users discover recent resources they may not have otherwise considered.

Incorporating fuzzy logic into resource recommendation systems can provide a more nuanced and personalized approach to resource discovery, leading to a more engaging and relevant user experience. The ability to handle uncertainty, personalize recommendations, and generate more relevant results can result in increased usage and satisfaction for library users.

D.   Role of recommender system chatbots to assist users in their search for resources and materials in intelligent libraries

Recommender system chatbots can assist users in their search for resources and materials in intelligent libraries in several ways (Almansor & Hussain, 2021; Anand et al., 2020; Ansari et al., 2021; Asemi & Ko, 2021; Coba et al., 2022; Collins et al., 2018; Huang et al., 2002; Jomsri, 2018; Koliarakis et al., 2022; Landbot, 2021; Nguyen et al., 2021; Paullier et al., 2020; Rhanoui et al., 2022; Rubei, Di Ruscio, et al., 2022; Rubei, Di Sipio, et al., 2022; Simovic, 2018; Tejeda-Lorente et al., 2018; Trichur Narayanan, 2021; Vaishali Kaushal & Rajan Yadav, 2022, 2022):

- Interactive search: Chatbots can interact with users in a conversational manner, helping them to clarify their search criteria and providing recommendations based on their preferences and past behavior.
- Personalized recommendations: By utilizing recommender systems and machine learning algorithms, chatbots can provide personalized recommendations to users based on their past behavior, preferences, and reading history.
- Quick access to information: Chatbots can provide quick and easy access to information, such as the availability of a particular book, its location, and due dates.
- Improved search results: Chatbots can use natural language processing (NLP) and other advanced technologies to understand and interpret user queries, providing more accurate and relevant search results.
- 24/7 Assistance: Chatbots can help 24/7, allowing users to access information and resources even outside of normal library hours.

Incorporating recommender system chatbots into intelligent libraries can provide a more convenient, personalized, and efficient user experience. The ability to provide quick and easy access to information, as well as personalized recommendations, can help users find the resources and materials they need more quickly and effectively. Also, 24/7 assistance can help increase user engagement and satisfaction with the library.

E.   Comparison intelligent libraries with traditional libraries in terms of the use of emerging technology-based services and systems

Intelligent libraries differ from traditional libraries in several keyways in terms of the use of emerging technology-based services and systems (Koliarakis et al., 2022; Trichur Narayanan, 2021; Wilson, 2021, 2021; X. Xie, 2018):

- Personalized Experience: Intelligent libraries use technology, such as recommender systems, to provide a personalized experience for users, based on their preferences, behavior, and reading history. Traditional libraries may not have the same level of personalization, relying more on static classification systems and manual recommendations.
- Automated Systems: Intelligent libraries often use automated systems, such as chatbots, to assist users with their information needs, whereas traditional libraries may rely more on human interaction, such as in-person reference services.
- Integration with Technology: Intelligent libraries are often integrated with a variety of technology-based services and systems, such as e-book platforms, online databases, and digital content management systems, allowing for seamless access to a wide range of resources and materials. Traditional libraries may have more limited access to technology-based services and systems and may have separate systems for several types of resources and materials.
- Data-Driven Decision Making: Intelligent libraries use data analysis and machine learning algorithms to make informed decisions about collection development, resource allocation, and user engagement, whereas traditional libraries may rely more on intuition and experience.
- User-Centered Design: Intelligent libraries are designed with the user experience in mind, using user feedback and data analysis to continually improve and refine their services and systems. Traditional libraries may have a more traditional, institutional approach to service design and delivery.

In general, intelligent libraries are characterized by a more sophisticated and integrated use of technology-based services and systems, leading to a more personalized, automated, and data-driven user experience. This can result in improved user engagement, satisfaction, and resource discovery, compared to traditional libraries.

F.   Potential challenges and limitations of using emerging technology-based services and systems in intelligent libraries

While the use of emerging technology-based services and systems can bring many benefits to intelligent libraries, there are also potential challenges and limitations to consider (Ali et al., 2020; Asemi et al., 2021; Coba et al., 2022; Y. Gao, 2018; Jamil, 2021; X. Su & Chen, 2022; Vekaria et al., 2021):

• Cost: Implementing and maintaining technology-based services and systems can be expensive, requiring significant investment in hardware, software, and personnel. This can be a challenge for libraries, particularly those with limited budgets.
• Technical Complexity: The use of technology-based services and systems can also bring technical complexity, requiring specialized skills and knowledge to manage and maintain them. This can pose a challenge for libraries, particularly those with limited technical resources.
• Privacy Concerns: The use of technology-based services and systems can raise privacy concerns, particularly regarding the collection and use of user data. Libraries need to ensure that they have appropriate privacy policies and procedures in place to address these concerns.
• User Adoption: The success of technology-based services and systems in libraries often depends on user adoption, and users may not be familiar or comfortable with the technology. Libraries need to ensure that they provide adequate training and support to help users get the most out of technology.
• Integration with Other Systems: The integration of technology-based services and systems with existing library systems and processes can also pose a challenge. Libraries need to ensure that the systems are compatible and can work together effectively to provide a seamless user experience.
• Limitations of Technology: While technology can enhance the user experience and improve resource discovery, it is not a panacea, and there may be limitations to its effectiveness. For example, the accuracy of recommender systems can be limited by the quality and quantity of data available, and users may still need assistance in finding and evaluating resources.

In summary, while the use of emerging technology-based services and systems can bring many benefits to intelligent libraries, there are also potential challenges and limitations that need to be carefully considered and addressed. Libraries need to weigh the costs and benefits of technology-based services and systems and implement them in a thoughtful and strategic manner to ensure their success.

G.   Role of intelligent libraries to integrate and utilize emerging technology-based services and systems to improve user experience and resource discovery

Intelligent libraries can effectively integrate and utilize emerging technology-based services and systems to improve user experience and resource discovery by following these best practices (Rubei, Di Sipio, et al., 2022; Soleimanzade et al., 2019; Wilson, 2021, 2021; J. Zhang et al., 2018):

- Conduct a Needs Assessment: Before implementing technology-based services and systems, libraries should conduct a thorough needs assessment to determine the specific needs and goals of their users, and how technology can best meet those needs. This will help to ensure that the technology is aligned with the library's mission and goals.
- Engage Users in the Process: User engagement and feedback is critical to the success of technology-based services and systems in libraries. Libraries should involve users in the design and testing of the technology and gather feedback on the user experience to continually improve and refine the services and systems.
- Use Data to Drive Decisions: The use of data and analytics can be a powerful tool for improving user experience and resource discovery in libraries. Libraries should use data to make informed decisions about the implementation and use of technology-based services and systems, and continually monitor and evaluate their effectiveness.
- Ensure Privacy and Security: Privacy and security are key concerns in the use of technology-based services and systems, particularly regarding the collection and use of user data. Libraries should have strong privacy policies and procedures in place and ensure that the technology is secure and protected from unauthorized access.
- Collaborate with Partners: Libraries can benefit from collaborating with other institutions, organizations, and vendors in the implementation and use of technology-based services and systems. This can help to share expertise, resources, and best practices, and increase the impact and reach of the technology.
- Provide Adequate Training and Support: Users need adequate training and support to make the most of technology-based services and systems, and to overcome any obstacles or limitations. Libraries should provide training and support for users and ensure that the technology is accessible and user-friendly.

Intelligent libraries can effectively integrate and utilize emerging technology-based services and systems to improve user experience and resource discovery by following these best practices and being strategic, user-centered, and data-driven in their approach. By doing so, libraries can create a more engaging and effective environment for users and support their information needs in new and innovative ways.

H.   Best practices for implementing and maintaining emerging technology-based services and systems in intelligent libraries

Here are some best practices for implementing and maintaining emerging technology-based services and systems in intelligent libraries (Asemi & Asemi, 2018; Xu, 2022):

- Start with a Plan: Develop a clear and comprehensive plan for the implementation and maintenance of technology-based services and systems, including goals, budget, timeline, and resources required.
- Involve Stakeholders: Engage stakeholders, including users, library staff, and technology vendors, in the planning and implementation process to ensure that the technology meets their needs and expectations.
- Assess Technical Requirements: Assess the technical requirements of the technology-based services and systems, including hardware, software, and network infrastructure, to ensure that the library has the necessary resources and capabilities to support them.
- Ensure Data Management: Ensure that data management systems are in place to effectively store, manage, and analyze data generated by technology-based services and systems.
- Prioritize User Experience: Prioritize the user experience in the design and implementation of technology-based services and systems, and ensure that they are accessible, user-friendly, and meet the needs of the library's diverse user population.
- Provide Adequate Training and Support: Provide adequate training and support for library staff and users to effectively use and maintain technology-based services and systems and ensure that support is available when needed.
- Continuously Evaluate and Improve: Continuously evaluate and improve technology-based services and systems, using data and user feedback to inform decisions and ensure that they remain relevant and effective over time.
- Maintain a Flexible Approach: Maintain a flexible approach to the implementation and maintenance of technology-based services and systems and be prepared to adjust and adapt as the technology evolves and user needs change.

Implementing and maintaining emerging technology-based services and systems in intelligent libraries requires a strategic and collaborative approach, involving all stakeholders, ensuring adequate resources and support, and continuously evaluating and improving the technology to ensure that it meets the needs of users and supports the mission of the library.

I.   Role of user preferences and Behaviors in the effectiveness of emerging technology-based services and systems in intelligent libraries

User preferences and behaviors play a significant role in the effectiveness of emerging technology-based services and systems in intelligent libraries. The following are some ways in which user preferences and behaviors impact the effectiveness of technology-based services and systems (Ansari et al., 2021; Shin, 2021; Trichur Narayanan, 2021; Wilson, 2021, 2021):

- Adoption and Usage: User preferences and behaviors impact the adoption and usage of technology-based services and systems. For example, users who are comfortable with technology and have experience using similar services and systems are more likely to adopt and effectively use the technology. However, users less familiar with technology or with different preferences may be less likely to use it.
- User Experience: User preferences and behaviors also impact the user experience of technology-based services and systems. For example, users who prefer a simple and intuitive interface will have a better experience with technology-based services and systems that are designed to be easy to use, while users who prefer a more complex interface may prefer more feature-rich technology-based services and systems.
- Resource Discovery: User preferences and behaviors also impact the effectiveness of technology-based services and systems in helping users discover recent resources and materials. For example, users who prefer to browse and explore new materials may be more likely to use technology-based services and systems that provide recommendations and allow them to easily discover recent resources, while users who prefer to search for specific materials may prefer technology-based services and systems that allow them to easily search and access specific resources.
- Data Privacy: User preferences and behaviors with regards to data privacy also play a role in the effectiveness of technology-based services and systems. For example, users who are concerned about their data privacy may be less likely to use technology-based services and systems that collect and store

their data or may prefer technology-based services and systems that provide strong privacy protections.

User preferences and behaviors significantly impact the effectiveness of emerging technology-based services and systems in intelligent libraries. It is important for libraries to understand their users' preferences and behaviors, and to design and implement technology-based services and systems that meet the needs and preferences of their users to ensure that they are effective and widely adopted.

J.   Success evaluation by intelligent libraries for implementation of emerging technology-based services and systems

Intelligent libraries evaluate the success and impact of their implementation of emerging technology-based services and systems by using various metrics and methods. Some common evaluation methods include (Ansari et al., 2021; Asemi & Ko, 2021; Cheng & Hou, 2022; Y. Gao, 2020; Liu, 2022; Rubei, Di Sipio, et al., 2022, 2022; Wilson, 2021):

- User Feedback: User feedback is one of the most important methods for evaluating the success and impact of technology-based services and systems in intelligent libraries. Surveys, focus groups, and interviews with users can provide valuable insights into their experiences with the technology and help libraries identify areas for improvement.
- Usage Data: Usage data, such as the number of searches, downloads, and resource recommendations, can help libraries understand the extent to which users are utilizing technology-based services and systems, and how they are impacting the discovery of resources and materials.
- Resource Impact: Libraries can evaluate the impact of technology-based services and systems on the discovery of resources and materials by comparing usage data before and after the implementation of the technology. For example, they can compare the number of resource recommendations, or the number of unique resources accessed before and after the implementation of technology-based services and systems.
- User Satisfaction: User satisfaction is another important metric for evaluating the success and impact of technology-based services and systems in intelligent libraries. Surveys and interviews can help libraries understand the extent to which users are satisfied with the technology and identify areas for improvement.
- Return on Investment (ROI): Return on investment (ROI) is another metric that can be used to evaluate the success and impact of technology-based

services and systems in intelligent libraries. This metric measures the budgetary impact of the technology, including the costs of implementation and maintenance, and the benefits in terms of increased resource discovery, user satisfaction, and other factors.

- Technical Performance: Technical performance, such as response time, uptime, and reliability, is another important metric for evaluating the success and impact of technology-based services and systems in intelligent libraries. Technical performance metrics help libraries understand the extent to which the technology is meeting the needs of users and delivering a positive user experience.

Evaluating the success and impact of the implementation of emerging technology-based services and systems in intelligent libraries requires a multi-faceted approach that considers a range of metrics and evaluation methods, including user feedback, usage data, resource impact, user satisfaction, ROI, and technical performance. This information can help libraries understand the impact of technology on the discovery of resources and materials, the user experience, and the financial viability of the technology, and make informed decisions about future investments in technology.

CONCLUSION

Emerging technology-based services and systems in intelligent libraries are gaining popularity and are being implemented in many libraries around the world. These technologies include recommender systems, which use data on a user's past borrowings and search history to suggest relevant materials, and fuzzy logic, which allows for more flexible and nuanced decision-making in the library's operations. One example of this is the use of a recommender system chatbot, which can provide personalized recommendations to users through a chat interface. The implementation of emerging technology-based services and systems, specifically intelligent libraries, recommender systems, fuzzy logic, and recommender system chatbots, has the potential to enhance user experience and resource discovery in the field of library and information science. However, it is important to carefully consider the potential challenges and limitations and implement best practices for the successful integration and utilization of these technologies. Further research is needed to fully understand the impact and success of these technologies in real-world applications and to identify any areas for improvement. The rise of intelligent libraries utilizing emerging technologies such as recommender systems and fuzzy logic chatbots is transforming the way people access and discover resources in the field of library and information science. These technologies bring a new level of convenience and personalization to the user experience by making recommendations based on individual preferences and behaviors. The use of fuzzy logic in intelligent libraries helps to improve resource recommendations and enhance the overall user experience. However, the successful implementation and utilization of these technologies in libraries requires careful consideration of potential challenges and limitations, as well as adherence to best practices for implementation and maintenance. The results of research in this field have shown that these technologies can improve the effectiveness and efficiency of library services. As technology continues to evolve and improve, it is likely that the use of intelligent services and systems utilizing emerging technologies will continue to revolutionize the library experience for users. The implementation of these technologies has several implications for practice in the field. For one, they allow libraries to offer more personalized and efficient services to their patrons, as they can use data to tailor their recommendations and operations to the specific needs of each user. Additionally, they can help to reduce the workload of the library staff by automating certain tasks and allowing them to focus on more complex and value-added activities. In terms of future research, there are several areas that could be explored. For example, there is a need to further study these technologies' effectiveness in terms of their impact on user satisfaction and engagement. Additionally, there is a need to explore the ethical implications of using data to personalize services, particularly regarding issues of privacy and security. Finally, there is a need to examine the potential for these technologies to be integrated with other emerging technologies, such as artificial intelligence and machine learning, to further enhance their capabilities and impact.

REFERENCES

Ali, M. Y., Naeem, S. B., & Bhatti, R. (2020). Artificial intelligence tools and perspectives of university librarians: An overview. *Business Information Review*, *37*(3), 116–124. doi:10.1177/0266382120952016

Almansor, E. H., & Hussain, F. K. (2021). Fuzzy Prediction Model to Measure Chatbot Quality of Service. *2021 IEEE International Conference on Fuzzy Systems (FUZZ-IEEE),* 1–4. 10.1109/FUZZ45933.2021.9494346

An, Y., & Yan, Y. (2022). Intelligent retrieval method of library document information based on hidden topic mining. *WEB INTELLIGENCE*, *20*(2), 93–102. doi:10.3233/WEB-210484

Anand, R., & Beel, J. (2020). *Auto-Surprise: An Automated Recommender-System (AutoRecSys) Library with Tree of Parzens Estimator (TPE) Optimization.* doi:10.1145/3383313.3411467

Ansari, N., Vakilimofrad, H., Mansoorizadeh, M., & Amin, M. (2021). Using data mining techniques to predict user's behavior and create recommender systems in libraries and information centers. *Global Knowledge Memory and Communication*, *70*(6–7), 538–557. doi:10.1108/GKMC-04-2020-0058

Asemi, A., & Asemi, A. (2018). Artificial Intelligence (AI) application in Library Systems in Iran: A taxonomy study. *Library Philosophy and Practice (e-Journal).* https://digitalcommons.unl.edu/libphilprac/1840

Asemi, A., & Ko, A. (2021). A Novel Combined Business Recommender System Model Using Customer Investment Service Feedback. *34th Bled EConference Digital Support from Crisis to Progressive Change: Conference Proceedings*, 223–237. 10.18690/978-961-286-485-9.17

Asemi, A., Ko, A., & Nowkarizi, M. (2021). Intelligent libraries: A review on expert systems, artificial intelligence, and robots. *Library Hi Tech*, *39*(2), 412–434. doi:10.1108/LHT-02-2020-0038

Bak, A., Kozik, V., Walczak, M., Fraczyk, J., Kaminski, Z., Kolesinska, B., Smolinski, A., & Jampilek, J. (2018). Towards Intelligent Drug Design System: Application of Artificial Dipeptide Receptor Library in QSAR-Oriented Studies. *Molecules (Basel, Switzerland)*, *23*(8), 1964. Advance online publication. doi:10.3390/molecules23081964 PMID:30082652

Chen, M., McNab, A., & Zhang, W. (2021, September 21). How to improve the university library intelligent knowledge service: A system dynamics model. *Journal of Information Science*. Advance online publication. doi:10.1177/01655515211042240

Chen, M., & Shen, C. (2019). The correlation analysis between the service quality of intelligent library and the behavioral intention of users. *The Electronic Library*, *38*(1), 95–112. doi:10.1108/EL-07-2019-0163

Cheng, T., & Hou, H. (2022). Intelligent Educational Evaluation of Research Performance between Digital Library and Open Government Data (vol 12, 791, 2022). *Applied Sciences-Basel*, *12*(21). Advance online publication. doi:10.3390/app122110925

Coba, L., Confalonieri, R., & Zanker, M. (2022). RecoXplainer: A Library for Development and Offline Evaluation of Explainable Recommender Systems. *IEEE Computational Intelligence Magazine*, *17*(1), 46–58. doi:10.1109/MCI.2021.3129958

Collins, A., Tkaczyk, D., Aizawa, A., & Beel, J. (2018). Position Bias in Recommender Systems for Digital Libraries. doi:10.1007/978-3-319-78105-1_37

Cox, A., Pinfield, S., & Rutter, S. (2019). The intelligent library Thought leaders' views on the likely impact of artificial intelligence on academic libraries. *Library Hi Tech*, *37*(3), 418–435. doi:10.1108/LHT-08-2018-0105

Cruz-Cunha, M. M., Miranda, I. M., & Gonçalves, P. (2013). What is Fuzzy Logic? In *Handbook of Research on ICTs and Management Systems for Improving Efficiency in Healthcare and Social Care* (Vols. 1–2). IGI Global. doi:10.4018/978-1-4666-3990-4

Gade, A., & Angal, Y. (2018). Real-Time Intelligent NI myRIO-Based Library Management Robotic System Using LabVIEW. doi:10.1007/978-981-10-5520-1_36

Gao, Y. (2018). Implementation of an Intelligent Library System Based on WSN and RFID. *International Journal of Online Engineering*, *14*(5), 211–224. doi:10.3991/ijoe.v14i05.8601

Gao, Y. (2020). Intelligent library knowledge innovation service system based on multimedia technology. *Personal and Ubiquitous Computing*, *24*(3), 333–345. doi:10.100700779-019-01269-2

Gao, Z., & Zou, G. (2021). Intelligent Data Mining of Computer-Aided Extension Residential Building Design Based on Algorithm Library. *Complexity*, *2021*, 1–9. Advance online publication. doi:10.1155/2021/6690746

Gu, B., & Tanoue, K. (2022). Research on Library Space Layout and Intelligent Optimization Oriented to Readers' Needs. *Mathematical Problems in Engineering*, *2022*, 1–12. Advance online publication. doi:10.1155/2022/4426091

Huang, Z., Chung, W., Ong, T.-H., & Chen, H. (2002). A graph-based recommender system for digital library. *Proceedings of the 2nd ACM/IEEE-CS Joint Conference on Digital Libraries*, 65–73. 10.1145/544220.544231

Iantovics, L., Rotar, C., & Nechita, E. (2018). Intelligent University Library Information Systems to Support Students Efficient Learning. doi:10.1007/978-3-030-04224-0_17

Jamil, H. (2021). Architecture of an Intelligent Personal Health Library for Improved Health Outcomes. doi:10.1109/ICDH52753.2021.00012

Jiang, S., & Yang, X. (2022). Analysis of the Influence of Library Information on the Utilization of Regional Environmental and Ecological Resources: From the Perspective of Intelligent Adaptive Learning. *Journal of Environmental and Public Health*, *2022*, 1–12. Advance online publication. doi:10.1155/2022/1110105 PMID:36213048

Jing, Z. (2019). *Logical Construction and Realization of Intelligent Service in Ai Library*. doi:10.1109/ICSGEA.2019.00069

Jomsri, P. (2018). FUCL mining technique for book recommender system in library service. doi:10.1016/j.promfg.2018.03.081

Kaushal, V., & Yadav, R. (2022). The Role of Chatbots in Academic Libraries: An Experience-based Perspective. *Journal of the Australian Library and Information Association*, *71*(3), 215–232. doi:10.1080/24750158.2022.2106403

Koliarakis, A., Krouska, A., Troussas, C., & Sgouropoulou, C. (2022). *Modified collaborative filtering for hybrid recommender systems and personalized search: The case of digital library*. doi:10.1109/SMAP56125.2022.9942020

Landbot. (2021). *Product Recommendation Chatbot: No-Code Tutorial*. Landbot.Io. https://landbot.io/blog/product-recommendation-chatbot-for-e commerce

Liu, Y. (2022). Investigating users' willingness of acceptance for background music service in intelligent library. *Library Hi Tech*, *40*(1), 33–44. doi:10.1108/LHT-02-2019-0052

Miller, R. A., & Shortliffe, E. H. (2022). Corrigendum to: The roles of the US National Library of Medicine and Donald A.B. Lindberg in revolutionizing biomedical and health informatics. *Journal of the American Medical Informatics Association : JAMIA*, *29*(5), 1025. doi:10.1093/jamia/ocac026 PMID:35226056

Nguyen, P., Di Ruscio, D., Di Rocco, J., Di Sipio, C., & Di Penta, M. (2021). *Adversarial Machine Learning: On the Resilience of Third-party Library Recommender Systems*. doi:10.1145/3463274.3463809

Nie, Y. (2018). The Way to Construct the Intelligent Library-Taking Nanyang College Library as an example. Academic Press.

Palo, T., & Tähtinen, J. (2013). Networked business model development for emerging technology-based services. *Industrial Marketing Management*, *42*(5), 773–782. doi:10.1016/j.indmarman.2013.05.015

Pang, L. (2022). Intelligent Big Information Retrieval of Smart Library Based on Graph Neural Network (GNN) Algorithm. *Computational Intelligence and Neuroscience, 2022*, 1–12. Advance online publication. doi:10.1155/2022/1475069 PMID:35875784

Paullier, A., & Sotelo, R. (2020). *A Recommender Systems' algorithm evaluation using the Lenskit library and MovieLens databases.* doi:10.1109/BMSB49480.2020.9379914

Qian, Y. (2022). *The Semantic Framework of Library Intelligent Question Answering System Based on Exploratory Search Behavior.* doi:10.1109/CCAI55564.2022.9807737

Rhanoui, M., Mikram, M., Yousfi, S., Kasmi, A., & Zoubeidi, N. (2022). A hybrid recommender system for patron driven library acquisition and weeding. *Journal of King Saud University-Computer and Information Sciences, 34*(6), 2809–2819. doi:10.1016/j.jksuci.2020.10.017

Rodriguez, S., & Mune, C. (2022). Uncoding library chatbots: Deploying a new virtual reference tool at the San Jose State University library. *RSR. Reference Services Review, 50*(3/4), 392–405. doi:10.1108/RSR-05-2022-0020

Rubei, R., Di Ruscio, D., Di Sipio, C., Di Rocco, J., & Nguyen, P. (2022). Providing upgrade plans for third-party libraries: A recommender system using migration graphs. *Applied Intelligence, 52*(10), 12000–12015. doi:10.100710489-021-02911-4

Rubei, R., Di Sipio, C., Di Rocco, J., Di Ruscio, D., & Nguyen, P. (2022). *Endowing third-party libraries recommender systems with explicit user feedback mechanisms.* doi:10.1109/SANER53432.2022.00099

Schuemie, M. J., van der Straaten, P., Krijn, M., & van der Mast, C. A. P. G. (2004, July 5). *Research on Presence in Virtual Reality: A Survey (world).* Http://Www.Liebertpub.Com/Cpb

Shang, S., Yu, Z., Geng, A., Xu, X., Ma, H., & Wang, G. (2022). Intelligent Optimization Method of Resource Recommendation Service of Mobile Library Based on Digital Twin Technology. *Computational Intelligence and Neuroscience, 2022*, 1–10. Advance online publication. doi:10.1155/2022/3582719 PMID:36065374

Shen, Y. (2018). *Digital Library Systems in Intelligent Infrastructure for Human-Centered Communities: Qualitative Research.* doi:10.1145/3197026.3203894

Shen, Y. (2019). Emerging scenarios of data infrastructure and novel concepts of digital libraries in intelligent infrastructure for human-centred communities: A qualitative research. *Journal of Information Science*, *45*(5), 691–704. doi:10.1177/0165551518811459

Shin, D. (2021). The effects of explainability and causability on perception, trust, and acceptance: Implications for explainable AI. *International Journal of Human-Computer Studies*, *146*, 102551. doi:10.1016/j.ijhcs.2020.102551

Simovic, A. (2018). A Big Data smart library recommender system for an educational institution. *Library Hi Tech*, *36*(3), 498–523. doi:10.1108/LHT-06-2017-0131

Soleimanzade, N., Asemi, A., Cheshmehsohrabi, M., & Shabani, A. (2019). The Scientific Information Exchange General Model at Digital Library Context: Internet of Things. *Library Philosophy and Practice*, *2019*, 21–38.

Su, H. (2021). Design of the online platform of intelligent library based on machine learning and image recognition. *Microprocessors and Microsystems*, *82*, 103851. Advance online publication. doi:10.1016/j.micpro.2021.103851

Su, X., & Chen, N. (2022). Intelligent Information Service System of Smart Library Based on Virtual Reality and Eye Movement Technology. *Scientific Programming*, *2022*, 1–12. Advance online publication. doi:10.1155/2022/9174756

Sun, H. (2020). Research on interest reading recommendation method of intelligent library based on big data technology. *Web Intelligence*, *18*(2), 121–131. doi:10.3233/WEB-200434

Sun, L., Li, Y., & Lu, Y. (2022). Construction of Cloud Library Intelligent Service Platform Relying on Artificial Neural Network. *Mobile Information Systems*, *2022*, 1–11. Advance online publication. doi:10.1155/2022/6259127

Tejeda-Lorente, A., Bernabe-Moreno, J., Porcel, C., & Herrera-Viedma, E. (2018). Using Bibliometrics and Fuzzy Linguistic Modeling to Deal with Cold Start in Recommender Systems for Digital Libraries. doi:10.1007/978-3-319-66827-7_36

Trichur Narayanan, R. (2021). Recommender System: Personalizing User Experience or Scientifically Deceiving Users? *2021 the 5th International Conference on Information System and Data Mining,* 138–144. 10.1145/3471287.3471303

Vekaria, K., Calyam, P., Sivarathri, S. S., Wang, S., Zhang, Y., Pandey, A., Chen, C., Xu, D., Joshi, T., & Nair, S. (2021). Recommender-as-a-service with chatbot guided domain-science knowledge discovery in a science gateway. *Concurrency and Computation*, *33*(19), e6080. doi:10.1002/cpe.6080 PMID:35495546

Wang, H., & Ding, J. (2022). Development Strategy of Intelligent Digital Library without Human Service in the Era of "Internet plus". *Computational Intelligence and Neuroscience*, *2022*, 1–11. Advance online publication. doi:10.1155/2022/7892738 PMID:36262604

Wang, Y., He, H., & Liu, T. (2019). *Intelligent Library Information Service Terminal Based on BDS and Wireless Communication*. Academic Press.

Wang, Y., & Wei, Y. (2019). *Research on Library functional layout based on Intelligent occupying system*. doi:10.1088/1742-6596/1176/3/032010

Wilson, S. (2021). User Experience Desires Personalization from Academic Library Websites. *School of Information Student Research Journal*, *11*(1). Advance online publication. doi:10.31979/2575-2499.110109

Xie, K., Liu, Z., Fu, L., & Liang, B. (2019). Internet of Things-based intelligent evacuation protocol in libraries. *Library Hi Tech*, *38*(1), 145–163. doi:10.1108/LHT-11-2017-0250

Xie, X. (2018). Study on the Intelligent Terminal Innovation of Personalized Active Service of Mobile Library-A Case Study of Zhejiang University of Media and Communications. doi:10.23977smme.2018.62224

Xu, K. (2022). Intelligent Library Service and Management Based on IoT Assistance and Text Recommendation. *Journal of Sensors*, *2022*, 1–10. Advance online publication. doi:10.1155/2022/3660135

Zhang, J., Zhang, Y., & Wu, X. (2018). *Research of Intelligent Library Based on RFID Technology*. doi:10.1109/ITME.2018.00129

Zhang, W. (2019). Research on the Construction of Community Library Intelligent Service System from the Perspective of Smart City. doi:10.2991schd-19.2019.1

Zhang, Y., & Yan, B. (2019). MIL-61 and Eu3+@MIL-61 as Signal Transducers to Construct an Intelligent Boolean Logical Library Based on Visualized Luminescent Metal-Organic Frameworks. *ACS Applied Materials & Interfaces*, *11*(22), 20125–20133. doi:10.1021/acsami.9b00179 PMID:31088052

Chapter 5

The Pivotal Role of the Internet of Things in Library Innovation:
A Step Towards Shifting Landscape of Libraries

Javaid Ahmad Wani
ⓘ https://orcid.org/0000-0003-2968-375X
University of Kashmir, India

Arshia Ayoub
University of Kashmir, India

ABSTRACT

In libraries, the internet of things (IoT) has enormous promise. It is not the sensor on the object, but it does have the capacity for electronic tracking and data exchange. This has created a plethora of new opportunities to improve the efficiency of libraries and, as a result, the user experience of various services. IoT has played a critical role in transforming libraries into Smart places by improving services such as "collection management," "instruction," "data security," and so on. It can also allow real-time global connection of a large library system. In this context, the chapter looks to explain IoT and its numerous technologies. The study additionally indicates possible library areas for implementation and how they affect library effectiveness in terms of patrons, operations, and technological innovation. This chapter will serve as a road map for scholars, practitioners, and readers interested in IoT, technosphere, and tech habitat.

DOI: 10.4018/978-1-6684-8671-9.ch005

INTRODUCTION

The Internet of Things (IoT) is regarded as a remarkable emerging technology, ranking above artificial intelligence, robots, and other comparable technological advancements (Burrus, 2014; Nord et al., 2019). IoT has unveiled new technology prospects by drastically expanding its influence area from "business and industry to home, health care, and knowledge management" (Al-athwari & Hossain, 2022; Khanna, & Kaur, 2020). The concept of the "Internet of Things" is 16 years old, but the concept of a linked gadget is far older. Earlier, this approach was named "pervasive computing". Kevin Ashton came up with the term "Internet of Things" in 1999 while engaged in an initiative at "Procter & Gamble" (P&G) to improve "supply chain management" by connecting "radio frequency identification" (RFID) data to the Internet (Ashton,2009; Kumar,2018, Rajan, et al., 2022). Since its beginnings, several researchers have described and defined it (Ali et al., 2015; Ben-Daya et al., 2017; Gigli & Koo, 2011; Huang et al., 2016; Lee & Lee, 2015; Lund et al., 2014; Madakam et al., 2015; Ornes, 2016), yet there is no conventional description. IoT is defined by Nord et al. (2019) as a system that consists of uniquely identified endpoints or "things" that record and distribute data. According to Ghasempour (2019), IoT is the interconnection of people and things in any environment, with anybody and anything utilising any system or platform. As a result, the basic essence of all the prevalent definitions may be summed as IoT is a word that depicts items engaging via the Internet (Alkhariji et al., 2023; Khanna, & Kaur, 2020), enabling almost limitless potential and interconnections.

Architecture

IoT communication architectures not only allow devices to connect to the internet but also allow them to communicate with one another autonomously (Petersen, et al. 2014). Several reputable organisations and working groups (ITU, IEEE, Cisco, and ETSI) have created IoT frameworks based on their technical specifications, corporate and service structures, and so forth. To date, however, none of them has been able to deliver the standardised structure (Gupta & Quamara, 2018). Initially, the acknowledged IoT architecture consisted of three layers: "perception, network, and application" (Bandyopadhyay, et al., 2013; Buyya et al., 2009; Miao et al., 2010; Sharma, 2015; Tilak, 2002; Yan et al., 2014). This 3-layer design became unsustainable as IoT growth progressed and 5-layer architecture emerged (Figure 1). The job of the "perception Layer" is to identify entities and gather information about them using embedded sensors. Depending on the entity's identifying mechanism, these devices can be Smart cards, biometrics, or indeed heat cameras (Anne et al., 2016). Similarly, the "application layer" is responsible for providing global oversight for

Figure 1. IoT architecture
Source: Figure by Author(s)

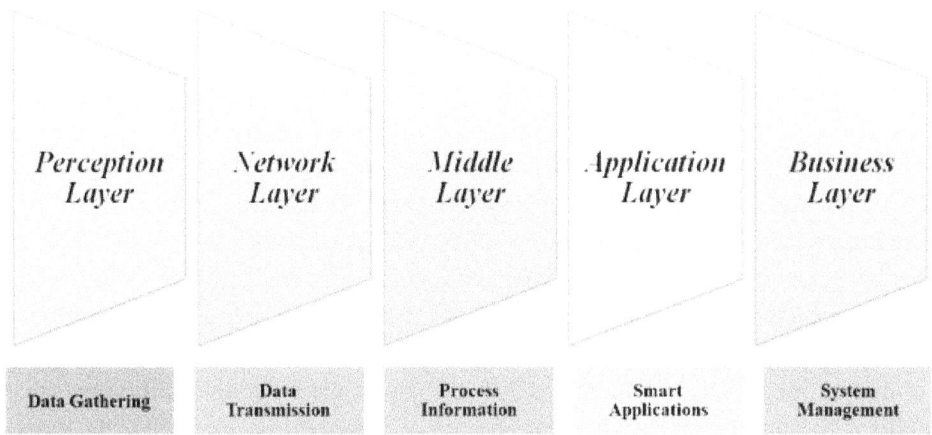

the whole programme according to the data of the item handled at the "middleware layer". It also makes IoT systems more sophisticated, authorized, and reliable.

The processing layer's major role is to assess information obtained from the network layer and make judgments based on the findings of pervasive computation Mouha (2021). The transport layer is responsible for transferring sensor data from the perception layer to the processing layer via networks such as *"WiFi, LAN, LTE, RFID, and Bluetooth"*. Ultimately, the business layer visualises data and analytics from the application layer and applies this information to upcoming initiatives and objectives (Chopra, et al., 2019; Miao et al., 2010; Said & Masud, 2013).

Application

IoT technology has proliferated in numerous fields, thus influencing every area of human lives (Lee & Lee, 2015; Wortmann & Fluchter, 2015). The broad spectrum of applications provided by IoT has been enumerated in various studies (Atzori et al., 2010; Bartje, 2016; Chopra et al., 2019; Khan et al., 2012; Lee & Lee, 2015; Lueth, 2015) that broadly encompass "healthcare", "smart environment" (home, office, plant), "personal" and "social domain". Even though IoT is recent technology the existing body of research confirms its value to the business, consumers, and the government. This is because of the potential of IoT-enabled technologies in enhancing "client interaction", making "economic choices", "object tracking", and management. (Nord et al., 2019).

IOT AND LIBRARIES

Librarianship as a significant and vital field has become an expedient scenario to be aided by the latest technology which led to the development of the concept *Smart library*. Smart library involves the employment of the latest information technology tools that enable to delivery of intelligent services and management techniques. The smart library applies "natural language processing" (NLP), "deep learning" (DL), "recommender systems", "machine vision", "smart acquisition" and "IoT-based approaches" (Anoop, & Ubale, 2020, Bagal, & Saindane, 2019; Choi, & Joo, 2019; Determe, et al., 2022; Jiang et al., 2021; Lin, et al., 2019). Of these, IoT technology is the most promising because it can be applied to most library services and facilitates the relationships between patrons, and information resources. In his study, Liang (2018) claimed that IoT has enormous potential in libraries since it allows for automated tracking and sharing of information. Its apex is the smart completion of all elements by the library, with no manual interference, to produce

Figure 2.
Source: Khan et al. (2022)

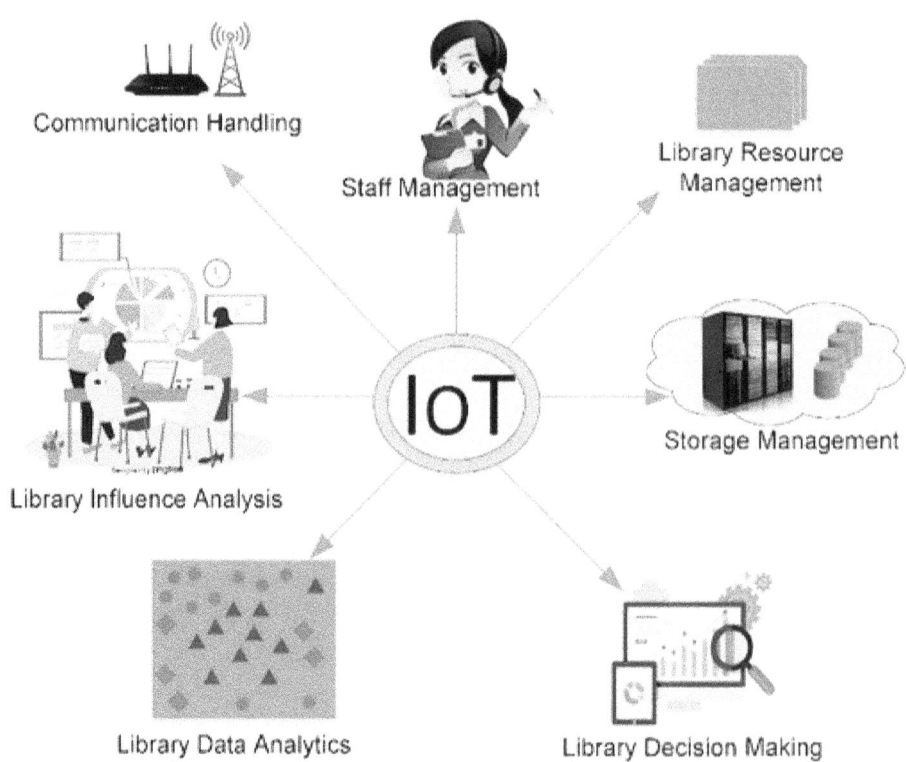

a "smart equilibrium" (Tian, 2020). As a result Khan et al., (2022) highlighted that libraries are incorporating IoT technologies into various library activities to make libraries more efficient and helpful to their customers (see Figure 2).

Moreover, before implementing IoT projects requires careful consideration of costs. This includes initial infrastructure setup, IoT sensor expenses, data plans, and integration with existing systems. Ongoing costs involve maintenance, support, and cybersecurity measures. Planning for scalability and regulatory compliance is crucial. Thorough research, expert consultation, and pilot projects are recommended to make informed decisions and assess feasibility before full deployment in libraries.

The implementation of IoT in libraries has a substantial impact on staff training and professional development. Library staff must acquire new technical skills, data analytics capabilities, and knowledge of privacy and security measures. They need to adapt to changes in workflows and be prepared to offer user assistance for IoT-enabled services. Continuous learning and a willingness to embrace innovation are crucial for maximizing the benefits of IoT in libraries.

PROMINENT IOT TECHNOLOGIES IN LIBRARIES

While IoT has many applications across various sectors, libraries have also begun to explore the potential of this technology to enhance their operations and services. From collection management to security and environmental monitoring, IoT offers a range of possibilities that can help academic libraries improve their efficiency and effectiveness in serving their patrons. The following sections will explore some of the key IoT devices applied in libraries and the potential benefits that this technology can offer to librarians and users alike.

Radio Frequency Identification (RFID)

RFID is a technology that uses radio waves to identify and track objects, including library materials such as books, media, and other items. The literature showed that RFID technology has several significant benefits for libraries (Li, 2022; Malipatil, et al., 2020; Timoshenko, 2020; Thamizhmaran, 2021). RFID technology can improve the efficiency of library operations by automating many of the manual tasks associated with managing library materials (Li, 2022). With RFID, librarians can quickly and easily locate items, check them in and out, and process returns. RFID can also enhance the security of library collections (Thamizhmaran, 2021). RFID tags can be used to trigger alarms if items are removed from the library without being checked out, and to track the location of materials within the library to prevent theft (Li, 2022). RFID technology can also increase accessibility for library users

(Thamizhmaran, 2021). With RFID-enabled self-checkout machines, patrons can easily check out items without the need for assistance from library staff. Additionally, RFID tags can be used to enable the automatic sorting and retrieval of materials, making it easier for patrons to locate the items they need (Malipatil, et al., 2020). RFID technology can help libraries better manage their collections by providing real-time data on item usage and location (Li, 2022). This data can help librarians make informed decisions about collection development and resource allocation. Overall, RFID technology can have a significant impact on the efficiency, security, accessibility, and management of library collections. As such, it has become an increasingly important tool for librarians seeking to enhance their operations and services in today's rapidly changing digital landscape (Timoshenko, 2020).

Near-Field Communication (NFC)

Near-field communication (NFC) is a short-range wireless technology that allows devices to communicate with each other when they are brought into proximity (typically within a few centimetres). NFC is a subset of Radio Frequency Identification (RFID) technology and operates at a frequency of 13.56 MHz. NFC is used in a variety of applications, including mobile payments, access control, and data transfer. In the context of libraries, NFC technology can be used in several ways (Engel, & Lie, 2022; Liao, & Shieh, 2015; Singh, 2020; Viet, 2019). NFC technology can be used to enable contactless checkout of library materials. Patrons can simply tap their NFC-enabled library card or mobile device on an NFC reader to check out items, without the need for physical contact or manual scanning of barcodes. It can also be used to enable location-based services in libraries (Singh, 2020). For example, NFC tags can be placed on bookshelves or other areas of the library, and when a patron taps their NFC-enabled device on the tag, they can be directed to relevant resources or receive recommendations based on their location. Further, NFC technology can also be used to enable access control for restricted library areas, such as staff-only rooms or special collections (Liao, & Shieh, 2015). By tapping their NFC-enabled library card or mobile device on an NFC reader, authorized users can gain access to these areas. Furthermore, NFC can be used to enable easy sharing of information between library patrons and staff. For example, NFC tags can be placed on library displays or exhibits, and patrons can tap their NFC-enabled devices on the tags to access additional information or multimedia content (Engel, & Lie, 2022). Overall, NFC technology offers a range of possibilities for enhancing library services and operations, particularly in the areas of contactless checkout, location-based services, access control, and information sharing.

Wireless Sensor Networks (WSNs)

Wireless Sensor Networks (WSNs) can be useful in libraries to monitor various environmental conditions such as temperature, humidity, light levels, air quality, and noise levels. This information can be used to optimize the library environment for the comfort of the users and preservation of the materials (Begum, & Nandury, 2023; Behera, et al., 2019; Hu, et al., 2022; Panahi, & Bayılmış, 2023). Here is a possible architecture for a WSN in a library:

1. Sensor Nodes: These are the devices that collect data about the environment. They could be temperature, humidity, light, air quality or noise sensors. The number of sensor nodes needed would depend on the size and layout of the library (Hu, et al., 2022).
2. Gateway: This is the device that collects data from the sensor nodes and forwards it to the central server. The gateway could be a laptop, Raspberry Pi or any other device capable of receiving and transmitting data (Begum, & Nandury, 2023).
3. Central Server: This is the device that stores and analyzes the data collected from the sensor nodes. The server could be a cloud-based server or an on-premise server located in the library (Panahi, & Bayılmış, 2023).
4. User Interface: This is the interface that allows library staff to view and analyze the data collected by the sensor nodes. The user interface could be a web-based dashboard or a mobile application (Hu, et al., 2022).
5. Alert System: This is a feature that sends alerts to library staff in case of any environmental condition that goes beyond a predefined threshold (Begum, & Nandury, 2023). For example, if the temperature or humidity goes beyond a certain level, the staff could be notified to take appropriate action.

The WSN architecture described above could be used to monitor the library environment and make data-driven decisions about how to optimize the environment for the comfort of the users and the preservation of the materials. For example, if the data shows that the temperature is consistently high in a certain area of the library, the staff could take action to improve the air conditioning or ventilation in that area (Panahi, & Bayılmış, 2023).

Bluetooth Low Energy (BLE)

Bluetooth Low Energy (BLE) is a wireless technology designed for low-power, short-range communication between devices. It is a subset of the Bluetooth standard and was introduced with Bluetooth 4.0. BLE is designed for use in devices that require low energy consumption, such as sensors, health monitors, and other IoT devices (Antevski, et al., 2016). BLE offers a range of benefits over traditional Bluetooth, including longer battery life, smaller size, and lower cost. BLE is designed to operate on low power, making it ideal for use in devices that require long battery life. BLE devices can operate on a single coin cell battery for years (Davidson & Piche, 2016). BLE has a shorter range than traditional Bluetooth, making it ideal for use in devices that need to communicate over short distances, such as wearables and sensors (Antevski, et al., 2016). BLE has a lower data rate than traditional Bluetooth, but this is sufficient for most IoT devices (Mendoza-Silva, et al., 2019). BLE is backwards compatible with *Bluetooth*, allowing devices to communicate with each other (Wang, 2019). BLE offers security features to protect against unauthorized access and data theft (Wang, 2019). BLE uses profiles to define the functionality of a device, allowing devices to communicate with each other (Mendoza-Silva, et al., 2019). BLE supports mesh networking, allowing devices to communicate with each other in a network without the need for a central hub (Jeon, et al., 2018). BLE is inexpensive to implement, making it ideal for use in low-cost devices (Wang, 2019). Therefore, BLE is a versatile wireless technology that offers a range of benefits over traditional Bluetooth. It is designed for use in low-power, short-range applications and is ideal for use in IoT devices.

Wireless Fidelity (Wi-Fi)

Wireless Fidelity, commonly known as Wi-Fi, is a technology that allows electronic devices to connect to the internet or communicate with one another wirelessly using radio waves (Determe, et al., 2022). Wi-Fi is a trademark of the Wi-Fi Alliance, which is an organization that promotes the adoption and certification of Wi-Fi technology. Wi-Fi technology works by transmitting data through radio waves over a wireless network (Goddard, 2020). Wi-Fi networks can be set up in a variety of settings, including homes, businesses, schools, airports, and public spaces. Wi-Fi technology has become a ubiquitous feature of modern life, enabling people to access the internet and connect without the need for physical wires or cables (Park, 2019). However, Wi-Fi technology can be vulnerable to security threats if proper precautions are not taken, such as using strong passwords and encryption protocols (Determe, et al., 2022; Goddard, 2020; Park, 2019).

Wireless Fidelity (Wi-Fi) has become an essential component of the Internet of Things (IoT) in libraries, providing seamless connectivity for both students and staff. Libraries have an extensive collection of digital resources, such as e-books, e-journals, and research databases (Park, 2019; Determe, et al., 2022). Wi-Fi technology allows students and staff to access these resources from their own devices without the need for physical access to the library. Smart libraries utilize IoT technology to optimize operations, including occupancy monitoring, temperature control, and book tracking. Wi-Fi technology can be used to support these IoT devices, enabling them to communicate and function seamlessly (Goddard, 2020). With Wi-Fi technology, students and staff can access library resources and services from their mobile devices. This includes mobile-friendly library catalogues, e-books, and other digital resources. Wi-Fi technology allows for collaboration among students and staff, promoting innovative learning and research activities (Determe, et al., 2022). Students can form study groups and work together on projects, accessing digital resources and sharing files seamlessly (Park, 2019). Wi-Fi technology has become an indispensable tool in university libraries, enabling students and staff to access digital resources, collaborate on projects, and optimize library operations through IoT technology.

Long-Range Wide-Area Network (LoRaWAN)

Long-Range Wide-Area Network (LoRaWAN) is a low-power wireless communication technology designed to connect IoT devices over long distances, typically several kilometres or more (Almuhaya, et al., 2022). Libraries invest a lot of money in books, technology equipment, and other materials, and LoRaWAN technology can be used to track these assets more effectively (Kufakunesu, et al., 2020). By placing IoT sensors on library resources, staff can monitor their location and status remotely, reducing the risk of loss or theft. Libraries can use LoRaWAN technology to create smart shelves that automatically detect when books are removed or returned (Minhaj, et al., 2023). This allows staff to track which books are in use and which are available, making it easier for students and staff to find the resources they need (Garrido-Hidalgo, et al., 2023). LoRaWAN sensors can be used to monitor temperature, humidity, and other environmental factors in the library. This can help staff to maintain optimal conditions for preserving library resources, reducing the risk of damage from factors such as moisture or mould (Noura, et al., 2020). Libraries can use LoRaWAN technology to monitor how different areas of the library are being used (Minhaj, et al., 2023). By tracking occupancy levels and usage patterns, staff can optimize library space more effectively, ensuring that users have access to the resources they need without overcrowding or underutilizing certain areas (Kufakunesu, et al., 2020). LoRaWAN technology can be used to enhance library security by enabling IoT sensors that detect unauthorized access, intrusions, and other security breaches. This allows staff to respond quickly and effectively to security threats, reducing the risk of theft or damage to library resources (Garrido-Hidalgo, et al., 2023). LoRaWAN technology is important to monitor occupancy levels in different areas of the library (Noura, et al., 2020). This data is being used to optimize library space more effectively, ensuring that students have access to the resources they need without overcrowding certain areas (Minhaj, et al., 2023). Further, it can be used to track the location of borrowed items and monitor their condition. This has helped to reduce losses and improve the borrowing experience for members. Overall, LoRaWAN technology has the potential to transform the way that libraries operate, providing valuable insights into library usage patterns, enhancing security, and optimizing library resources and services. By utilizing LoRaWAN technology, libraries can improve the user experience, reduce costs, and better meet the needs of their users (Kufakunesu, et al., 2020).

Robotic Systems

Robotic systems are rapidly transforming the way libraries operate. These technologies have the potential to enhance the user experience, improve the efficiency of library operations, and provide valuable insights into library usage patterns (Prabhu, et al., 2021). Robotic systems are becoming increasingly common in libraries, particularly for tasks such as book sorting and shelving (Angal, & Gade, 2017). These systems are typically comprised of autonomous robots that can navigate through the library, identifying and sorting books based on their location and category (De Sarkar, 2022). Robotic systems are particularly useful in large libraries where manual book sorting can be time-consuming and labour-intensive (Bajeh, et al., 2021). By using robotic systems, libraries can reduce the amount of time and effort required for book sorting, freeing up staff to focus on other tasks (Prabhu, et al., 2021). One of the main benefits of robotic systems in libraries is their ability to integrate with IoT devices. For example, sensors can be placed on library shelves to detect when books are removed or returned (De Sarkar, 2022). This data can be sent to the robotic system, which can use to identify and sort books more efficiently (Tella, 2020). Robotic systems can also be integrated with other IoT devices, such as smart lighting and environmental monitoring systems, (Tella, 2020) to optimize library operations and enhance the user experience (Bajeh, et al., 2021). IoT devices are also being used in libraries to provide real-time data on library usage patterns (Gupta et al., 2020). For example, sensors can be placed on library doors to detect when students enter and leave the library (Prabhu, et al., 2021). This data can be used to track library usage patterns, such as peak usage times and popular areas of the library (Tella, 2020). By analyzing this data, libraries can optimize library space more effectively, ensuring that students have access to the resources they need without overcrowding certain areas (De Sarkar, 2022). Another area where IoT is being used in university libraries is asset tracking. Libraries invest a significant amount of money in books, technology equipment, and other materials, and IoT-enabled asset-tracking systems can help libraries keep track of these resources more effectively (Gupta et al., 2020). By placing sensors on library resources, staff can monitor their location and status remotely, reducing the risk of loss or theft (Prabhu, et al., 2021). One of the main advantages of using IoT and robotic systems in libraries is their ability to enhance the user experience (Bajeh, et al., 2021). Robotic systems and IoT devices, provide valuable insights into library usage patterns (Angal, & Gade, 2017), enhancing the user experience, and improving the efficiency of library operations (De Sarkar, 2022). As these technologies continue to develop, we can expect to see even more innovative applications in the library setting, helping to make libraries more accessible, efficient, and user-friendly than ever before (Gupta et al., 2020).

APPLICATIONS OF IOT IN VARIOUS OPERATIONS OF LIBRARIES

Applications of the Internet of Things (IoT) in libraries are vast and diverse, spanning various operations and services. Some eye-catching key applications of IoT in different aspects of libraries are as follows:

Management of Resources

Every organisation requires resource management, and as such, a sustainability intake management system for libraries can lead to increased sustainability usage efficiency. Sustainability can be effectively scheduled based on practical needs by utilising IoT. The suggested framework comprises multi-source sensors, a sensor network, and a server, with sensors collecting data such as "temperature", "humidity", "personal information", and so on. The acquired data are synchronised with one another and with the server and are based on prior sensor data. The control centre installed on the server can perform smart planning for equipment consumption, for example, the light system can be dynamically scheduled based on time series data analysis (Yang, et al. 2019). Xue et al. (2021) advocated using IoT solutions to handle light shading in the library to maximise natural light utilisation efficiency. The reader can access the device control system and select smart mode. The sensors begin detecting the intensity and angles of light and adapt to the settings chosen by the readers.

RFID sensors can also monitor, evaluate, and manage these usage statistics (Daniel, et al. 2019; Maepa, et al., 2021). Users can use their mobile devices to "log in, reserve, scan, check-in, and cancel seats" in the reading room. Real-time data can be used to intelligently schedule based on the changing needs of space consumption status (Liu, et al. 2021; Zhou, 2019). In a similar example, the Internet of Things has been utilized to examine room occupancy with "access authorization" based on a *"facial recognition"* system (Upala & Wong, 2019). Another study offered a system that can help students organise all types of subject study groups spontaneously and collaboratively. This approach, which is based on the combination of "BLE beacons", "Wi-Fi", and "K-nearest neighbour" (KNN), is offered to aid in the orderly and efficient setup of library space usage (Antevski, et al., 2016).

Stock Verification/User Statistics

IoT may also be used to provide data on library resource utilisation, a map displaying the most heavily utilised locations of the library, and a rating of satisfaction with the user's experience. *"Deep learning aided computational RFIDs"* (CRFID) technology is used to recognize and track user reading actions such as picking up the book, surfing the title, skimming through the page, altering the location of the book, reading, and borrowing (Bai, et al. 2020). The importance of embracing these tools is that the data they collect can proactively assess and evaluate the user's behaviours and demands (Jung & Kim, 2020; Preuveneers & Berbers, 2008; Zhou et al., 2017).

Cataloguing

The "Internet of Things" can be a valuable tool in cataloguing and organizing library resources in libraries. IoT can be applied to cataloguing in libraries in different ways. IoT sensors can be integrated with RFID technology to track books and other materials as they are shelved or retrieved (Ozeer et al., 2019). This can help in the automation of the cataloguing process, enabling more efficient inventory management and tracking. This can enable faster and more accurate identification of resources, leading to more efficient cataloguing and retrieval (Schopfel, 2018). IoT sensors can be used to automatically scan books and other materials, and then generate tags or metadata for each resource. IoT sensors can be used to analyze user behaviour and preferences, and then generate personalized recommendations for books and other resources. This can enhance the user experience and encourage greater engagement with the library's resources (Ozeer et al., 2019; Schopfel, 2018). Moreover, Wi-Fi (wireless fidelity) has been widely used in navigation (Asemi et al., 2020) to help people find books. IoT technology can be used to intelligently deliver optimal navigation (Asemi et al., 2020) for the precise localization of resources (Cao et al., 2018) like books, journals, microfiche, etc.). Further, *"Bluetooth low energy"* (BLE) can also be used in navigation to discover specific locations, as well as social connectedness for student discussion and learning (Uttarwar, & Chong, 2017). The entire procedure can save time and require no direct human intervention, which is especially important during COVID-19 (Temiz, & Salelkar, 2020).

Circulation

With the IOT applying to libraries, self-service operations for the issue, return, library book renewal and library dues have become a new norm. RFID tags embedded in the library resources can be read even if they are covered by objects, and a stack of books can be read simultaneously (Dong-Ying Li et al., 2016) with high efficiency (Cui, et al., 2019). With advancements *"UHF band RFID system"* is now being introduced that provides a wider reading range and larger multi-reading capacity, which can be introduced into IOT System for "library materials management" with higher efficiency and convenience (Dong-Ying Li et al., 2016; Irankunda, 2021). The book sorting system is another significant tool for improving the efficiency of the circulation service. This has been achieved by the deployment of barcode and deep learning-based optical character recognition (OCR) (Shi et al., 2021). Book inventories can also be prepared by drone robots. Based on the "visual localization" and "optical character recognition" OCR (Irankunda, 2021), the tags on the books can be identified by the drone robot and further used for inventory service (Martinez-Martin, et al., 2021). This has improved the experience of users (Irankunda, 2021) in libraries.

Location Guide

IoT can provide a self-guided "virtual tour" of the library (Pera, 2015). Beacons like "wireless devices" can be installed in various sections of the library, when users visit a particular section, their mobile phone will play a video or audio providing details about that section (Bansal, et al., 2018). *Virtual library cards* can be provided access to resources of libraries using a mobile app. Pera (2015) mentioned that if a user accesses the library catalogue to locate the required resources, the library app installed on the user's mobile device will provide a map of the library guiding the user to the exact location of resources (Pujara & Satyanarayanab 2015).

Reference Service

IoT may also use client data to make personalised suggestions based on existing borrowing records, using real-time data (Bai, et al. 2020). When a researcher searches a database for material on the topic associated with his or her research interest, other related resources that may be of interest to them can be suggested (Bi et al., 2022). "RNN deep learning aided computational RFIDs" (CRFID) technology is used to detect and identify readers' reading behaviours, such as "picking up the book", browsing the title, skimming through the page, altering the location of the book, reading, and borrowing (Bi et al., 2022). The activity data are collected via

CRFID, and the collected data naturally have sequential properties and are ideally adapted for the "RNN-based technology" for machine learning, as well as providing constructive ideas for meeting the readers' needs (Bai, et al. 2020). Another novel technique to suggest is based on "cloud computing" and "recommender systems". This method not only helps readers to assess borrowed books, but it also recommends novels based on previous information that is stored online. Users may benefit from the entire system suite (Bi et al., 2022).

Preservation

Another interesting applicability of IoT is monitoring environmental conditions of rooms devoted to storing books and other documents, where it is important to steadily check "temperature, humidity, pressure, light, and pollution agents" (Perles, et al., 2018). Guaranteeing proper environmental conditions is strategic in these contexts, and it belongs to those care activities required to adequately maintain the persistent accumulation of knowledge (Cerf, 2019). IoT provides the provision of constantly measuring these parameters, to ensure optimal conditions, supporting the paper "conservation and preservation". Monti, et al. (2019) presented a novel preservation solution aiming at preserving the "environmental conditions". This solution can provide better preservation for the environmental conditions, with the help of a "multi-sensor and monitoring data visualization".

Services for Physically Special Users

The "Internet of Things" can potentially be utilised to give services to *"physically special users"*. Karthikeyan et al. (2021) developed a revolutionary library audio solution that uses OCR, deep learning, and ultrasonic sensors to assist blind persons in hearing and comprehending book content. The ultrasonic sensor is initially used to help figure out the range between the OCR equipment and the book. Following that, OCR is used in connection with deep learning to recognise the substance of the physical book and transform it into textual information. Finally, the written content is translated into the associated audio file and played over the headset (Karthikeyan et al., 2021). This can effectively assist blind persons in reading actual books in libraries.

CONCLUSION

In the age of the Internet of Things (IoT), libraries are likely to undergo significant transformations to remain relevant and vibrant hubs for information and community engagement (Ho et al., 2023). Libraries will increasingly become smart spaces, equipped with IoT sensors and devices that enhance the user experience. Smart kiosks, interactive displays, and mobile apps will provide seamless access to information, personalized recommendations, and real-time updates on events and resources (Igbinovia, 2021). IoT technologies will enable libraries to bridge the gap between physical and digital collections. RFID tags, QR codes, or other IoT-enabled tracking systems will streamline the borrowing and returning of materials, making it easier and quicker for patrons to access the resources they need. Libraries will leverage IoT data to offer personalized services to their patrons (Leorke et al., 2018). By analyzing user behaviour and preferences, libraries can deliver tailored recommendations, program suggestions, and targeted outreach initiatives, fostering a stronger sense of connection between the library and its community. IoT-generated data will play a crucial role in shaping library operations. Libraries will rely on data analytics to optimize resource allocation, space usage, and service offerings based on the changing needs and preferences of their users. IoT will enable libraries to foster community engagement through interactive displays, digital signage, and collaborative spaces (Ola, 2020). Libraries will host a diverse range of technology-driven events, workshops, and programs that cater to the interests and aspirations of their local communities. Libraries will prioritize the implementation of IoT-based security solutions to safeguard their collections and ensure the safety of their patrons. Smart surveillance, access control systems, and environmental monitoring will help prevent theft, vandalism, and potential safety hazards (Kaba, & Ramaiah, 2019). Libraries will embrace IoT solutions to minimize their environmental impact. Energy-efficient lighting, HVAC systems, and automated utilities management will contribute to reducing the library's carbon footprint and aligning with sustainability goals (Khan, et al., 2022). IoT-enabled devices will allow libraries to extend their reach beyond physical spaces. Virtual libraries, accessible through smart devices and IoT platforms, will offer digital resources, virtual reality experiences, and online collaborations, catering to patrons who prefer remote access to information and services. Libraries may form partnerships with tech companies, IoT providers, and educational institutions to stay at the forefront of technological innovations and ensure access to cutting-edge IoT solutions. Libraries will need to navigate the ethical implications of collecting and utilizing IoT-generated data. Ensuring data privacy, security, and transparency will be critical to maintaining patron trust and compliance with relevant regulations. As the IoT continues to advance, libraries will embrace these technologies to remain dynamic and adaptive institutions that meet the evolving needs of their communities (Taha Osman, et al., 2023; Igbinovia & Okuonghae, 2021). By strategically integrating IoT solutions, libraries can ensure their continued relevance as essential knowledge hubs in the digital age.

LIMITATIONS

Research on IoT in libraries may encounter several limitations that could influence the scope and depth of the study. The implementation of IoT in libraries can vary significantly across different institutions, and each library may have unique requirements and resources. As a result, findings from one library may not be easily generalizable to others. The adoption of IoT in libraries is a relatively recent development, and the research may be constrained by a limited historical dataset, making it challenging to assess long-term impacts. Access to a diverse range of libraries and their willingness to participate in the study could affect the representativeness of the sample. Moreover, certain libraries might be more inclined to adopt IoT, leading to potential selection bias. Collecting data on library users' interactions with IoT devices may raise privacy concerns. Addressing ethical considerations related to data collection and user consent could limit the type and amount of data that can be gathered. Detailed studies on IoT implementation may require considerable resources, which might be a limitation for researchers with restricted budgets or access to necessary technology. Longitudinal studies examining the impact of IoT over time might be limited by the availability of historical data on library operations and services. IoT implementations may involve sensitive data related to library operations or user behavior. Researchers must ensure data confidentiality and security while still conducting a comprehensive analysis. IoT technology is evolving rapidly, and the devices and systems used during the research may become outdated or replaced by newer versions, affecting the relevance of findings. The success of IoT in libraries largely depends on user acceptance. The research might be limited in its ability to fully gauge user attitudes, preferences, and barriers to adoption. The research might not be able to address the long-term sustainability and maintenance of IoT systems in libraries, which is critical for understanding their ongoing impact. External factors, such as changes in funding, policy, or technology trends, might influence the effectiveness and sustainability of IoT in libraries, which could be challenging to account for in the research. Despite these limitations, research on IoT in libraries offers valuable insights into the potential benefits, challenges, and opportunities for improving library services and user experiences. Researchers can address these limitations through thoughtful study design, careful data collection, and transparent reporting of their findings.

FUTURE RESEARCH

Future research on IoT in libraries should focus on longitudinal studies to understand long-term impacts, examine user acceptance and experiences, address privacy and security implications, and conduct cost-benefit analyses. Additionally, investigating integration, interoperability, sustainability, and best practices can optimize IoT implementations. Exploring IoT in specialized libraries, accessibility improvements, and smart spaces will contribute to a more comprehensive understanding of IoT's role in libraries and drive innovative approaches for community engagement and library services.

REFERENCES

Al-athwari, B., & Hossain, M. A. (2022). IoT Architecture: Challenges and Open Research Issues. In *Proceedings of 2nd International Conference on Smart Computing and Cyber Security.* Springer. 10.1007/978-981-16-9480-6_39

Ali, Z. H., Ali, H. A., & Badawy, M. M. (2015). Internet of Things (IoT): Definitions, challenges and recent research directions. *International Journal of Computer Applications*, *128*(1), 37–47. doi:10.5120/ijca2015906430

Alkhariji, L., De, S., Rana, O., & Perera, C. (2023). Semantics-based privacy by design for Internet of Things applications. *Future Generation Computer Systems*, *138*, 280–295. doi:10.1016/j.future.2022.08.013

Almuhaya, M. A., Jabbar, W. A., Sulaiman, N., & Abdulmalek, S. (2022). A survey on Lorawan technology: Recent trends, opportunities, simulation tools and future directions. *Electronics (Basel)*, *11*(1), 164. doi:10.3390/electronics11010164

Angal, Y., & Gade, A. (2017, February). Development of library management robotic system. In *2017 International Conference on Data Management, Analytics and Innovation (ICDMAI)* (pp. 254-258). IEEE. 10.1109/ICDMAI.2017.8073520

Anoop, A., & Ubale, N. A. (2020, August). Cloud Based Collaborative Filtering Algorithm for Library Book Recommendation System. In *2020 Third International Conference on Smart Systems and Inventive Technology (ICSSIT)* (pp. 695-703). IEEE. 10.1109/ICSSIT48917.2020.9214243

Antevski, K., Redondi, A. E., & Pitic, R. (2016, July). A hybrid BLE and Wi-Fi localization system for the creation of study groups in smart libraries. In *2016 9th IFIP wireless and mobile networking conference (WMNC)* (pp. 41-48). IEEE. 10.1109/WMNC.2016.7543928

Asemi, A., Ko, A., & Nowkarizi, M. (2020). Intelligent libraries: A review on expert systems, artificial intelligence, and robot. *Library Hi Tech*, *39*(2), 412–434. doi:10.1108/LHT-02-2020-0038

Ashton, K. (2009). That ''Internet of Things'' thing: In the Real World Things Matter More than Ideas. *RFiD Journal*. Retrieved from http://kevinjashton.com/2009/06/22/the-internet-of-things/

Bagal, D., & Saindane, P. (2019, July). Librany-A Face Recognition and QR Code Technology based Smart Library System. In *2019 International Conference on Communication and Electronics Systems (ICCES)* (pp. 253-258). IEEE. 10.1109/ICCES45898.2019.9002530

Bai, R., Zhao, J., Li, D., Lv, X., Wang, Q., & Zhu, B. (2020). RNN-based demand awareness in smart library using CRFID. *China Communications*, *17*(5), 284–294. doi:10.23919/JCC.2020.05.021

Bajeh, A. O., Mojeed, H. A., Ameen, A. O., Abikoye, O. C., Salihu, S. A., Abdulraheem, M., & Awotunde, J. B. (2021). Internet of robotic things: its domain, methodologies, and applications. In *Emergence of Cyber Physical System and IoT in Smart Automation and Robotics: Computer Engineering in Automation* (pp. 135–146). Springer International Publishing. doi:10.1007/978-3-030-66222-6_9

Bandyopadhyay, S., Balamuralidhar, P., & Pal, A. (2013). Interoperations among IoTstandards. *Journal of ICT Standardization*, *1*(2), 253–270. doi:10.13052/jicts2245-800X.12a9

Bansal, A., Arora, D., & Suri, A. (2018). Internet of things: Beginning of new era for libraries. *Library Philosophy and Practice*, 1.

Barile, L. (2011). Mobile technologies for libraries: A list of mobile applications and resources for development. *College & Research Libraries News*, *72*(4), 222–228. doi:10.5860/crln.72.4.8545

Begum, B. A., & Nandury, S. V. (2023). Data Aggregation Protocols for WSN and IoT Applications–A Comprehensive Survey. *Journal of King Saud University-Computer and Information Sciences*.

Behera, T. M., Mohapatra, S. K., Samal, U. C., Khan, M. S., Daneshmand, M., & Gandomi, A. H. (2019). I-SEP: An improved routing protocol for heterogeneous WSN for IoT-based environmental monitoring. *IEEE Internet of Things Journal*, *7*(1), 710–717. doi:10.1109/JIOT.2019.2940988

Ben-Daya, M., Hassini, E., & Bahroun, Z. (2017). Internet of Things and supply chainmanagement: A literature review. *International Journal of Production Research*, •••, 1–24. doi:10.1080/00207543.2017.1402140

Bi, S., Wang, C., Zhang, J., Huang, W., Wu, B., Gong, Y., & Ni, W. (2022). A Survey on Artificial Intelligence Aided Internet-of-Things Technologies in Emerging Smart Libraries. *Sensors (Basel)*, *22*(8), 2991. doi:10.339022082991 PMID:35458974

Brindley, L. (2006). Re-defining the library. *Library Hi Tech*, *24*(4), 484–495. doi:10.1108/07378830610715356

Burrus, D. (2014). *The Internet of Things is Far Bigger than Anyone Realizes*. Retrieved from https://www.wired.com/insights/2014/11/the-internet-of-things-bigger/

Buyya, R., Yeo, C. S., Venugopal, S., Broberg, J., & Brandic, I. (2009). Cloud computing and emerging IT platforms: Vision, hype, and reality for delivering computing as the 5th utility. *Future Generation Computer Systems*, *25*(6), 599–616. doi:10.1016/j.future.2008.12.001

Cao, G., Liang, M., & Li, X. (2018). How to make the library smart? The conceptualization of the smart library. *The Electronic Library*, *36*(5), 811–825. doi:10.1108/EL-11-2017-0248

Castiglione, J. (2008). Facilitating employee creativity in the library environment: An important managerial concern for library administrators. *Library Management*, *29*(3), 159–172. doi:10.1108/01435120810855296

Cerf, V. G. (2019). Libraries considered hazardous. *Communications of the ACM*, *62*(2), 5–5. doi:10.1145/3302508

Choi, Y., & Joo, S. (2019, June). Topic detection of online book reviews: preliminary results. In *2019 ACM/IEEE Joint Conference on Digital Libraries (JCDL)* (pp. 418-419). IEEE. 10.1109/JCDL.2019.00098

Chopra, K., Gupta, K., & Lambora, A. (2019, February). Future internet: The internet of things-a literature review. In *2019 International Conference on Machine Learning, Big Data, Cloud and Parallel Computing (COMITCon)* (pp. 135-139). IEEE. 10.1109/COMITCon.2019.8862269

Cui, L., Zhang, Z., Gao, N., Meng, Z., & Li, Z. (2019). Radio frequency identification and sensing techniques and their applications—A review of the state-of-the-art. *Sensors (Basel)*, *19*(18), 4012. doi:10.339019184012 PMID:31533321

Daniel, O. C., Ramsurrun, V., & Seeam, A. K. (2019, September). Smart library seat, occupant and occupancy information system, using pressure and RFID sensors. In *2019 Conference on Next Generation Computing Applications (NextComp)* (pp. 1-5). IEEE. 10.1109/NEXTCOMP.2019.8883610

Davidson, P., & Piché, R. (2016). A survey of selected indoor positioning methods for smartphones. *IEEE Communications Surveys and Tutorials*, *19*(2), 1347–1370. doi:10.1109/COMST.2016.2637663

De Sarkar, T. (2022). Internet of Things (IOT) and library services. *Library Hi Tech News*, *39*(9), 18–22. doi:10.1108/LHTN-06-2022-0079

Determe, J. F., Azzagnuni, S., Singh, U., Horlin, F., & De Doncker, P. (2022). Monitoring large crowds with WiFi: A privacy-preserving approach. *IEEE Systems Journal*, *16*(2), 2148–2159. doi:10.1109/JSYST.2021.3139756

Engel, M. M., & Lie, H. D. (2022). Library Self Service System Using Nfc And 2fa Google Authenticator. *Jurnal Teknik Informatika (Jutif)*, *3*(3), 753–761.

Garrido-Hidalgo, C., Roda-Sanchez, L., Ramírez, F. J., Fernández-Caballero, A., & Olivares, T. (2023). Efficient online resource allocation in large-scale LoRaWAN networks: A multi-agent approach. *Computer Networks*, *221*, 109525. doi:10.1016/j.comnet.2022.109525

Ghasempour, A. (2019). Internet of things in smart grid: Architecture, applications, services, key technologies, and challenges. *Inventions (Basel, Switzerland)*, *4*(1), 22. doi:10.3390/inventions4010022

Gigli, M., & Koo, S. (2011). Internet of Things: Services and applications categorization. *Advances in Internet of Things*, *1*(2), 27–31. doi:10.4236/ait.2011.12004

Goddard, J. (2020). Public Libraries Respond to the COVID-19 Pandemic, Creating a New Service Model. *Information Technology and Libraries*, *39*(4). Advance online publication. doi:10.6017/ital.v39i4.12847

Gopalakrishnan, K. (2020). Security vulnerabilities and issues of traditional wireless sensors networks in IoT. *Principles of internet of things (IoT) ecosystem: Insight paradigm*, 519-549.

Gupta, B. B., & Quamara, M. (2018). An overview of Internet of Things (IoT): Architectural aspects, challenges, and protocols. *Concurrency and Computation*, *e4946*(21). doi:10.1002/cpe.4946

Gupta, T., Tripathi, R., Shukla, M. K., & Mishra, S. (2020). Design and development of IoT based smart library using line follower robot. *Int. J. Emerg. Technol*, *11*(2), 1105–1109. https://d1wqtxts1xzle7.cloudfront.net/82644987/Design_20and_20Development_20of_20IOT_20Based_20Smart_20Library_20using_20Line_20Follower_20Robot_20Rohit_20Tripathi_202798-libre.pdf

Ho, C. Y., Chiu, D. K., & Ho, K. K. (2023). Green space development in academic libraries: a case study in Hong Kong. In *Global Perspectives on Sustainable Library Practices* (pp. 142–156). IGI Global.

Hu, B., Tang, W., & Xie, Q. (2022). A two-factor security authentication scheme for wireless sensor networks in IoT environments. *Neurocomputing*, *500*, 741–749. doi:10.1016/j.neucom.2022.05.099

Huang, X., Craig, P., Lin, H., & Yan, Z. (2016). SecIoT: A security framework for theInternet of Things. *Security and Communication Networks*, *9*(16), 3083–3094. doi:10.1002ec.1259

Igbinovia, M. O. (2021). Internet of things in libraries and focus on its adoption in developing countries. *Library Hi Tech News*, *38*(4), 13–17. doi:10.1108/LHTN-05-2021-0020

Igbinovia, M. O., & Okuonghae, O. (2021). Internet of Things in contemporary academic libraries: Application and challenges. *Library Hi Tech News*, *38*(5), 1–4. doi:10.1108/LHTN-05-2021-0019

Irankunda, D. (2021). *Development of radio frequency identification based library management and anti-theft system: a case of east African community region* [Doctoral dissertation]. NM-AIST.

Jeon, K. E., She, J., Soonsawad, P., & Ng, P. C. (2018). BLE beacons for internet of things applications: Survey, challenges, and opportunities. *IEEE Internet of Things Journal*, *5*(2), 811–828. doi:10.1109/JIOT.2017.2788449

Jiang, M., Hu, Y., Worthey, G., Dubnicek, R. C., Underwood, T., & Downie, J. S. (2021, September). Evaluating BERT's Encoding of Intrinsic Semantic Features of OCR'd Digital Library Collections. In *2021 ACM/IEEE Joint Conference on Digital Libraries (JCDL)* (pp. 308-309). IEEE. 10.1109/JCDL52503.2021.00045

Jung, S., & Kim, S. (2020). A Study of Promoting Method a traditional market by implementing RFID technology and 6W1H context awareness. *Journal of Convergence for Information Technology*, *10*(10), 9–14.

Kaba, A., & Ramaiah, C. K. (2019). The internet of things: opportunities and challenges for libraries. *Library Philosophy and Practice (e-Journal)*, 3704. Retrieved from https://digitalcommons.unl.edu/libphilprac/3704

Karthikeyan, D., Arumbu, V. P., Surendhirababu, K., Selvakumar, K., Divya, P., Suhasini, P., & Palanisamy, R. (2021). Sophisticated and modernized library running system with OCR algorithm using IoT. *Indonesian Journal of Electrical Engineering and Computer Science*, *24*(3), 1680–1691. doi:10.11591/ijeecs.v24.i3.pp1680-1691

Khan, A. U., Zhang, Z., Chohan, S. R., & Rafique, W. (2022). Factors fostering the success of IoT services in academic libraries: A study built to enhance the library performance. *Library Hi Tech*, *40*(6), 1976–1995. doi:10.1108/LHT-06-2021-0179

Khan, R., Khan, S. U., Zaheer, R., & Khan, S. (2012, December). Future internet: the internet of things architecture, possible applications and key challenges. In *2012 10th international conference on frontiers of information technology* (pp. 257-260). IEEE. 10.1109/FIT.2012.53

Khanna, A., & Kaur, S. (2020). Internet of things (IoT), applications and challenges: A comprehensive review. *Wireless Personal Communications*, *114*(2), 1687–1762. doi:10.100711277-020-07446-4

Kufakunesu, R., Hancke, G. P., & Abu-Mahfouz, A. M. (2020). A survey on adaptive data rate optimization in lorawan: Recent solutions and major challenges. *Sensors (Basel)*, *20*(18), 5044. doi:10.339020185044 PMID:32899454

Kumar, V. (2018). Application of internet of things for smart libraries: An overview. *International Journal of Multidisciplinary Education Research*, *7*(2). https://www.researchgate.net/publication/328475575_APPLICATIONS_OF_INTERNET_OF_THINGS_FOR_SMART_LIBRARIES_AN_OVERVIEW

Lee, I., & Lee, K. (2015). The Internet of Things (IoT): Applications, investments, andchallenges for enterprises. *Business Horizons*, *58*(4), 431–440. doi:10.1016/j.bushor.2015.03.008

Leorke, D., Wyatt, D., & McQuire, S. (2018). More than just a library: Public libraries in the smart city.*City. Cultura e Scuola*, *15*, 37–44. doi:10.1016/j.ccs.2018.05.002

Li, D.-Y., Xie, S.-D., Chen, R.-J., & Tan, H.-Z. (2016). Design of Internet of Things System for Library Materials Management using UHF RFID. *2016 IEEE International Conference on RFID Technology and Applications (RFID-TA)*, 44-48. 10.1109/RFID-TA.2016.7750755

Li, D. Y., Xie, S. D., Chen, R. J., & Tan, H. Z. (2016, September). Design of Internet of Things system for library materials management using UHF RFID. In 2016 IEEE international conference on RFID technology and applications (RFID-TA) (pp. 44-48). IEEE. doi:10.1109/RFID-TA.2016.7750755

Li, X. (2006). Library as incubating space for innovations: Practices, trends and skill sets. *Library Management*, *27*(6/7), 370–378. doi:10.1108/01435120610702369

Li, X. (2022). Application Analysis of RFID in Library Automation Management. *International Journal of e-Collaboration*, *18*(2), 1–10. doi:10.4018/IJeC.304036

Liang, X. (2018). Internet of Things and its applications in libraries: A literature review. *Library Hi Tech*, *38*(1), 67–77. doi:10.1108/LHT-01-2018-0014

Liao, P. K., & Shieh, J. C. (2015, July). The development of library mobile book-finding system based on NFC. In *2015 IIAI 4th International Congress on Advanced Applied Informatics* (pp. 148-153). IEEE. 10.1109/IIAI-AAI.2015.198

Lin, W. H., Chang, S. S., Li, P., Chiu, T. T., & Lou, S. J. (2019, May). Exploration of usage behavioral model construction for university library electronic resources from Deep Learning Multilayer perceptron. In *2019 IEEE International Conference on Consumer Electronics-Taiwan (ICCE-TW)* (pp. 1-2). IEEE. 10.1109/ICCE-TW46550.2019.8991756

Liu, Y., Ye, H., & Sun, H. (2021). Mobile phone library service: Seat management system based on WeChat. *Library Management*, *42*(6/7), 421–435. doi:10.1108/LM-09-2020-0132

Lund, D., MacGillivray, C., Turner, V., & Morales, M. (2014). Worldwide and Regional Internet of Things (IoT) 2014-2020 Forecast: A Virtuous Circle of Proven Value and Demand. *International Data Corporation (IDC)*. Retrieved from http://branden.biz/wp-content/uploads/2017/06/IoT-worldwide_regional_2014-2020-forecast.pdf

Madakam, S., Ramaswamy, R., & Tripathi, S. (2015). Internet of Things (IoT): A literature review. *Journal of Computer and Communications*, *3*(5), 164–173. doi:10.4236/jcc.2015.35021

Maepa, M. R., & Moeti, M. N. (2021, December). IoT-Based Smart Library Seat Occupancy and Reservation System using RFID and FSR Technologies for South African Universities of Technology. In *Proceedings of the International Conference on Artificial Intelligence and Its Applications* (pp. 1-8). 10.1145/3487923.3487933

Malathy, S., & Kantha, P. (2013). Application of mobile technologies to libraries. *DESIDOC Journal of Library and Information Technology*, *33*(5).

Malipatil, N., Roopashree, V., Gowda, R. S., Shobha, M. R., & Kumar, H. S. (2020). RFID based library management system. *Int. J. Res. Eng. Sci. Manag*, *3*(7), 112–115.

Martinez-Martin, E., Ferrer, E., Vasilev, I., & Del Pobil, A. P. (2021). The uji aerial librarian robot: A quadcopter for visual library inventory and book localisation. *Sensors (Basel)*, *21*(4), 1079. doi:10.339021041079 PMID:33557363

Mendoza-Silva, G. M., Matey-Sanz, M., Torres-Sospedra, J., & Huerta, J. (2019). BLE RSS measurements dataset for research on accurate indoor positioning. *Data*, *4*(1), 12. doi:10.3390/data4010012

Minhaj, S. U., Mahmood, A., Abedin, S. F., Hassan, S. A., Bhatti, M. T., Ali, S. H., & Gidlund, M. (2023). Intelligent Resource Allocation in LoRaWAN using Machine Learning Techniques. *IEEE Access : Practical Innovations, Open Solutions*, *11*, 10092–10106. doi:10.1109/ACCESS.2023.3240308

Monti, L., Mirri, S., Prandi, C., & Salomoni, P. (2019, December). Preservation in smart libraries: An experiment involving iot and indoor environmental sensing. In *2019 IEEE Global Communications Conference (GLOBECOM)* (pp. 1-6). IEEE. 10.1109/GLOBECOM38437.2019.9014149

Morgan, J. (2014). A Simple Explanation of 'The Internet of Things'. *Forbes*. Retrieved from https://www.forbes.com/sites/jacobmorgan/2014/05/13/simple-explanation-internet-things-that-anyone-can-understand/

Mouha, R. (2021). Internet of Things (IoT). *Journal of Data Analysis and Information Processing*, *9*(2), 77–101. doi:10.4236/jdaip.2021.92006

Negi, D. S. (2014). Using mobile technologies in libraries and information centers. *Library Hi Tech News*, *31*(5), 14–16. doi:10.1108/LHTN-05-2014-0034

Ngu, A. H. H., Gutierrez, M., Metsis, V., Nepal, S., & Sheng, M. Z. (2017). IoT Middleware: A Survey on Issues and Enabling technologies. *IEEE Internet of Things Journal*, *4*(1), 1–20. doi:10.1109/JIOT.2016.2615180

Nord, J. H., Koohang, A., & Paliszkiewicz, J. (2019). The Internet of Things: Review and theoretical framework. *Expert Systems with Applications*, *133*, 97–108. doi:10.1016/j.eswa.2019.05.014

Noura, H., Hatoum, T., Salman, O., Yaacoub, J. P., & Chehab, A. (2020). LoRaWAN security survey: Issues, threats and possible mitigation techniques. *Internet of Things*, *12*, 100303. doi:10.1016/j.iot.2020.100303

Ola, A. C. (2020). Funding and ICT use as determinants of sustainable library development. *Covenant Journal of Library and Information Science*, *3*(1), 32–49.

Ornes, S. (2016). The Internet of Things and the explosion of interconnectivity. *Proceedings of the National Academy of Sciences of the United States of America*, *113*(40), 11059–11060. doi:10.1073/pnas.1613921113 PMID:27702874

Ozeer, A., Sungkur, Y., & Nagowah, S. D. (2019, December). Turning a traditional library into a smart library. In *2019 International Conference on Computational Intelligence and Knowledge Economy (ICCIKE)* (pp. 352-358). IEEE. 10.1109/ICCIKE47802.2019.9004242

Panahi, U., & Bayılmış, C. (2023). Enabling secure data transmission for wireless sensor networks based IoT applications. *Ain Shams Engineering Journal*, *14*(2), 101866. doi:10.1016/j.asej.2022.101866

Park, S. (2019). Analyzing Library Space Use Patterns in a Public Library through Smartphone WiFi. *Jeongbo Gwan'ri Hag'hoeji*, *36*(1), 295–313.

Pera, M. (2015). *Libraries and the "Internet of Things". OCLC Symposium shows benefits, raises questions.* Available at: https://americanlibrariesmagazine.org/blogs/thescoop/libraries-and-the-internet-of-things/

Pera, M. (2015). *Libraries and the "Internet of Things": OCLC Symposium shows benefits, raises questions.* Retrieved from https://americanlibrariesmagazine.org/blogs/thescoop/libraries-and-the-internet-of-things/

Perles, A., Pérez-Marín, E., Mercado, R., Segrelles, J. D., Blanquer, I., Zarzo, M., & Garcia-Diego, F. J. (2018). An energy-efficient internet of things (IoT) architecture for preventive conservation of cultural heritage. *Future Generation Computer Systems*, *81*, 566–581. doi:10.1016/j.future.2017.06.030

Prabhu, V. S., Abinaya, R. M., Archana, G., & Aishwarya, R. (2021). IoT-based automatic library management robot. In *Advances in Smart System Technologies: Select Proceedings of ICFSST 2019* (pp. 567-575). Springer Singapore. 10.1007/978-981-15-5029-4_46

Preuveneers, D., & Berbers, Y. (2008). Internet of things: A context-awareness perspective. In *The Internet of things* (pp. 287–308). Auerbach Publications.

Pujara, S. M., & Satyanarayanab, K. V. (2015). Internet of Things and libraries. *Annals of Library and Information Studies*, *62*, 186–190.

Rajan, S. S., Esmail, M., & Musthafa, K. M. (2022). Repositioning Academic Libraries as a Hub of technology enhanced learning space: Innovations and Challenges. *Library Philosophy and Practice*, 1-14. Retrieved from https://media.proquest.com/media/hms/PFT/1/D0O6M?_s=KnuGNUzf73G%2BpYJcmo%2BmkOrneZ8%3D

Said, O., & Masud, M. (2013). Towards internet of things: Survey and future vision. *International Journal of Computer Networks*, *5*(1), 1–17.

Schopfel, J. (2018). Smart libraries. *Infrastructures*, *3*(4), 43. doi:10.3390/infrastructures3040043

Sharma, A. (2015). *The tech behind Internet of things*. Retrieved from https://www.pcquest.com/the-tech-behind-internet-things/

Sheng, X., & Sun, L. (2007). Developing knowledge innovation culture of libraries. *Library Management*, *28*(1/2), 36–52. doi:10.1108/01435120710723536

Shi, X., Tang, K., & Lu, H. (2021). Smart library book sorting application with intelligence computer vision technology. *Library Hi Tech*, *39*(1), 220–232. doi:10.1108/LHT-10-2019-0211

Singh, N. K. (2020). Near-field Communication (NFC). *Information Technology and Libraries*, *39*(2). Advance online publication. doi:10.6017/ital.v39i2.11811

Taha Osman, G. T. O., Hassan Mohammed, M., & Babiker Al Shafei, I. (2023). The Internet of Things in information institutions: Concept, use and challenges. *BSU-Journal of Pedagogy and Curriculum*, *2*(3), 153–172. doi:10.21608/bsujpc.2023.278442

Tella, A. (2020). Robots are coming to the libraries: Are librarians ready to accommodate them? *Library Hi Tech News*, *37*(8), 13–17. doi:10.1108/LHTN-05-2020-0047

Temiz, S., & Salelkar, L. P. (2020). Innovation during crisis: Exploring reaction of Swedish university libraries to COVID-19. *Digital Library Perspectives*, *36*(4), 365–375. doi:10.1108/DLP-05-2020-0029

Thamizhmaran, K. (2021). RFID for Library Management System. *Journal of Advancement in Communication System*, *4*(1).

Tian, L. (2020). Research on the construction of smart library based on the Internet of Things. *Journal of Library Science*, *42*(10), 101–104.

Tilak, S., Abu-Ghazaleh, N., & Heinzelman, W. (2002). A taxonomy of wireless microsensor network models. *ACM Mobile Computing and Communications Review*, *6*(2), 28–36. doi:10.1145/565702.565708

Timoshenko, I. (2020). RFID in Libraries: automatic identification and data collection technology for library documents. In Maintenance Management. IntechOpen. doi:10.5772/intechopen.82032

Upala, M., & Wong, W. K. (2019, April). IoT solution for smart library using facial recognition. *IOP Conference Series. Materials Science and Engineering*, *495*(1), 012030. doi:10.1088/1757-899X/495/1/012030

Uttarwar, M. L., & Chong, P. H. (2017). Bealib: A beacon enabled smart library system. *Wireless Sensor Network*, *9*(8), 302–310. doi:10.4236/wsn.2017.98017

Viet, D. H. (2019). *Near field communication enabled library* [Doctoral dissertation]. Vietnamese-German University. Retrieved from http://epub.vgu.edu.vn/handle/dlibvgu/545

Wang, C. S. (2019). An AR mobile navigation system integrating indoor positioning and content recommendation services. *World Wide Web (Bussum)*, *22*(3), 1241–1262. doi:10.100711280-018-0580-3

Wu, M., Lu, T. J., Ling, F. Y., Sun, J., & Du, H. Y. (2010, August). Research on the architecture of Internet of Things. In *2010 3rd international conference on advanced computer theory and engineering (ICACTE)* (Vol. 5, pp. V5-484). IEEE. 10.1109/ICACTE.2010.5579493

Xue, J., Wang, Y., & Wang, M. (2021). Smart Design of Portable Indoor Shading Device for Visual Comfort—A Case Study of a College Library. *Applied Sciences (Basel, Switzerland)*, *11*(22), 10644. doi:10.3390/app112210644

Yan, Z., Zhang, P., & Vasilakos, A. V. (2014). A survey on trust management forInternet of Things. *Journal of Network and Computer Applications*, *42*, 120–134. doi:10.1016/j.jnca.2014.01.014

Yang, C. J., Kang, H. B., Zhang, L., & Zhang, R. Y. (2019). A design of smart library energy consumption monitoring and management system based on IoT. In *Proceedings of the Fifth Euro-China Conference on Intelligent Data Analysis and Applications 5* (pp. 217-224). Springer International Publishing. 10.1007/978-3-030-03766-6_24

Zhou, D. (2019, October). Case Study on Seat Management of University Library Based on WeChat Public Number Client—Taking Jianghan University Library as an Example. In *2019 4th International Conference on Mechanical, Control and Computer Engineering (ICMCCE)* (pp. 630-6303). IEEE.

Zhou, L., Yan, S., & Zhu, X. (2017). Context-aware service model and evaluation research of smart library. *Research on Library Science*, *21*, 23–30.

KEY TERMS AND DEFINITIONS

BLE: It is a variation of the Bluetooth wireless standard designed for low power consumption. Its goal is to connect devices over relatively short range.

IoT: It refers to use of internet connected objects and system to obtain data gathered by embedded sensors, actuators in machines and other physical objects.

Library Operations: The major tasks performed in the librariesperformed in the library such as acquisition, cataloguing and classification, storage, retrieval, and service provision.

NFC: It is a short-range wireless connectivity technology that uses magnetic field induction to enable communication between devices.

RFID: It is a technology whereby digital data encoded in RFID tags or smart labels are captured by a reader via radio waves. RFID belongs to a group of technologies known as Automatic Identification and Data Capture (AIDC) that automatically identify object, collect data and save that data into computer.

Smart Library: The interconnection between different library devices, users and librarians in a wireless networking environment within a library will be called a smart library.

Wi-Fi: Wi-Fi is a wireless technology for networking that uses electromagnetic waves to transmit networks and allows high-speed data transfer over short distances.

Chapter 6
Chatbots Benefications Towards the Education Sector

Tanvi Jindal
Chitkara University, India

Ishika L. N. U.
Chitkara University, India

Parth Sharma
Chitkara University, India

Gurpreet Kaur
Chitkara University, India

ABSTRACT

One of the many promising developments in data innovation is chatbots. They were created as a brand-new user interface that would let users communicate with services merely through chat, replacing or enhancing the need for applications or website visits. Since digitalisation is pacing up around the globe, usage of chatbots is prevalent in many domains for the interface between customers and entities. However, there is still another industry where chatbots have a considerable amount of prospect: education. In the educational sector, chatbots are equipped but to a certain extent only. The aim of the study is to create awareness about potential of chatbots in learning processes and its various applications in the academic sector. Chatbots various verticals promote digitalisation in the future of education support systems enhancing the productivity and boosting growth as the education sector of any country is the foundation of its growing future.

DOI: 10.4018/978-1-6684-8671-9.ch006

INTRODUCTION

A chatbot is a piece of software that mimics human dialogue through text chats, voice commands or both. A chatbot, often known as a chatterbot, is an AI feature that can be integrated into and utilised with any significant messaging platform. Chatbots typically function in one of two ways: either by pattern recognition or with predetermined rules. Chatbots that follow predefined rules are, however, becoming a thing of the past as a result of improvements in AI technology. As technology continues to evolve, more companies are switching from using conventional platforms to doing business online. Businesses are bringing efficiency through technology by integrating AI techniques on their online platforms. Chatbots are an artificial intelligence approach that is expanding in both use and application. Virtual assistants like Google Assistant and Alexa are some instances of chatbot technology.

An automated programme known as a chatbot engages with clients much like a human would and is usually free to use. Customers can contact chatbots at any time of day or week; they are not restricted by time or place. Due to this, many firms that might not have the personnel or financial means to maintain staff working 24 hours a day are attracted to its implementation. Industry studies show that the COVID-19 significantly sped up chatbot deployment and user adoption on a global scale. The fact that every student in a classroom has a particular set of learning requirements and interests has long been recognised. As a result, everyone could benefit from a trained tutor's assistance. Ironically, even the costliest universities around the globe do not offer this kind of service. What is the best workable and affordable solution to this issue? Education-related chatbots. The majority of the time, basic courses can be delivered by chatbots. The idea is for chatbots to act as virtual counsellors and to adjust to the students' skill levels as they do so. They thereby adjust to the speed of their own learning. On the contrary side, recent proposals call for chatbots to act as vertical tutors and engage in conversation with every student. As a result, experts can assess them and determine which subjects they require assistance with.

Before the pandemic, the education sector already possessed considerable technical capabilities. In instance, data reveals that one of the top five industries adopting chatbots profitably is education. They are being utilized on all online platforms, including social media as well as corporate websites and applications. Schools are seeing satisfied children and happier staff as a result of tech-savvy pupils, parents and teachers having the opportunity to communicate with chatbots. Chatbots are becoming more and more popular in the education sector across a range of business applications, including online tutoring, student support, teacher's aide, managerial tool, evaluating and providing results.

With the use of technology, user is allowed to communicate with an artificial intelligence-based robot instead of humans (Molnár & Szuts 2018). Adoption of chatbots is essential in online classes with a large number of students since it makes it difficult for teachers to provide each student with one-on-one support (Winkler & Söllner 2018). Chatbots can help students learn in an educational setting by instantly supplying them with course material (Cunningham-Nelson et al. 2019), homework (Ismail & Ade-Ibijola 2019), practice questions (Sinha et al. 2020), and study materials (Mabunda 2020).

Chatbots are used to respond to thousands of commonly requested questions, some of which are repeated many times per day, in business sectors with customer care support. A fresh aspect is introduced to a company's website and the way its customer interacts with the business with these conversational agents (Rahim et al. 2022). With rising consumer expectations, it's critical to sustain an increasing trend in customer service by implementing chatbots as virtual agents (Adam et al. 2020). In the age of AI, chatbots are used more frequently in higher education institutions (HEIs) for educational goals by providing quick and tailored services to all individuals within the sector, including institutional employees and students (Sharafi et al. 2022). Leading universities and colleges have used AI-based chatbots for their college inquiry websites during the past few years (Rahim et al. 2022). These bots will work as 24/7 university advisers for a small percentage of the cost of hiring numerous human staff (Lee et al. 2021). Students will find it simpler to use this technology because they won't need to interact with the staff at the information desk (Rahim et al. 2022). Utilizing new technologies can improve academic and research outcomes (Aljami et al. 2018). Therefore, utilising various technology acquisition models and frameworks, extensive research has been done to identify the essential factors influencing the application of various innovations in the educational setting (Ma et al. 2017).

Depending on their AI level, chatbots have varying levels of comprehension. Chatbots gains information while enhancing its language abilities, learning how to tailor a message and developing its own communication and responding norms through subsequent experience. As a result, chatbots may respond to queries, offer solutions and resolve issues by comprehending the user's intents (Rahim et al. 2022).

REVIEW OF LITERATURE

The below stated information is regarding various discussions done in accordance to chatbot's implications and suitability in educational sector. Despite of its presence in other eminent sectors chatbot based on AI has a long way to develop in educational sector at different levels. The literature was retrieved by computerized database and hand search from reputed sources. For the study of research paper www.googlescholar is used, for the article (www.sciencedirect, www.spinger, www.proquest, etc.)

and for websites www.google is used. Past researches have already discussed the working and assessment of chatbot in educational settings stating various elements contributing in growth as well as stating both sides of the coin in different manner.

Kuhail et al. (2022) follows the guidelines described by Keele which includes further three steps. The objective of the study is to find out an organized summary to understand the utilization of chatbots in education sector by applying a variety of aspects. It shows that chatbots use a pre-arranged conversation pathway are assessed with investigation which results in an upgrade warning. The study predicted many educational chatbots access and how they allow learners across different spheres. In addition to future learning surveyed the consequences and benefits of chatbot in order to make learning more effective and efficient. Moreover, chatbots are flexible web platforms which give way to numerous devices such as laptops, mobile phones etc.

Rahim et al. (2022) collected data from post-graduate students via a study questionnaire by employing a purposive sampling approach where study states that in the field of information systems, the research on chatbot adoption in the setting of higher education institutions is yet relatively unexplored. Additionally, the majority of the current research on chatbot adoption in the HEI context is not very specialised in a student services solution viewpoint. This work seeks to discover characteristics that affect the efficacy of chatbot adoption in the HEI setting. This exploration uncovers that apparent trust is impacted by intelligence, plan, and morals. In the meantime, conduct aim is affected by apparent trust, execution hope, and propensity towards the utilization of chatbot applications in the HEI setting. Finally, the discoveries of this study can be useful to the HEI understudy administrations unit and can be an aide towards efficiency and promoting system in serving the understudies better.

Merel et al. (2022) focused to improve and elevate student's progress through chatbots in education sector. For this, inspections are done to take the view of teachers and collecting the evidences which will help in improving chatbot process. The study winded up by saying that there are various elements which contribute to the growth of chatbots in educational sector which will lead to preferred knowledgeable conclusions. Moreover, chatbots shed lights on the efficient ways of enhancing a student's personality.

Aldahne (2022) aims to determine the status of ai usage in the field of education wherein observed that chatbots usage provides assistance as we all have been gone through covid 19 pandemic which affected the usage of chatbots in a positive manner to enhance the productivity of students and teachers.

Bowman (2022) intended and conducted research regarding chatbots has led to a success among first year of higher education by creating a link between them and their respective institutions. This has led a promotion in student engagement and enhanced their skills. The study winded up by saying that there are different aspects of creating student-led online societies and the author also gives recommendations on how they can improve the learning of students in the near future.

Leonardi and Torchiano (2022) with aim of study to develop chatbots in such a way so that it generates a proper answer based on knowledge. This will help to test in real world scenarios. The study summarized by saying that chatbots makes the answers very effective and efficient and the answers came with full accuracy which helps in building a load of neural network.

Almada et al. (2022) summarized by saying that the proactive chatbot shows sound accuracy and more effective and efficient learning supports than other chatbots. This is so because the new chatbot framework uses knowledge from the PS2CLH model. The objective of the study is to introduce a new chatbot structure which will integrate the profile of students and to enhance chatbot elements to improve the interaction of students.

Kumar (2021) conducted study from second year bachelor of education when divided into two groups and exposed to same context with target to determine that chatbots are mainly considered for bookish purposes and are accomplished to supply customized learning. Moreover, educational chatbots are considered as a body of knowledge and gives a practical impact on the teaching-learning process. The study concluded that chatbots are still dealing with a number of consequences but still it is coping up with these situations and has become an efficient teaching tool in the education sector. Moreover, chatbots lead to effective and efficient encouragement learning conclusions which helps in practical knowledge. Mathew (2021) to check the consideration of chatbots as future electronic transformation in the context of computer education in primary and secondary schools, this study aims to investigate the idea of an artificial intelligence and natural language processing (NLP) based intelligent tutoring system (ITS) and reached to point that a learning assistant is one of an ITS's components that enables students to ask for help whenever they do.

Sophia and Jacob (2021): The focus of this study is on the covid period that during the covid pandemic, chatbots provide a human-like interface which helps in solving the problems of students and males learning easier. Frangoudes et al. (2021) goal of the study is to explore the uses of chatbots in Medical and Healthcare and how they are developed in that fields. In addition, focused on areas of computerized patients in medical education. The study concluded by saying that chatbots are flexible in the field of medical and healthcare education but the efficiency is not properly tested. Also, a few examples have been found but it can be improved through future research. Lin and Mubarok (2021) conducted study in Taiwan regarding overcome the challenges of English as Foreign Language (EFL) faced by students through chatbots with the advancement of Artificial Intelligence. The study winded up by saying that the chatbot approach helps the students to learn English as a foreign language and improves their performance. Moreover, chatbots are also useful for researchers who pretends to conduct AI-support flipped classrooms in learning language.

Yang and Evans (2020) proposed a prototype by taking into account three case studies in the fields of academic simulation, software training, and help desk support while the purpose of study to create awareness about potential of chatbots in education sector and explaining its prototypes which are currently developing. The study anticipated the development of chatbots in educational domain by creating awareness among chatbot software developers so that the potential of chatbots providing personalized services and accomplishing pedagogical need is discovered and availed in near future. Mendoza et al. (2020) with aim of study as to determine chatbot as an extensive extra school tool. Chatbots work as a platform between the teachers and students and provides a text-based user observation. In addition, chatbots mark various intercommunications which further helped in predicting chatbots have become a crucial part of academic studies and is a beneficial tool in a teaching-learning course by supplying data and findings. They mostly capture favourable consequences from the evaluated members.

Perez (2020) to find out the number of chatbots that are involved in education sector and examined the type of technology used to reveal the outcomes through using chatbots. Chatbots are considered as service agent in educational sector, through the study it's been observed that chatbots are based on guidelines of PRISMA function. Chat bots can be useful and can be considered as human tutor. Thomas (2020) focuses on the chatbots and its history concluding it as a virtual technique that are diminishing monotonous duties of educator, which are researched and been exploited. The solution provided by chatbots in the sector of education are contingently responded, which acts as time saver.

Tamayo and Herrero (2020) focuses on the reason behind the adoption, advantage, usage, modification for chat bots and role of teachers in implementation of these technology in education sector. The author suggested that chatbots are the future of educational support system, but these bots require some changes to be more effective and reliant. Wildfred (2020) through the research objectives and problems faced by the usage of chatbots are been recognised and suggestions on issues were discussed concluding that introduction of chatbots in the modern world acted as boom, as chatbots have potential to solve the query without any time-lapse issue with the internet facility. Plunzokia (2020) analysed the usage of al technology in education by drawing the line between online education in Russia and suggested that chatbots can become the best source of information provider and mentor for the students staying far away from respective institution. Dokulina and Gumanova (2020) point of the study is to tell that chatbots makes easier to learn a foreign language by providing uncomplicated communications just like humans. This has created a new learning scenario for foreign language learners. The study summarized by saying that chatbots are one of the basic components of easy and smart learning. This is making India a digital country and learning an easy tool. Bahja et al. (2020) focus of the study is on the covid period that during the pandemic, the learning capacity of the students went down which was a great loss to the economy. That's why, the focus of chatbots in education sector has been increased till now to make learning an interesting way. The study concluded by saying that chatbot learning is an effective and efficient tool for developing a learner's personality. In addition, a user-centric framework for designing chatbot was also developed.

Ouatu and Gifu (2020) with goal of study to introduce chatbots as a conventional learning process and to develop them more. The author concluded by saying that chatbots is an artificial intelligence tutor which do not get angry on explaining the same problem again and again. Moreover, they function as a personal therapist. Sandhu (2019) applied an empirical research design and quantitative method through which factors affecting usage of chatbots had been covered to enhance the learning techniques of students. A survey had been conducted in this paper for telling the educational institutes and web developers to know about the issues faced by chatbots users. The study suggested that in the era of technology, chatbots based on AI are in trend in all sectors, but these chatbots are not that much developed in educational sector due to change in trends (offline to online) due to the pandemic. using chatbots productivity, communication, effectiveness and efficient usage of time can be performed. Hobert and Sebastian (2019) with aim to make a blueprint which shows a variety of multidimensional research areas and using natural and simple language in chatbots. Then, it will be easy for the users of that information to understand the concepts clearly. The study concluded by saying that firstly, the author reviewed current state in art literature of educational representative and then they use the concluded results to give an advice for changing these educational agents, i.e., chatbots.

Rose and Sofie (2018) applied qualitative approach and strategy used to state that there is limited usage of chatbots by the persons involved in the educational sector. Therefore, chatbots can be utilized to enhance the learning experience but it all depends on its implementation. The study concluded chatbots has shown interesting potential for both learning and management in educational sector. This study is based only on chatbots under xml derived under aiml. Ahmad 2018 focuses on the various type of technique used in chatbot's designing as well as conversational trend between human and computer has been in trend from past few years, from enormous natural language processing technique. With the constant development of chatbots, author stated that chatbots can be constrain for human and it tool to optimize productivity. Winkler (2018) to fulfil the gap in the past research that were conducted and chatbots can be considered as intermediate for the learner and teacher as it provides 24/7 service. Best source for higher education.

Benotti et al. (2014) conducted a study with the purpose of relying on finding chatbots potential for imparting fundamental knowledge and engaging students which has not been assessed yet. Experimenting using a chatbot software platform aiming to promote interaction while teaching topics like variables to school students. The study suggested that with the use of chatbots within the conventional learning methods resulted into higher retention of concepts and girl interest among learners which demonstrates student involvement. For decades, chatbot has been used in different scenarios and can be utilised for educating students and inculcating better learning.

All the above esteemed authors are talking about various aspects of chatbots functioning and performance mentioned in reputed sources. The studies not only discussed the advantages, implications and working but another end of sword as well. Every possible related factor and topic were covered by following different research methodologies. The studies carried out by authors anticipated the pathway of chatbots interface. Explained how chatbots implication lack in high educational institutions on the other hand can be used for various purposes besides a human tutor in class like for admission process, doubt engine, counselling and many more. In order to improve chatbot process its necessary to bring the problems faced in attention of software developers so they could enhance it to fulfil pedagogical needs in more better way that now. During the pandemic in 2020, chatbots came in trend and further increased in a positive manner acting as a service agent while helping people in distant learning. It helps people by accomplishing tasks in efficient and effective manner. Its advantages doesn't end here but continues with its various operations like accurate solutions, helps in developing skills, increased interest and focus in class. By acting like a human interface can also help in learning foreign languages. While AI based technology are acing in other significant industries still have a far way to travel to develop completely in educational sector which ironically provides and prepares the human resource for any other industry. Without human intervention everything is not possible but to some extent it can be operated.

RESEARCH GAP

Human came too far in the field of innovation and modernization through human's research and development. The chatbot business built on AI which is one of the leading and supporting industries in the field of innovation. Different instances discovered sufficient situations in the study that can be improved to increase the benefits of chatbots in the educational sector. Poor infrastructure development, leading the biggest impact on chatbot use as there is a lack of public awareness of chatbots and its usage. However, the need to raise awareness among society about how chatbots can benefit people if used effectively and efficiently still exists. Therefore, chatbots can be considered as linkage tool between teachers and students.

There are both positive as well as negative side of the subject matter similarly chat bots are having both sides but it depends on its implication and usage to make it as boom not destructive matter.

OBJECTIVE OF STUDY

As chatbots are the forthcoming future in the contemporary world, a study on chatbots and its implications in the field of education was done. Both students and teachers can benefit from chatbots, which give students free interactive links about their studies and futures. As already seen in the pandemic situation, where the entire education sector was affected and disrupted, chatbots came in as a boom and helped the entire sector move forward and succeed, chatbots helped students to learn about the use of online learning techniques. Human can think of chatbots as helping hands that assist students in asking questions, maintaining their position, and openly proving it.

RESEARCH METHODOLOGY

In the present study, a comprehensive collation of the research findings and suggestions by several other researchers. The literature was retrieved by computerized database and hand search. Number of research papers, articles, website are used and their references are cited along with them.

INCLUSION OF DATA

The research thus validates that the data were obtained from reputed sources. Additionally, because these databases index articles from other significant databases including Science Direct, Emerald, ProQuest, Wiley, Springer, and many more, they are useful for generalization purposes. On the other hand, the information should originate from an even more dependable source in order to convey the insights and future directions. The data was narrowed down in many earlier studies in order to offer the findings based on subjective opinion. However, 28 papers from reputable journals and authors were objectively chosen for the current investigation. As a result, data were meticulously gathered from dependable sources through keyword

Figure 1. Flowchart of the process of selection of the studies
Source: Author's compilation

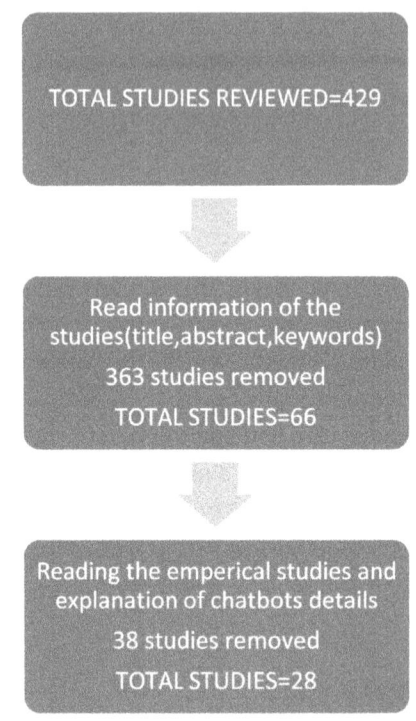

Figure 2. Number of citations conducted by recent research papers
Source: Author's compilation

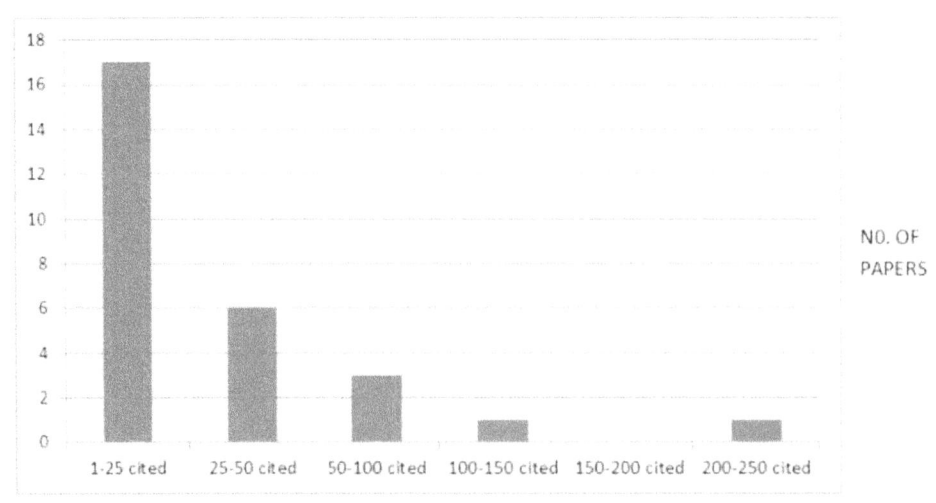

Figure 3. Showing number of research papers taken in accordance of publishing years
Source: Author's compilation

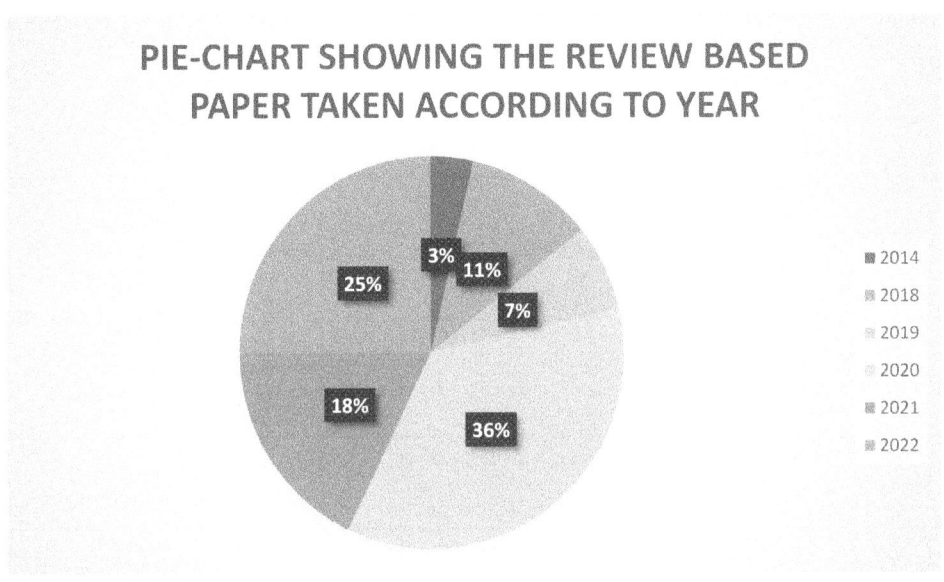

Figure 4. Depicts the countries on map where researches were conducted
Source: Author's compilation

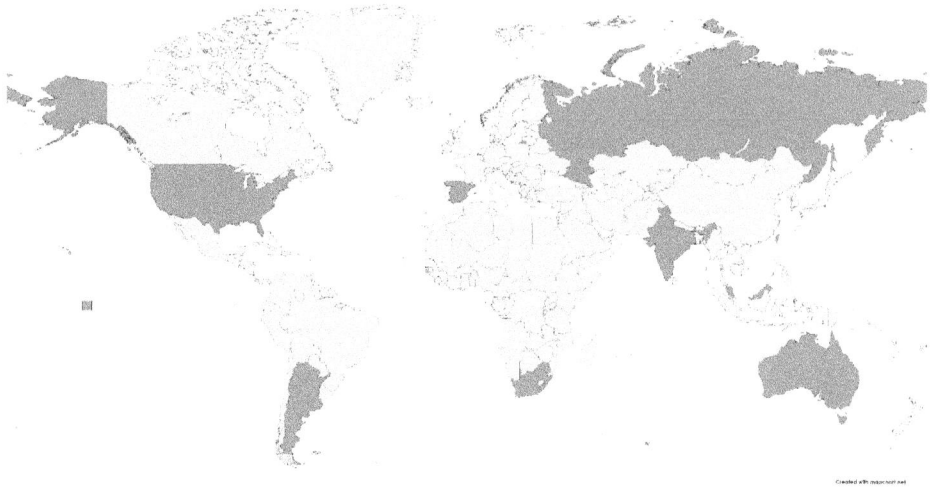

Table 1. research conducted countries with journal and title name in accordance to map

Name of Countries	Name of the Journal	Title of Paper
Argentina	Iticse'14: Proceedings of the 2014 conference on innovation & technology in computer science education	Engaging high school students using Chatbots
Malaysia	Sustainability	AI-Based Chatbots Adoption Model for Higher-Education Institutions: A Hybrid PLS-SEM Neural network modeling Approach
Australia		Adoption of AI-Chatbots to enhance student learning experience in Higher education in India
Malaysia	Learning and collaboration Technologies. Human and Technology Ecosystems	Supporting Student-Teacher Interaction Through A Chatbots
Spain	Rediscovering the use of chatbots in education: A systematic literature review	computer application in engineering learning
Malaysia	International Journal of Computer Applications	Review of Chatbots Design Techniques
south Africa	computer and education: artificial intelligence	Chatbots applications in education: A systematic review
India	International Journal of Trend in Scientific Research and Development (IJTSRD)	Critical Literature Review on Chatbots in Education
Chicago	Academy of Management Annual Meeting (AOM)	unleashing the potential of chat bots in education sector a state of art analysis
Karnataka	International Journal of Electrical and Computer Engineering (IJECE)	NLP-based personal learning assistant for school education
Russia	Engineering Management of Communication and Technology (EMCTECH), International Conference	Technologies of Artificial Intelligence in Educational Management
Taiwan	Journal article	Learning analytics for Investigating the Mind Map-Guided AI Chatbot Approach in an EFL Flipped Speaking Classroom

Source: Author's Compilation

searches in order to ensure that they are coming from a rich data source and to retain the study's sense of objectivity.

DISCUSSION

When you think of industries that use technology frequently, the education sector may not come to mind right away. However, the COVID-19 pandemic made the employment of technology in education a must. The top five industries now include education, which not only uses chatbots but also benefits financially from them.

There are some of the technological advancements that chatbots can bring to the world of online education. It picks up on the pupil's emotional states, which chatbots can use to adjust their language in their answer when they recognise them. Provides individualised instruction that adjusts to the needs and demands of the individual student. When communicating or addressing questions pertaining to a course, this offers a more direct orientation.

Since chatbots instantly respond to frequently asked questions from the pupils with responses that have been previously predesigned, it enables the teacher to spend less time organising and carrying out activities. This time saved can be used for group supervision and motivating, as well as research or projects that are still due for the course.

Effectively gather and interpret data when analysing student evaluations and progress. The application of artificial intelligence has the effect of assisting students in efficiently allocating their time and assigning work in accordance with their goals. The chatbot with its pattern recognition is focused on interaction and formation; it does not take the student's resources, language or location into account. It can be considered something like a "transformation of learning".

CONCLUSION

A variety of commercial applications, such as online tutoring, student assistance, teacher's help, management tool, analysing and giving results, are making chatbots more and more ordinary in the education sector. Even though there are many technical breakthroughs that chatbots can contribute to the realm of online education, such as the ability to discover a person's emotional states and adapt their language in the answer when they identify them, they still have a pocket-sized role in the education sector.

The usage of chatbots in the education sector is still booming. Alexa and Google Assistant are two examples of virtual assistants that use chatbot technology. Chatbots are one of the many potential data innovation trends. Customers are not restricted by location or time; they may communicate with chatbots at any moment during the day or week. Most of the time, chatbots can give fundamental training. By providing students with course content immediately, chatbots can assist learning in a classroom context. In the era of AI, higher education institutions (HEIs) are using chatbots more often to achieve educational goals by offering speedy and individualized services to everyone in the sector, including institutional staff and students.

Since many users are unaware, chatbots today have interactive ways to serve information through graphical interactivity or graphical widgets. They no longer just provide response in the form of text or voice commands. Therefore, it can be concluded that chatbots could be lifesavers for those shy language learners who find it difficult to learn communicative skills of a new language through talking to a realistic person and prefer to practice independently on their computers or smartphones. These devices won't judge them and will give them time to think of what to say before uttering a word or a phrase. Chatbots create a strong basis for a new educational scenario based on the concept of micro learning and an adaptable learning environment when combined with the visual content of educational blogs posted on social media sites like Instagram. Although a real educator cannot be replaced in this scenario, it can aid in the student's learning process and give individuals who would not otherwise be able to get foreign language instruction due to their location or way of life convenient approach to it.

SUGGESTIONS

A well-designed and operated chatbot might be a tool to increase user engagement and offer a positive user experience between a human and the field being serviced. The creation of chatbots should be well planned out, and selecting the right platforms and technologies is crucial as it may increase the efficacy and efficiency of the chatbots.

FUTURE SCOPE OF STUDY

The study will assist students in comprehending the study material. This has truly aided the development of online learning and distance learning for all. Chatbots are undoubtedly the future of education, but will they eventually replace all teachers?

We've heard that education chatbots cover all bases while also being low-maintenance. Even so, chatbots are not human and lack the EQ and capabilities that only humans have. It is obvious that teachers do more than just teach; they also help students grow into better people.

Chatbots are considered as valuable during the difficult times of covid pandemic and it has also helped a number of students but still it can be improved in many ways.

REFERENCES

Abbas, N., Whitfield, J., Atwell, E., Bowman, H., Pickard, T., & Walker, A. (2022). Online chat and chatbots to enhance mature student engagement in higher education. *International Journal of Lifelong Education*, *41*(3), 1–19. doi:10.1080/02601370.2022.2066213

Adam, M., Wessel, M., & Benlian, A. (2021). AI-based chatbots in customer service and their effects on user compliance. *Electronic Markets*, *31*(2), 427–445. doi:10.100712525-020-00414-7

Ahmad, N. A., Che, M. H., Zainal, A., Abd Rauf, M. F., & Adnan, Z. (2018). Review of chatbots design techniques. *International Journal of Computer Applications*, *181*(8), 7–10. doi:10.5120/ijca2018917606

Al-Sharafi, M. A., Al-Emran, M., Iranmanesh, M., Al-Qaysi, N., Iahad, N. A., & Arpaci, I. (2022). Understanding the impact of knowledge management factors on the sustainable use of AI-based chatbots for educational purposes using a hybrid SEM-ANN approach. *Interactive Learning Environments*, 1–20. doi:10.1080/10494820.2022.2075014

Alajmi, Q. A., Kamaludin, A., Arshah, R. A., & Al-Sharafi, M. A. (2018). The effectiveness of cloud-based e-learning towards quality of academic services: An Omanis' expert view. *International Journal of Advanced Computer Science and Applications*, *9*(4). Advance online publication. doi:10.14569/IJACSA.2018.090425

AlDhaen, F. (2022). The Use of Artificial Intelligence in Higher Education–Systematic Review. *COVID-19 Challenges to University Information Technology Governance*, 269-285.

Almada, A., Yu, Q., & Patel, P. (2023). Proactive chatbot framework based on the PS2CLH model: an AI-Deep Learning chatbot assistant for students. In *Proceedings of SAI Intelligent Systems Conference* (pp. 751-770). Springer. 10.1007/978-3-031-16072-1_54

Bahja, M., Hammad, R., & Butt, G. (2020, July). A user-centric framework for educational chatbots design and development. In *International Conference on Human-Computer Interaction* (pp. 32-43). Springer. 10.1007/978-3-030-60117-1_3

Benotti, L., Martínez, M. C., & Schapachnik, F. (2014, June). Engaging high school students using chatbots. In *Proceedings of the 2014 conference on Innovation & technology in computer science education* (pp. 63-68). 10.1145/2591708.2591728

Cunningham-Nelson, S., Boles, W., Trouton, L., & Margerison, E. (2019). A review of chatbots in education: practical steps forward. In *30th Annual Conference for the Australasian Association for Engineering Education (AAEE 2019): Educators Becoming Agents of Change: Innovate, Integrate, Motivate* (pp. 299-306). Engineers Australia.

Dokukina, I., & Gumanova, J. (2020). The rise of chatbots–new personal assistants in foreign language learning. *Procedia Computer Science*, *169*, 542–546. doi:10.1016/j.procs.2020.02.212

Frangoudes, F., Hadjiaros, M., Schiza, E. C., Matsangidou, M., Tsivitanidou, O., & Neokleous, K. (2021, July). An Overview of the Use of Chatbots in Medical and Healthcare Education. In *International Conference on Human-Computer Interaction* (pp. 170-184). Springer. 10.1007/978-3-030-77943-6_11

Hobert, S. (2019). How are you, chatbot? evaluating chatbots in educational settings–results of a literature review. *DELFI 2019*. https://www.chatcompose.com/chatbot-learning.html https://www.investopedia.com/terms/c/chatbot.asp https://yellow.ai/chatbots/use-cases-of-chatbots-in-education industry/#:~:text=Chatbots%20act%20 as%20a%20data,all%20their%20students%20at%20onc

Ismail, M., & Ade-Ibijola, A. (2019, November). Lecturer's apprentice: A chatbot for assisting novice programmers. In *2019 International Multidisciplinary Information Technology and Engineering Conference (IMITEC)* (pp. 1-8). IEEE. 10.1109/IMITEC45504.2019.9015857

Kuhail, M. A., Alturki, N., Alramlawi, S., & Alhejori, K. (2022). Interacting with educational chatbots: A systematic review. *Education and Information Technologies*, 1–46.

Kumar, J. A. (2021). Educational chatbots for project-based learning: Investigating learning outcomes for a team-based design course. *International Journal of Educational Technology in Higher Education*, *18*(1), 1–28. doi:10.118641239-021-00302-w PMID:34926790

Lee, C. T., Pan, L. Y., & Hsieh, S. H. (2021). Artificial intelligent chatbots as brand promoters: A two-stage structural equation modelling-artificial neural network approach. *Internet Research*.

Leonardi, S., & Torchiano, M. (2023). Educational Chatbot to Support Question Answering on Slack. In *International Conference in Methodologies and intelligent Systems for Technology Enhanced Learning* (pp. 20-25). Springer. 10.1007/978-3-031-20617-7_4

Lin, C. J., & Mubarok, H. (2021). Learning analytics for investigating the mind map-guided AI Chatbot approach in an EFL flipped speaking classroom. *Journal of Educational Technology & Society*, *24*(4), 16–35.

Ma, Y. J., Gam, H. J., & Banning, J. (2017). Perceived ease of use and usefulness of sustainability labels on apparel products: Application of the technology acceptance model. *Fashion and Textiles*, *4*(1), 1–20. doi:10.118640691-017-0093-1

Mabunda, K. (2020). *An Intelligent Chatbot for Guiding Visitors and Locating Venues* [Doctoral dissertation]. University of Johannesburg.

Mathew, A. N., & Paulose, J. (2021). NLP-based personal learning assistant for school education. *International Journal of Electrical & Computer Engineering, 11*(5).

Mendoza, S., Hernández-León, M., Sánchez-Adame, L. M., Rodríguez, J., Decouchant, D., & Meneses-Viveros, A. (2020, July). Supporting student-teacher interaction through a chatbot. In *International conference on human-computer interaction* (pp. 93-107). Springer.

Merelo, J. J., Castillo, P. A., Mora, A. M., Barranco, F., Abbas, N., Guillén, A., & Tsivitanidou, O. (2022). Exploring the role of chatbots and messaging applications in higher education: a teacher's perspective. In *International Conference on Human-Computer Interaction* (pp. 205-223). Springer. 10.1007/978-3-031-05675-8_16

Mohd Rahim, N. I., & Iahad, N., Yusof, A. F., & Al-Sharafi, M. (2022). AI-Based chatbots adoption model for higher-education institutions: A hybrid PLS-SEM-Neural network modelling approach. *Sustainability*, *14*(19), 12726. doi:10.3390u141912726

Molnár, G., & Szuts, Z. (2018, September). The role of chatbots in formal education. In *2018 IEEE 16th International Symposium on Intelligent Systems and Informatics (SISY)* (pp. 197-202). IEEE. 10.1109/SISY.2018.8524609

Okonkwo, C. W., & Ade-Ibijola, A. (2021). Chatbots applications in education: A systematic review. *Computers and Education: Artificial Intelligence*, *2*, 100033. doi:10.1016/j.caeai.2021.100033

Ouatu, B. I., & Gifu, D. (2021). Chatbot, the Future of Learning? In *Ludic, Co-design and Tools Supporting Smart Learning Ecosystems and Smart Education* (pp. 263–268). Springer. doi:10.1007/978-981-15-7383-5_23

Pérez, J. Q., Daradoumis, T., & Puig, J. M. M. (2020). Rediscovering the use of chatbots in education: A systematic literature review. *Computer Applications in Engineering Education, 28*(6), 1549–1565. doi:10.1002/cae.22326

Pluzhnikova, N. N. (2020, October). Technologies of artificial intelligence in educational management. In *2020 International Conference on Engineering Management of Communication and Technology (EMCTECH)* (pp. 1-6). IEEE. 10.1109/EMCTECH49634.2020.9261561

Roos, S. (2018). Chatbots in education: A passing trend or a valuable pedagogical tool? IEEE.

Sinha, S., Basak, S., Dey, Y., & Mondal, A. (2020). An educational Chatbot for answering queries. In *Emerging technology in modelling and graphics* (pp. 55–60). Springer. doi:10.1007/978-981-13-7403-6_7

Sophia, J. J., & Jacob, T. P. (2021, August). Edubot-a chatbot for education in covid-19 pandemic and vqabot comparison. In *2021 Second International Conference on Electronics and Sustainable Communication Systems (ICESC)* (pp. 1707-1714). IEEE. 10.1109/ICESC51422.2021.9532611

Tamayo, P. A., Herrero, A., Martín, J., Navarro, C., & Tránchez, J. M. (2020). Design of a chatbot as a distance learning assistant. *Open Praxis, 12*(1), 145–153. doi:10.5944/openpraxis.12.1.1063

Winkler, R., & Söllner, M. (2018). Unleashing the potential of chatbots in education: A state-of-the-art analysis. *Academy of Management Annual Meeting (AOM).* 10.5465/AMBPP.2018.15903abstract

Yang, S., & Evans, C. (2019, November). Opportunities and challenges in using AI chatbots in higher education. In *Proceedings of the 2019 3rd International Conference on Education and E-Learning* (pp. 79-83). 10.1145/3371647.3371659

Chapter 7
Adoption of Digital Art NFTs in Hong Kong

Sze Wing Wong
The University of Hong Kong, Hong Kong

Mimi Mei Wa Chan
The University of Hong Kong, Hong Kong

Dickson K. W. Chiu
https://orcid.org/0000-0002-7926-9568
The University of Hong Kong, Hong Kong

ABSTRACT

The most popular uses of blockchain technology in the art market are related to non-fungible tokens (NFTs). This chapter explores the adoption of NFT in the digital art market and its future development. The authors explore NFT adoption in the digital art market with the five pillars of the digital entrepreneurship model, including knowledge base, business environment, finance, technology, and culture. Their scarcity and utilities determine the value of NFTs. Collectors and artists should also be aware of the benefits and drawbacks of blockchain technology and take appropriate steps to guarantee their rights are protected. This chapter provides a fundamental review of the current development and outlook of NFT and opens new opportunities for future study. Scant research focuses on the present condition and adoption of NFTs within the Hong Kong digital art market. This review offers a much-needed exploration and understanding, particularly beneficial for potential investors and participants seeking a comprehensive insight into NFT adoption.

DOI: 10.4018/978-1-6684-8671-9.ch007

INTRODUCTION

Blockchain technology, initially renowned for its role in cryptocurrency, has taken giant strides in various sectors (Fang et al., 2022; Au et al., 2022). Blockchain's application has stretched beyond cryptocurrencies like Bitcoin and has recently seeped into various industries, including the digital art market, with the introduction of non-fungible tokens (NFTs). Recently, the advent of NFTs has dramatically transformed the dynamics of the digital art market, inviting both awe and scrutiny.

Conceptually, NFTs are unique cryptographic assets on a blockchain and have become increasingly popular, especially in digital art. They provide a means of asserting ownership over a unique piece of digital content, and they've sparked a major shift in how we perceive the value and ownership of digital creations. These tokens are inseparably linked to digital items, such as works of art, and cannot be interchanged like traditional fungible tokens, such as cryptocurrencies. Each NFT represents a distinct set of characteristics that set it apart from any other token, thereby encapsulating uniqueness and validating ownership and authenticity (Cointelegraph, 2021).

Historically, the concept of NFTs emerged from the cryptographic design of the Ethereum blockchain, with the introduction of ERC-721, a free and open standard that describes how to build non-fungible tokens on the Ethereum blockchain. In 2017, the introduction of CryptoKitties, a digital game built on Ethereum's blockchain, marked the early development of NFTs. Since then, the NFT market has grown exponentially and gained unprecedented popularity in 2021 when digital artist Mike Winkelmann, known as Beeple, sold an NFT of his artwork for $69 million (Sharma, 2023).

The significance of NFTs lies in their potential to democratize the art industry. They provide digital artists with an innovative platform for monetizing their work and obtaining a fair share of the profits from their art. NFTs also bring transparency and traceability to the often-opaque art market, as each token's history of ownership is openly verifiable on the blockchain. Furthermore, they have the power to redefine the concept of ownership in the digital realm, where duplication and plagiarism have previously made it challenging to own and trade digital assets uniquely.

Applying the Digital Entrepreneurship Model (Bogdanowicz, 2015), this chapter presents a detailed analysis of the NFT market and examines how Hong Kong can position itself in this rapidly evolving digital landscape. It delves into the application of blockchain technology in the Hong Kong digital art market, offering a nuanced understanding of the technology and its potential impact on the market. This exploratory study is based on comprehensive literature reviews and analysis of relevant tech sites and media publications.

The art industry, like many others, has begun to experiment with smart contracts to establish ownership and usage rights and address the challenge of proving the delivery of digital artworks (Abbate et al., 2022). However, the applications of blockchain in the digital art market in Hong Kong remain largely under-researched and under-exploited. This chapter seeks to address this gap, presenting insights that would be particularly valuable to investors unfamiliar with NFTs. Thus, the research questions guiding this study are:

RQ1: How is blockchain technology used in digital art?

RQ2: How will Hong Kong play a crucial role in developing digital art?

LITERATURE REVIEW

The emergence of blockchain technology has had a transformative impact on various industries, and the realm of digital art is no exception. This literature review aims to delve further into the intricacies of blockchain technology, its development, and its relevance to the world of digital art.

Blockchain Technology

Blockchain is a groundbreaking development in financial technology (FinTech), initially proposed as a distributed ledger system for Bitcoin in 2008 by the pseudonymous person or group Satoshi Nakamoto. The technology behind blockchain has the potential to bring transformational changes to the FinTech industry by improving security, reducing costs, increasing efficiency, and enabling transparency. Since its inception, blockchain has attracted interest and investment from various sectors, including banking, insurance (Beckman, 2021), and healthcare. With its decentralized, tamper-proof, and secure database, blockchain is expected to revolutionize how transactions are conducted and recorded across industries globally. As it is still in its early stages of development, the potential application of blockchain technology continues to expand, promising new opportunities for innovation and growth across various sectors (Nakamoto, 2008).

Blockchain technology is a decentralized and immutable ledger system that enables secure and transparent transactions. It functions as a distributed database, where records, or blocks, are cryptographically linked together in a chronological chain. This design ensures the integrity and immutability of data, making it resistant to tampering or fraud. The decentralized nature of blockchain eliminates the need for intermediaries and enables peer-to-peer transactions, empowering individuals to have greater control over their digital assets (Warkentin & Orgeron, 2020). Blockchain technology not only helps safeguard against the double-spending problem by maintaining a shared record that continually reflects changes in asset ownership but also creates a ledger of transactions that documents every exchange (Abbate et al., 2022).

Numerous iterations and improvements have been made to blockchain protocols, leading to the development of alternative platforms, such as Ethereum, that support the execution of smart contracts and the creation of NFTs. This fusion has eliminated the need for significant manual verification and validation of information, thereby improving the quality and efficiency of transactions. By utilizing blockchain technology, the transparency and security of transactions are enhanced, reducing costs and boosting efficiency (Seebacher & Schüritz, 2017). Furthermore, incorporating smart contracts into the blockchain technology allows automatic transactions based on predetermined conditions, ensuring a smooth and efficient process (Liu & Ye, 2021).

Opportunities and Challenges in the Art Industries

The art industry is a unique ecosystem with distinct features. The demand for artistic creations is unpredictable; artists' pride is deeply associated with their works; there is a vertical differentiation due to varying levels of talent, originality, and proficiency among artists; and creative outputs are time-sensitive but durable (O'Dair, 2018; Jiang et al., 2019; Lo et al., 2021). Nevertheless, the industry faces significant challenges. The valuation of art is subjective, with no predefined criteria to evaluate the quality of the artwork, and the concept of what qualifies as art is perpetually debatable (Abbate et al., 2022). Furthermore, the art industry suffers from a lack of complete transparency. With private sales accounting for nearly half of all market transactions, crucial sales information is often not publicly accessible (Coslor, 2016). With such challenges, blockchain technology emerges as a potentially transformative solution, enhancing transparency and reducing administrative burdens (Patrickson, 2021).

According to Private Museum (n.d.), the advent of NFTs has profoundly transformed the art industry, democratizing it in unprecedented ways. The art ecosystem has become more inclusive and diverse by enabling artists, particularly those traditionally marginalized, to create and sell their digital art via NFT marketplaces. Furthermore, the intermediary roles of galleries and auction houses are being redefined as artists can sell their work directly to collectors, ensuring a higher profit margin and cultivating more personal artist-collector relationships. Another significant impact of NFTs is their redefinition of art ownership and authenticity. Thanks to the transparent, immutable ledger of ownership provided by blockchain records, proving the authenticity and ownership of digital art has become less challenging than in traditional art scenarios. Lastly, NFTs have enhanced the accessibility and affordability of art. The concept of fractional ownership in NFTs allows a wider audience to invest in art and support artists without the need for full ownership of an art piece. This phenomenon has lowered the barriers to entry in the art market, making it more accessible to a larger demographic. In the context of digital art, the true potential of blockchain technology began to unfold with the introduction of NFTs. The emergence of NFTs has opened exciting possibilities in digital art and collectibles, offering unique opportunities for creators and collectors. NFTs empower creators to maintain ownership of their works while enabling collectors to invest in them. NFTs also provide an alternative revenue stream that bypasses traditional intermediaries like galleries and auction houses.

However, the advent of NFTs in the art industry has prompted a range of considerations and challenges. Sustainability is a major concern due to the high energy consumption associated with NFT creation and trading, raising environmental considerations (Salisu, n.d.). Additionally, the potential for fraud and counterfeiting within the ecosystem risks market trust (Private Museum, n.d.). From a legal standpoint, the complex issues of intellectual property rights, copyright, and varied ownership rights in the digital realm present substantial hurdles, necessitating clear and refined regulations around blockchain technology. The significant environmental impact, largely due to Ethereum's energy-intensive proof-of-work consensus algorithm, is another critical issue, calling for a shift towards more sustainable blockchain practices. The rise of NFT art has also sparked social and ethical debates regarding inclusivity, diversity, market accessibility, and the risk of fraud. Interoperability and standardization present additional challenges as the NFT market grows, with the need for industry-wide standards and cross-chain compatibility becoming increasingly pressing. Security and privacy considerations inherent to the decentralized nature of blockchain require robust solutions to ensure data integrity and user privacy. Despite these challenges, the NFT art market is showing promise with increasing acceptance, innovative uses, and wider adoption. However, adoption should consider potential pitfalls, such as environmental concerns, regulatory uncertainty, and market volatility, to shape a prosperous future trajectory.

Relevance to Digital Art

Blockchain technology has revolutionized the digital art landscape by addressing critical issues such as provenance, ownership, and authenticity. In traditional art markets, establishing the authenticity and provenance of artworks has often been challenging, leading to concerns about forgeries and disputes. By leveraging blockchain technology, NFTs provide an immutable and transparent record of ownership and transaction history, alleviating these concerns. This feature has empowered artists and collectors alike, ensuring the integrity and value of digital artworks. Blockchain technology has also democratized the art market by removing intermediaries and facilitating direct interactions between artists and collectors.

Online marketplaces powered by blockchains, such as OpenSea and Rarible, have emerged as platforms for artists to showcase and sell their digital artworks, reaching a global audience without the need for traditional art galleries or auction houses (Fortnow et al., 2021). This has significantly reduced barriers to entry, enabling emerging artists to gain exposure and recognition for their work.

Among the various applications of blockchain technology in the art market, non-fungible tokens and traceability of transactions have garnered significant attention. NFTs, a distinct category of tradable assets on the blockchain, enable monetizing digital or physical assets (Abbate et al., 2022). This advancement in digital art uses distributed, decentralized systems to establish a sustainable and verifiable online ecosystem (Chan, 2021). Creating an NFT, or tokenization, involves registering the data that identifies the artwork as unique and preserving crucial information about the artwork, such as the artist's name, creation date, category, estimated price, and other details asserting its authenticity. Upon purchase, collectors obtain the NFT, guaranteeing the piece's ownership and authenticity (Tempone, 2021).

Furthermore, tokenization introduces the possibility of fractional ownership, widening the pool of potential collectors and democratizing the ownership of investment-grade artworks (Abbate et al., 2022). The intricacies and evolution of blockchain technology and the emergence of NFTs offer compelling prospects for the art industry. Despite the current challenges, these innovations promise improved transparency, enhanced efficiency, and more equitable access to the art mark.

ANALYSES

Five Pillars of Digital Entrepreneurship Model

The Digital Entrepreneurship Model comprises five critical pillars: knowledge base, business environment, finance, technology, and culture (Bogdanowicz, 2015). This model forms the basis for digital enterprises to harness cutting-edge information technologies, such as social media (Cheung et al., 2023; Chung et al., 2020; Jiang et al., 2023; Lei et al., 2021; Li, Chiu, Kafeza, & Ho, 2023; Xie, Chiu, & Ho, 2023; Xie, Wong, Chiu, & Lei, 2023; I.H.S. Wong et al., 2023; J. Wong et al., 2023; Yang et al., 2022), Big Data (Deng & Chiu, 2023; Gao et al., 2021; Wang et al., 2022a; 2022b), mobile computing (Chan et al., 2022; Ni et al., 2022; Ezeamuzie et al., 2022; Wai et al., 2019; Yip et al., 2021; Zhang et al., 2021), and cloud computing (Feng et al., 2016; Zhuang et al., 2014). These technologies significantly augment and innovate business activities while enhancing the precision of business intelligence (Swati et al., 2019). Digital entrepreneurship pivots on exploiting the full potential of digital technology to amplify entrepreneurial opportunities and drive business

Table 1. Coding for findings according to the five pillars of the digital entrepreneurship model

Themes	Coding
Knowledge base	Unique intellectual propertyScarcity in the digital worldCommunity members and social networks are the sources of NFT knowledge
Business Environment	NFT MarketplacesGateway for collectors and artists to trade NFTUser interface should be clean and easy to useMake connections between physical and digital
Finance	NFT's value expected to rise for profitAlternative way to support artistsLower the cost of art transactionsCreate a more equal and transparent environment
Technology	Gaps of innovation and digitalization in the art sectorInsufficient IT knowledgeUnderstanding the benefit and drawbacks of NFT is necessary
Culture	Serve as marketing tools in traditional industryNFT exhibitions take place at major eventsMuseums are under pressure because of challenging established methods of collecting art

growth (A.K-k. Wong & Chiu, 2023; Tse et al., 2022; Sun et al., 2022). Within the Digital Entrepreneurship Model framework, our analysis investigates the integration and impact of blockchain technology within the digital art market, providing an in-depth exploration within the context of these five foundational pillars. The themes and codes of our findings shown in Table 1 support the lens of the five pillars of the Digital Entrepreneurship Model effectively analyzed blockchain technology, particularly NFTs, in the digital art market.

Knowledge Base

As we address the first pillar of the Digital Entrepreneurship Model, the knowledge base, our focus narrows to the intricacies of purchasing an NFT. Upon buying an NFT, an investor often becomes the owner of the underlying asset and any attached rights. However, it is a rarity that they secure exclusive ownership of the underlying intellectual property (Mo, 2021). According to the definition of "non-fungible," non-fungible tokens cannot be converted into another form or traded for their equivalent. Consequently, each NFT is one of a kind and cannot be replicated, contributing to the already existing scarcity in the digital world (Ross, 2022). Although the exponential growth and popularity of NFTs have occurred worldwide, most of the demand comes from Central and Southern Asia as well as North America. People are becoming more reliant on technology and the internet due to the rising digitalization of society (Ding et al., 2021; Lo & Chiu, 2015; Lo et al., 2015, 2017, 2019; Suen et al., 2020; Sun et al., 2022; Tse et al., 2022; Wang et al., 2016; K.C. Wong & Chiu, 2023; Wu et al., 2023; Zuo et al., 2023), which is particularly prevalent in an era marked by more restrictions and self-isolation during the pandemic (Chan et al., 2022; Dai et al., 2023; Li, Xie, Chiu, & Ho, 2023; Sung & Chiu, 2022; Yi & Chiu, 2023; Mo, 2021). Most people learn about NFTs and the extent of their knowledge regarding trading in NFTs. Community members, auction houses, Sotheby's, and social networks were knowledge sources of NFT. The first case of tokenization occurred on June 20, 2018, when the global art blockchain market, Maecenas, selected Andy Warhol's 14 Small Electric Chairs for the world's first cryptocurrency art auction. During this auction, buyers paid with cryptocurrency to purchase fractional-ownership digital certificates of this painting. More recently, in October 2020, Christie's was responsible for organizing the sale of the Barney Ellsworth collection. Since this was the first auction sale documented in a blockchain database (Abbate et al., 2022), most investors started investing in NFT recently.

Business Environment

Pivoting to the second pillar of the Digital Entrepreneurship Model, the business environment, we delve into the NFT marketplace and the trading platforms. An NFT marketplace is the gateway for collectors to participate in purchasing and selling NFTs (Schachter, 2022). Opensea, Rarible, and Foundation are some of the major platforms. In particular, OpenSea is the first and largest NFT marketplace, and it has gained a significant lead over its rivals in recent months. Its most recent fundraising round, which was spearheaded by venture capital companies Paradigm and Coatue Management, resulted in the company having a valuation of $13.3 billion (Abrams, 2022). OpenSea is the most popular marketplace to trade NFT. Opeasea was easy for people new to NFTs to understand, and the user interface was clean.

Finance

For the third pillar of the Digital Entrepreneurship Model, finance, we focus on NFTs' value. Like any other speculative asset, NFTs can be purchased with the expectation that their value will rise so that they can be sold for a profit (Clark, 2022). The value of NFTs is determined by their scarcity and utilities (Ali, n.d.). In particular, the Bored Ape Yacht Club is one of the most successful NFT projects ever. Millions of dollars have been spent on the most expensive BAYCs. The availability is constrained to scarcity because there are only 10,000 Bored Apes. In addition, even within the same collection, some Bored Ape avatars have a higher rarity than others. Each Ape is a one-of-a-kind creation resulting from a unique random combination of 170 characteristics. These characteristics include background color, earrings, expression, headdress, and attire (Erinfolami, 2022), contributing to the inflated price of certain Bored Apes. For utilities, the value of artwork created using Bored Ape comes from the fact that it can be used to create a digital identity and the economic usage rights that come along with it (Erinfolami, 2022). A BAYC NFT can also be used as a membership card for the online "Yacht Club," providing holders with additional perks and privileges (Bitstamp, 2022).

Technology

Addressing the fourth pillar, technology, we take on the task of bridging the knowledge gap that exists in the sphere of innovation and digitalization in the art industry. The advent of NFTs in the art world has been accompanied by various technological issues, even as they gain popularity. In the art industry, there is a serious lack of understanding of the information entered into the blockchain (Abbate et al., 2022). NFTs are regarded heavily in more and more conventional areas of the art world, but as their popularity grows, so do their technological issues. The blockchain works like a digital ledger that keeps track of an item's complete transaction history, from its first owner to its most recent one. Once smart contracts are in place, they cannot be changed or deleted. A smart contract is a computer program or transaction protocol for automatically carrying out, controlling, and documenting trade terms, especially digital rights. If the information on the blockchain and the ledger are also wrong, it is almost impossible to fix a blockchain ledger (Lo, 2022). Lack of transparency and knowledge asymmetry impact art market transactions (Abbate et al., 2022). Therefore, collectors and artists should understand the benefits and drawbacks of blockchain technology and take appropriate steps to ensure their rights are appropriately protected.

Culture

Lastly, when discussing the fifth pillar, culture, we consider the impact of NFTs on traditional and cultural industries. In addition to the growth and evolution of NFTs, it is vital to cultivate a culture that emphasizes the acceptance of NFTs in the community. Since the introduction of NFT, many businesses operating in traditional areas have started to accept NFT and use NFT as a marketing and public relations tool (Hashkey, n.d.). In Hong Kong, NFC exhibitions have occurred at major events and locations, including the Digital Art Fair (DAFA), K11, Fine Art Asia, and the M+ Museum of visual culture. NFTs are primarily utilized as collections, but in the coming years, they will have greater utility and be linked to numerous traditional industries. Museums are pressured to keep up with the latest digital developments, viewed as threats to institutions, challenging established art collection methods (Mplus, n.d.).

DISCUSSION

Application of Blockchain Technology in the Digital Art Market (RQ1)

With its decentralized and immutable nature, blockchain technology has emerged as a transformative force within the realm of digital art, bringing forth many opportunities and novel paradigms. This transformation's core is the groundbreaking NFT, a distinct digital asset powered by blockchain technology that has revolutionized how art is created, bought, and sold (Fortnow et al., 2021).

The blockchain's key features of transparency and immutability underpin the potential solution to a long-standing challenge in the art market - the verification of authenticity and provenance. For years, collectors and investors have grappled with the daunting task of tracing an artwork's lineage, especially in cases where the original artist is deceased. With blockchain technology, each artwork can have a unique digital signature permanently recorded on the blockchain, thereby providing an unalterable record of its origin and ensuring its authenticity (Botz, n.d.). This not only enhances trust and confidence in transactions but also adds significant value to the artwork.

Additionally, the rise of NFTs, a direct outcome of blockchain technology, is shaping the future contours of the art industry. These unique tokens imbue digital assets with the irreplaceable quality of physical ones, thereby challenging traditional notions of ownership and value. NFTs stand to democratize the art industry, mitigating the power asymmetry between artists and intermediaries like galleries and auction houses. Artists can now sell their work directly to the public, retain more profits, and earn royalties on future sales, thereby ushering in a new era of equitable and transparent art commerce (Abbate et al., 2022). The democratization of the art marketplace, catalyzed by NFTs, presents both opportunities and challenges. While it enables more artists to market their work, attain recognition, and secure compensation, it simultaneously emphasizes the importance of caution and informed decision-making. This is particularly pertinent to NFT transactions, where the open marketplace demands both awareness and due diligence for successful and safe participation (Dsouza, 2023).

Furthermore, blockchain technology heralds the emergence of peer-to-peer art markets, effectively breaking down the walls erected by intermediaries, which is much more facilitated by contemporary automated transactions and workflows (Chan & Chiu, 2022; Chin & Chiu, 2023; S.M. Li et al., 2023; Lin et al., 2022). This paradigm is shifting to transform how we interact with and consume art, making it a more inclusive and accessible domain. The innovative fusion of art and technology represented by NFTs is thus disrupting the traditional art marketplace, seeding a fertile ground for novel art forms and collection methods to flourish.

Hong Kong's Crucial Role in NFT Art Development (RQ2)

Positioned at the epicenter of finance and art trading, Hong Kong plays a pivotal role in accelerating the adoption and development of NFT Art. Advocates of blockchain-based assets, particularly cryptocurrencies, and NFTs, are progressively recognizing Hong Kong's strategic importance in shaping the future of the digital landscape (Shen, 2022).

According to Statista (n.d.), the NFT market in Hong Kong is experiencing substantial growth, with projections indicating that revenue could reach US$6,038.00k in 2023. Furthermore, the sector is expected to maintain this upward trajectory with an impressive annual growth rate (CAGR) of 18.46% from 2023 to 2027. This would result in a total market revenue of approximately US$11,890.00k by 2027. Statista also estimates that the average revenue per user within Hong Kong's NFT market will be around US$270.40 in 2023. This suggests that while the market is expanding, it still caters to a relatively niche group of users. The number of users in Hong Kong's NFT market is also expected to increase over the coming years, with Statista predicting it will reach around 26.26k users by 2027. These figures suggest that Hong Kong's NFT market is experiencing significant growth and has a committed user base.

HashKey Group, one of the leading digital asset management groups, offers an excellent example of the NFTs market in Hong Kong (HashKey Group, n.d.). In late 2021, HashKey made headlines by releasing a commemorative NFT for Hong Kong FinTech Week. The NFT was designed to tribute Hong Kong's iconic skyline and has highlighted the city's significance as an international financial hub and burgeoning center for fintech innovation. The creation and distribution of this NFT signify several important trends in the Hong Kong NFT market:

Digital Celebrations of Culture and Progress: NFTs can serve as digital representations of significant cultural and technological milestones. The HashKey NFT was not just a piece of digital art but a celebration of Hong Kong's fintech developments, indicating the growing recognition and adoption of NFTs in the cultural and financial sectors.

Corporate Involvement: The involvement of HashKey, a major player in digital asset management, shows that established corporations in Hong Kong are engaging with NFTs. This indicates the increasing institutional acceptance of NFTs, which can drive market growth and development.

Innovative Marketing Strategies: HashKey's creation of a commemorative NFT represents a novel marketing strategy. Companies in Hong Kong are leveraging

NFTs as a unique and innovative way to engage with their audiences and promote events or products.

Fintech Innovation: The release of this NFT during Hong Kong FinTech Week signifies the innovative use of fintech in the city. It showcases how NFTs and blockchain technology are becoming integral parts of the financial technology landscape in Hong Kong.

In summary, this case demonstrates the dynamic and growing nature of the NFT market in Hong Kong and the innovative ways in which organizations are beginning to integrate NFTs into their strategies. It's a strong indicator of the growing significance of NFTs in the cultural, financial, and technological sectors in Hong Kong.

Hong Kong's robust legal framework, aligned with international business law standards, ensures a reliable and secure business environment, attracting a wealth of international art companies dealing in tangible assets. Moreover, the city's advantageous tax policies, especially those regarding importing and exporting art and collectibles, further cement its reputation as a global art trading hub (Chan, 2021). Hong Kong's forward-looking approach towards NFTs and digital innovation, coupled with its thriving art scene, makes it a conducive incubator for the mainstream acceptance of NFTs in the art sector. As traditional businesses begin embracing NFTs as marketing and public relations tools, Hong Kong is emerging as a cultural pioneer in promoting the adoption of NFTs in conventional industries (HashKey Group, n.d.; Lam & Li, 2023). Furthermore, the city's strategic geographical location, bridging the East and the West, positions it as a significant player in the global NFT market. With its advanced infrastructure and commitment to integrating blockchain technologies, Hong Kong stands to lead the digital revolution in art, paving the way for the greater acceptance and utility of NFTs in the broader economy.

Suggestions

A strategic approach that embraces collaboration and innovation is paramount in navigating the ever-evolving landscape of blockchain technology and NFT art. Artists, art organizations, investors, and other key stakeholders should consider their unique capacities and resources to create and capture value within this burgeoning digital realm. This involves recognizing the opportunities arising from establishing new markets across diverse sectors and understanding the innovative features and functionalities that NFT art brings to the table.

However, the journey is not without its challenges. The volatile nature of cryptocurrencies, as evidenced by recent market crashes resulting in substantial losses for investors, underlines the risks inherent in the blockchain ecosystem. One striking instance is the sudden depreciation of Apecoin, a cryptocurrency launched by Yuga Labs, the creator of the renowned NFT collection, Bored Ape Yacht Club.

The coin's value plummeted to about one-third of its initial value in late April 2022 (Shen, 2022). Such scenarios highlight the need for careful and informed decision-making when dealing with blockchain-based assets.

While blockchain technology is instrumental in establishing the authenticity of digital artworks, it also raises complex legal questions. As NFTs gain traction, the accompanying legal implications become increasingly prominent. For example, given their rapidly growing popularity, the Hong Kong Securities and Futures Commission (SFC) recently warned investors about the risks associated with NFTs. Although unique copies of digital assets like photographs and artwork are currently exempt from regulation under the SFC's guidelines (Sun, 2022), this could change as the industry evolves.

To navigate these complexities, stakeholders must actively foster partnerships and build reliable networks with artists, institutions, and regulatory bodies. Sustained dialog and collaboration between these entities will facilitate a phased, coordinated implementation of blockchain technology within the art industry, ensuring that all participants adapt and evolve with the digital revolution (Abbate et al., 2022).

Furthermore, stakeholders should consider implementing educational initiatives to improve understanding and awareness of NFTs and blockchain technology. Such initiatives could help demystify the technology, equip artists and collectors with the knowledge to navigate the NFT art market confidently, and potentially drive greater adoption and acceptance of NFTs in the broader community.

In conclusion, the growth and acceptance of blockchain technology and NFTs in the art industry require a holistic approach involving strategic planning, risk management, regulatory awareness, education, and, most importantly, collaboration among all stakeholders. The digital art revolution has begun, and a thoughtful, well-considered approach can harness its potential to redefine the art world.

CONCLUSION

Blockchain technology harbors significant potential for revolutionizing the digital art market. This chapter takes a comprehensive look at the adoption of Non-Fungible Tokens (NFTs), particularly within Hong Kong's unique position in the art world. One of the main reasons behind the sweeping success of blockchain technology is its decentralized nature. Implementing NFTs can empower artists to assert ownership over their works unequivocally and share this ownership within a decentralized database framework. There is a prevailing sentiment among artists that incorporating blockchain technology within the art sector will pave the way for a more equitable and transparent marketplace.

Although NFT art is still in its nascent stage, there is an expectation for an upward trend in its adoption. This will likely be seen not only in the number of traditional organizations embracing blockchain technology but also in the volume of NFT assets making their way into established industries. Taken together, NFTs will likely continue their march toward becoming mainstream.

Furthermore, we are actively investigating the role of social networks in promoting arts and culture (Mak et al., 2022; Leung et al., 2022; Fong et al., 2020; Zhang et al., 2015). Our research also encompasses understanding the impact of the COVID-19 pandemic on the art market and exhibitions, particularly in the context of the extensive lockdowns of public and cultural facilities (Huang et al., 2021, 2022, 2023; Yu, Lam, & Chiu, 2023; Yu, Chiu, & Chiu, 2023; Meng et al., 2023).

In addition, our interests span the promotion of reading and other information services via social media (Chan et al., 2020; Fong et al., 2020; Cheng et al., 2020; Lam et al., 2019, 2023; Jiang et al., 2023). This includes using social media analytics to glean reading preferences and evaluate reading promotion effectiveness (Deng & Chiu, 2022; He et al., 2022; Wang et al., 2022). Moreover, we are keenly exploring the digital transformation of reading and reference materials (Sun et al., 2022; Tse et al., 2022; Chan and Chiu, 2023). The goal is to integrate these various strands of investigation to yield a holistic understanding of the evolving landscape of digital arts, reading, and culture.

REFERENCES

Abbate, T., Vecco, M., Vermiglio, C., Zarone, V., & Perano, M. (2022). Blockchain and art market: Resistance or adoption? *Consumption Markets & Culture*, *25*(2), 105–123. doi:10.1080/10253866.2021.2019026

Abrams, Z. (2022). *Beyond OpenSea: Our guide to to the biggest NFT marketplaces.* https://www.businessofbusiness.com/articles/plenty-of-fish-besides-opensea-here-is-our-guide-to-the-most-popular-nft-marketplaces/

Ali, G. (n.d.). *Everything about NFT - According to a Hong Kong NFT* artist. Retrieved from https://www.aligstudios.com/everything-about-nft

Au, C. H., Law, K. M. Y., Chiu, D. K. W., & Ho, K. K. W. (2022). Investigating mainstreaming strategies of hot cryptocurrencies-wallet. *Proceedings of Australasian Conference on Information Systems (ACIS), 1.* https://aisel.aisnet.org/acis2022/1

Beckman, M. (2021). *The Comprehensive Guide: NFTS, Digital Artwork, Blockchain Technology*. Skyhorse Pubishing.

Bitstamp. (2022). *What is the Bored Ape Yacht Club? (BAYC)*. https://www.bitstamp. net/learn/web3/what-is-the-bored-ape-yacht-club-bayc/

Bogdanowicz, M. (2015). *Digital Entrepreneurship Barriers and Drivers*. JRC Technical Report European Commission.

Botz, A. (n.d.). *Is blockchain the future of art?* https://artbasel.com/news/blockchain-artworld-cryptocurrency-cryptokitties

Chan, L. (2021). *Virtual Masterpieces: The Dawn of Digital Art*. https://research. hktdc.com/en/article/ODQ1OTg4MjUx

Chan, M. M. W., & Chiu, D. K. W. (2022). Alert Driven Customer Relationship Management in Online Travel Agencies: Event-Condition-Actions rules and Key Performance Indicators. In A. Naim & S. Kautish (Eds.), *Building a Brand Image Through Electronic Customer Relationship Management*. IGI Global. doi:10.4018/978-1-6684-5386-5.ch012

Chan, M. M. W., & Chiu, D. K. W. (2022). Alert Driven Customer Relationship Management in Online Travel Agencies: Event-Condition-Actions rules and Key Performance Indicators. In A. Naim & S. Kautish (Eds.), *Building a Brand Image Through Electronic Customer Relationship Management* (pp. 250–268). IGI Global. doi:10.4018/978-1-6684-5386-5.ch012

Chan, T. T. W., Lam, A. H. C., & Chiu, D. K. W. (2020). From Facebook to Instagram: Exploring user engagement in an academic library. *Journal of Academic Librarianship*, *46*(6), 102229. doi:10.1016/j.acalib.2020.102229 PMID:34173399

Chan, V. H. Y., & Chiu, D. K. W. (2023). Integrating the 6C's Motivation into Reading Promotion Curriculum for Disadvantaged Communities with Technology Tools: A Case Study of Reading Dreams Foundation in Rural China. In A. Etim & J. Etim (Eds.), *Adoption and Use of Technology Tools and Services by Economically Disadvantaged Communities: Implications for Growth and Sustainability*. IGI Global.

Chan, V. H. Y., Ho, K. K. W., & Chiu, D. K. W. (2022). Mediating effects on the relationship between perceived service quality and public library app loyalty during the COVID-19 era. *Journal of Retailing and Consumer Services*, *67*, 102960. doi:10.1016/j.jretconser.2022.102960

Cheung, V. S. Y., Lo, J. C. Y., Chiu, D. K. W., & Ho, K. K. W. (2023). Predicting Facebook's influence on travel products marketing based on the AIDA model. *Information Discovery and Delivery*, *51*(1), 66–73. doi:10.1108/IDD-10-2021-0117

Chin, G. Y. L., & Chiu, D. K. W. (2023). RFID-based Robotic Process Automation for Smart Museums with an Alert-driven Approach. In R. Tailor (Ed.), *Application and Adoption of Robotic Process Automation for Smart Cities*. IGI Global.

Chung, C., Chiu, D. K. W., Ho, K. K. W., & Au, C. H. (2020). Applying social media to environmental education: Is it more impactful than traditional media? *Information Discovery and Delivery*, *48*(4), 255–266. doi:10.1108/IDD-04-2020-0047

Clark, M. (2022). *People are spending millions on NFTs. What? Why?* The Verge.

Cointelegraph. (2021). *How to create an NFT: A guide to creating a non-fungible token*. https://cointelegraph.com/nonfungible-tokens-for-beginners/how-to-create-an-nft

Coslor, E. (2016). Transparency in an opaque market: Evaluative frictions between "thick" valuation and "thin" price data in the art market. *Accounting, Organizations and Society*, *50*, 13–26. doi:10.1016/j.aos.2016.03.001

Dai, C., & Chiu, D. K. W. (2023). Impact of COVID-19 on reading behaviors and preferences: Investigating high school students and parents with the 5E instructional model. *Library Hi Tech*. Advance online publication. doi:10.1108/LHT-10-2022-0472

Deng, S., & Chiu, D. K. W. (2023). Analyzing Hong Kong Philharmonic Orchestra's Facebook Community Engagement with the Honeycomb Model. In M. Dennis & J. Halbert (Eds.), *Community Engagement in the Online Space*. IGI Global. doi:10.4018/978-1-6684-5190-8.ch003

Ding, S. J., Lam, E. T. H., Chiu, D. K. W., Lung, M. M., & Ho, K. K. W. (2021). Changes in reading behavior of periodicals on mobile devices: A comparative study. *Journal of Librarianship and Information Science*, *53*(2), 233–244. doi:10.1177/0961000620938119

Dong, G., Chiu, D. K. W., Huang, P.-S., Lung, M. M., Ho, K. K. W., & Geng, Y. (2021). Relationships between research supervisors and students from coursework-based Master's degrees: Information usage under social media. *Information Discovery and Delivery*, *49*(4), 319–327. doi:10.1108/IDD-08-2020-0100

Dsouza, L. (2023). *Are NFTs the Future for the Art World?* Motiva. https://www.motiva.art/blog/are-nfts-the-future-for-the-art-world/

Dukic, Z., Chiu, D. K. W., & Lo, P. (2015). How useful are smartphones for learning? Perceptions and practices of Library and Information Science students from Hong Kong and Japan. *Library Hi Tech*, *33*(4), 545–561. doi:10.1108/LHT-02-2015-0015

Erinfolami, K. (2022, May 7). *Bored ape yacht club: What is it & why are they so expensive?* https://bitkan.com/learn/what-is-bored-ape-and-what-makes-bored-ape-expensive-8695

Ezeamuzie, N. M., Rhim, A. H. R., Chiu, D. K. W., & Lung, M. M. W. (2022). Mobile Technology Usage by Foreign Domestic Helpers: Exploring Gender Differences. *Library Hi Tech.* Advance online publication. doi:10.1108/LHT-07-2022-0350

Fan, K. Y. K., Lo, P., Ho, K. K. W., So, S., Chiu, D. K. W., & Ko, K. H. T. (2020). Exploring the mobile learning needs amongst performing arts students. *Information Discovery and Delivery*, *48*(2), 103–112. doi:10.1108/IDD-12-2019-0085

Fang, J., Chiu, D. K. W., & Ho, K. K. W. (2021). Exploring Cryptocurrency Sentimental with Clustering Text Mining on Social Media. In Z. Sun (Ed.), *Handbook of Research on Intelligent Analytics with Multi-Industry Applications* (pp. 157–171). IGI Global. doi:10.4018/978-1-7998-4963-6.ch007

Feng, Z., Chiu, D., Peng, R., Gong, P., He, K., & Huang, Y. (2016). Facilitating Cloud Process Family Co-evolution by Reusable Process Plug-in: An Open-source Prototype. *IEEE Transactions on Services Computing*, *10*(6), 854–867. doi:10.1109/TSC.2015.2504973

Fong, K. C. H., Au, C. H., Lam, E. T. H., & Chiu, D. K. W. (2020). Social network services for academic libraries: A study based on social capital and social proof. *Journal of Academic Librarianship*, *46*(1), 102091. doi:10.1016/j.acalib.2019.102091

Fortnow, M., Terry, Q., & Nguyen, K. (2021). *The NFT Handbook: How to Create, Sell and Buy Non-Fungible Tokens.* Ascent Audio.

Gao, W., Lam, K. M., Chiu, D. K. W., & Ho, K. K. W. (2021). A Big data Analysis of the Factors Influencing Movie Box Office in China. In Z. Sun (Ed.), *Handbook of Research on Intelligent Analytics with Multi-Industry Applications* (pp. 232–249). IGI Global. doi:10.4018/978-1-7998-4963-6.ch011

Hashkey Group. (n.d.). *Combing NFT Summer: What are the future directions of NFT worth looking forward to?* https://www.hashkey.com/en/insights/combing-nft-summer-what-are-the-future-directions-of-nft-worth-looking-forward-to.html

HashKey Group. (n.d.). *HashKey Releases Commemorative NFT for Hong Kong FinTech Week.* https://www.hashkey.com/en/insights/hashkey-releases-commemorative-nft-for-hong-kong-fintech-week.html

He, Z., Chiu, D. K. W., & Ho, K. K. W. (2022). Weibo Analysis on Chinese Cultural Knowledge for Gaming. In Z. Sun (Ed.), *Handbook of Research on Foundations and Applications of Intelligent Business Analytics* (pp. 320–349). doi:10.4018/978-1-7998-9016-4.ch015

Huang, P. S., Paulino, Y. C., So, S., Chiu, D. K. W., & Ho, K. K. W. (2021). Editorial. *Library Hi Tech, 39*(3), 693–695. doi:10.1108/LHT-09-2021-324

Huang, P. S., Paulino, Y. C., So, S., Chiu, D. K. W., & Ho, K. K. W. (2022). Guest editorial: COVID-19 Pandemic and Health Informatics Part 2. *Library Hi Tech, 40*(2), 281–285. doi:10.1108/LHT-04-2022-447

Jiang, T., Lo, P., Cheuk, M. K., Chiu, D. K. W., Chu, M. Y., Zhang, X., Zhou, Q., Liu, Q., Tang, J., Zhang, X., Sun, X., Ye, Z., Yang, M., & Lam, S. K. (2019). 文化新語:兩岸四地傑出圖書館、檔案館及博物館傑出工作者訪談 [New Cultural Dialog: Interviews with Outstanding Librarians, Archivists, and Curators in Greater China]. Systech Publications.

Jiang, X., Chiu, D. K. W., & Chan, C. T. (2023). Application of the AIDA model in social media promotion and community engagement for small cultural organizations: A case study of the Choi Chang Sau Qin Society. In M. Dennis & J. Halbert (Eds.), *Community Engagement in the Online Space* (pp. 48–70). IGI Global. doi:10.4018/978-1-6684-5190-8.ch004

Lam, A. H. C., Ho, K. K. W., & Chiu, D. K. W. (2023). Instagram for student learning and library promotions? A quantitative study using the 5E Instructional Model. *Aslib Journal of Information Management, 75*(1), 112–130. doi:10.1108/AJIM-12-2021-0389

Lam, E. T. H., Au, C. H., & Chiu, D. K. W. (2019). Analyzing the use of Facebook among university libraries in Hong Kong. *Journal of Academic Librarianship, 45*(3), 175–183. doi:10.1016/j.acalib.2019.02.007

Lam, S. Y., & Li, Y. (2023, May 30). *By embracing NFTs and the metaverse, Hong Kong can remodel its identity and global role.* South China Morning Post. https://www.scmp.com/comment/letters/article/3222159/embracing-nfts-and-metaverse-hong-kong-can-remodel-its-identity-and-global-role

Lei, S. Y., Chiu, D. K. W., Lung, M. M., & Chan, C. T. (2021). Exploring the aids of social media for musical instrument education. *International Journal of Music Education, 39*(2), 187–201. doi:10.1177/0255761420986217

Leung, T.N., Hui, Y.M., Luk, C.K.L., Chiu, D.K.W., & Kevin, K.K.W. (2022). Evaluating Facebook as aids for learning Japanese: learners' perspectives, *Library Hi Tech*. . doi:10.1108/LHT-11-2021-0400

Li, S., Chiu, D. K. W., & Ho, K. K. W. (2023). Social Media Analytics for Non-governmental Organizations: A Case Study of Hong Kong Next Generation Arts. In Z. Sun (Ed.), *Driving Socioeconomic Development with Big Data: Theories, Technologies, and Applications*. IGI Global. doi:10.4018/978-1-6684-5959-1.ch013

Li, S., Chiu, D. K. W., Kafeza, E., & Ho, K. K. W. (2023). Social Media Analytics for Non-governmental Organizations: A Case Study of Hong Kong Next Generation Arts. In Z. Sun (Ed.), *Driving Socioeconomic Development with Big Data: Theories, Technologies, and Applications* (pp. 277–295). IGI Global. doi:10.4018/978-1-6684-5959-1.ch013

Li, S. M., Lam, A. H. C., & Chiu, D. K. W. (2023). Digital transformation of ticketing services: A value chain analysis of POPTICKET in Hong Kong. In J. D. Santos & I. V. Pereira (Eds.), *Management and Marketing for Improved Retail Competitiveness and Performance*. IGI. Global.

Lin, C.-H., Chiu, D.K.W., Lam, K. T. (2022). Hong Kong Academic Librarians' Attitudes Towards Robotic Process Automation. *Library Hi Tech*. . doi:10.1108/LHT-03-2022-0141

Liu, N., & Ye, Z. (2021). Empirical research on the blockchain adoption - based on TAM. *Applied Economics*, *53*(37), 4263–4275. doi:10.1080/00036846.2021.1898535

Lo, P., Chan, H. H. Y., Tang, A. W. M., Chiu, D. K. W., Cho, A., Ho, K. K. W., See-To, E., & He, J. (2019). Visualising and Revitalising Traditional Chinese Martial Arts – Visitors' Engagement and Learning Experience at the 300 Years of Hakka KungFu. *Library Hi Tech*, *37*(2), 273–292. doi:10.1108/LHT-05-2018-0071

Lo, P., & Chiu, D. K. W. (2015). Enhanced and Changing Roles of School Librarians under the Digital Age. *New Library World*, *116*(11/12), 696–710. doi:10.1108/NLW-05-2015-0037

Lo, P., Cho, A., Law, B. K. K., Chiu, D. K. W., & Allard, B. (2017). Progressive Trends in Electronic Resources Management among Academic Libraries in Hong Kong. *Library Collections, Acquisitions & Technical Services*, *40*(1-2), 28–37. doi:10.1080/14649055.2017.1291243

Lo, P., Hsu, W. E., Wu, S. H. S., Travis, J., & Chiu, D. K. W. (2021). *Creating a Global Cultural City via Public Participation in the Arts: Conversations with Hong Kong's Leading Arts and Cultural Administrators*. Nova Science Publishers.

Lo, P., Yu, K., & Chiu, D. K. W. (2015). A Research Agenda for the Enhancing Roles and Practice of School Librarians in Hong Kong's 21st Century Learning Environments. *School Libraries Worldwide*, *21*(1), 19–37. doi:10.29173lw6881

Lo, Z. (2022). *Hong Kong NFT experts decode why the blockchain isn't as safe as you think*. https://www.tatlerasia.com/culture/arts/hong-kong-art-tech-screens-guru-new-blockchain-solution-interview

Mak, M. Y. C., Poon, A. Y. M., & Chiu, D. K. W. (2022). Using Social Media as Learning Aids and Preservation: Chinese Martial Arts in Hong Kong. In S. Papadakis & A. Kapaniaris (Eds.), *The Digital Folklore of Cyberculture and Digital Humanities* (pp. 171–185). IGI Global. doi:10.4018/978-1-6684-4461-0.ch010

Meng, Y., Chu, M. Y., & Chiu, D. K. W. (2023). The impact of COVID-19 on museums in the digital era: Practices and Challenges in Hong Kong. *Library Hi Tech*, *41*(1), 130–151. doi:10.1108/LHT-05-2022-0273

Mo, Q. (2021). *How NFTs are transforming culture, society and financial investments*. https://www.asiaglobalonline.hku.hk/how-nfts-are-transforming-culture-society-and-financial-investments

Mplus. (n.d.). *Museums and digital culture: A Field of productive tension*. https://www.mplus.org.hk/en/events/museums-and-digital-culture-a-field-of-productive-tension/

Nakamoto, S. (2008). Bitcoin: A peer-to-peer electronic cash system. *Decentralized Business Review*, 21260.

Ni, J., Chiu, D. K. W., & Ho, K. K. W. (2022). Information search behavior among Chinese self-drive tourists in the smartphone era. *Information Discovery and Delivery*, *50*(3), 285–296. doi:10.1108/IDD-05-2020-0054

Patrickson, B. (2021). What do blockchain technologies imply for digital creative industries? *Creativity and Innovation Management*, *30*(3), 585–595. doi:10.1111/caim.12456

Private Museum. (n.d.). *Art NFTs: a Revolution in the Art Industry*. https://www.privatemuseum.art/blog/guide/art-nfts-revolution-art-industry

Ross, L. (2022). *How to make and sell NFTs*. https://www.benzinga.com/money/how-to-make-your-own-nft

Salisu, M. S. (n.d.). *Non-Fungible Tokens (Nfts): Future of Digital Art & Collectibles*. LinkedIn. https://www.linkedin.com/pulse/non-fungible-tokens-nfts-future-digital-art-m-seun-salisu-

Schachter, K. (2022). *10 best NFT marketplaces.* https://www.fool.com/the-ascent/cryptocurrency/nft-marketplaces/

Seebacher, S., & Schüritz, R. (2017). Blockchain Technology as an Enabler of Service Systems: A Structured Literature Review. *Exploring Service Science*, *279*, 12–23.

Sharma, R. (2023). *Non-Fungible Token (NFT): What It Means and How It Works.* Investopedia. https://www.investopedia.com/non-fungible-tokens-nft-5115211

Shen, X. (2022). *Hong Kong NFT gamers among the world's most enthusiastic.* https://www.scmp.com/tech/tech-trends/article/3179694/hong-kong-nft-gamers-among-worlds-most-enthusiastic-only-behind

Statista. (n.d.). *NFT - Hong Kong | Statista Market Forecast.* https://www.statista.com/outlook/dmo/fintech/digital-assets/nft/hong-kong

Suen, R. L. T., Tang, J., & Chiu, D. K. W. (2020). Virtual reality services in academic libraries: Deployment experience in Hong Kong. *The Electronic Library*, *38*(4), 843–858. doi:10.1108/EL-05-2020-0116

Sun, X., Chiu, D. K. W., & Chan, C. T. (2022). Recent Digitalization Development of Buddhist Libraries: A Comparative Case Study. In S. Papadakis & A. Kapaniaris (Eds.), *The Digital Folklore of Cyberculture and Digital Humanities* (pp. 251–266). IGI Global. doi:10.4018/978-1-6684-4461-0.ch014

Sun, Z. (2022). *Hong Kong's securities and futures commission warns of non-fungible token risks.* https://cointelegraph.com/news/hong-kong-s-securities-and-futures-commission-warn-of-nonfungible-token-risks

Sung, Y. Y. C., & Chiu, D. K. W. (2022). E-book or print book: Parents' Current View in Hong Kong. *Library Hi Tech*, *40*(5), 1289–1304. doi:10.1108/LHT-09-2020-0230

Swati, B., Namrata, P., Shikha, S., & Shivani, S. (2019). Digital Entrepreneurship And The Bottom of The Pyramid- A Conceptual Framework. *International Journal on Recent Trends in Business and Tourism*, *3*(1), 33–42.

Tempone, D. (2021). *How Can Artists Tokenize Their Work and Enter the New Crypto Art Market?* https://www.domestika.org/en/blog/7273-how-can-artists-tokenize-their-work-and-enter-the-new-crypto-art-market

Tse, H. L., Chiu, D. K., & Lam, A. H. (2022). From Reading Promotion to Digital Literacy: An Analysis of Digitalizing Mobile Library Services With the 5E Instructional Model. In A. Almeida & S. Esteves (Eds.), *Modern Reading Practices and Collaboration Between Schools, Family, and Community* (pp. 239–256). IGI Global. doi:10.4018/978-1-7998-9750-7.ch011

Wai, I. S. H., Ng, S. S. Y., Chiu, D. K. W., Ho, K. K., & Lo, P. (2018). Exploring undergraduate students' usage pattern of mobile apps for education. *Journal of Librarianship and Information Science, 50*(1), 34–47. doi:10.1177/0961000616662699

Wang, B., Chiu, D. K. W., & Ho, K. K. W. (2022a). A Comparison of Deep Learning Models in Time Series Forecasting of Web Traffic Data from Kaggle. In Z. Sun (Ed.), *Handbook of Refromch on Foundations and Applications of Intelligent Business Analytics* (pp. 301–319). IGI Global. doi:10.4018/978-1-7998-9016-4.ch014

Wang, J., Deng, S., Chiu, D. K. W., & Chan, C. T. (2022b). Social Network Customer Relationship Management for Orchestras: A Case Study on Hong Kong Philharmonic Orchestra. In N. B. Ammari (Ed.), *Social Customer Relationship Management (Social-CRM) in the Era of Web 4.0* (pp. 250–268). IGI Global. doi:10.4018/978-1-7998-9553-4.ch012

Wang, P., Chiu, D. K. W., Ho, K. K., & Lo, P. (2016). Why read it on your mobile device? Change in reading habit of electronic magazines for university students. *Journal of Academic Librarianship, 42*(6), 664–669. doi:10.1016/j.acalib.2016.08.007

Warkentin, M., & Orgeron, C. (2020). Using the Security Triad to Assess Blockchain Technology in Public Sector Applications. *International Journal of Information Management, 52*, 102090–102098. doi:10.1016/j.ijinfomgt.2020.102090

Wong, A. K. K., & Chiu, D. K. W. (2023). Digital Transformation of Museum Conservation Practices: A Value Chain Analysis of Public Museums in Hong Kong. In R. Pettinger, B. B. Gupta, A. Roja, & D. Cozmiuc (Eds.), *Handbook of Research on the Digital Transformation Digitalization Solutions for Social and Economic Needs*. IGI Global. doi:10.4018/978-1-6684-4102-2.ch010

Wong, I. H. S., Fan, E. C. H., Chiu, D. K. W., & Ho, K. K. W. (2023). Internet Celebrities' Influence on Youths' Diet Behaviors: A Gender Study on Social Media. *Aslib Journal of Information Management*. Advance online publication. doi:10.1108/AJIM-11-2022-0495

Wong, J., Chiu, D. K. W., Leung, T. N., & Kevin, K. K. W. (2023). Exploring the Associations of Addiction as a Motive for Using Facebook with Social Capital Perceptions. *Online Information Review, 47*(2), 283–298. doi:10.1108/OIR-06-2021-0300

Wong, K. C., & Chiu, D. K. W. (2023). Promoting The Use of Electronic Resources of International Schools in Hong Kong: The Case Study of ESF King George V School. In E. Meletiadou (Ed.), *Handbook of Research on Redesigning Teaching, Learning, and Assessment in the Digital Era*. IGI Global. doi:10.4018/978-1-6684-8292-6.ch007

Wu, M., Lam, A. H. C., & Chiu, D. K. W. (2023) Transforming and Promoting Reference Services with Digital Technologies: A Case Study on Hong Kong Baptist University Library. In B. Holland (Ed.), Handbook of Research on Advancements of Contactless Technology and Service Innovation in Library and Information Science (pp. 128-145). IGI Global. doi:10.4018/978-1-6684-7693-2.ch007

Xie, Z., Chiu, D. K. W., & Ho, K. K. W. (2023). The Role of Social Media as Aids for Accounting Education and Knowledge Sharing: Learning Effectiveness and Knowledge Management Perspectives in Mainland China. *Journal of the Knowledge Economy*. Advance online publication. doi:10.100713132-023-01262-4

Xie, Z., Wong, G. K. W., Chiu, D. K. W., & Lei, J. (2023). Bridging K-12 Mathematics and Computational Thinking in the Scratch Community: A Big Data Analysis. *IT Professional*, *25*(2), 64–70. doi:10.1109/MITP.2023.3243393

Yang, Z., Zhou, Q., Chiu, D. K. W., & Wang, Y. (2022). Exploring the factors influencing continuance usage intention of academic social network sites. *Online Information Review*, *46*(7), 1225–1241. doi:10.1108/OIR-01-2021-0015

Yi, Y., & Chiu, D.K.W. (2023). Public information needs during the COVID-19 outbreak: A qualitative study in mainland China. *Library Hi Tech*. doi:10.1108/LHT-08-2022-0398

Yip, K. H. T., Chiu, D. K. W., Ho, K. K. W., & Lo, P. (2021). Adoption of Mobile Library Apps as Learning Tools in Higher Education: A Tale between Hong Kong and Japan. *Online Information Review*, *45*(2), 389–405. doi:10.1108/OIR-07-2020-0287

Yu, H. H. K., Chiu, D. K. W., & Chan, C. T. (2023). Resilience of symphony orchestras to challenges in the COVID-19 era: Analyzing the Hong Kong Philharmonic Orchestra with Porter's five force model. In W. Aloulou (Ed.), *Handbook of Research on Entrepreneurship and Organizational Resilience During Unprecedented Times* (pp. 586–601). IGI Global.

Yu, H. Y., Tsoi, Y. Y., Rhim, A. H. R., Chiu, D. K. W., & Lung, M. M. W. (2022). Changes in habits of electronic news usage on mobile devices in university students: A comparative survey. *Library Hi Tech*, *40*(5), 1322–1336. doi:10.1108/LHT-03-2021-0085

Yu, P. Y., Lam, E. T. H., & Chiu, D. K. W. (2023). Operation management of academic libraries in Hong Kong under COVID-19. *Library Hi Tech*, *41*(1), 108–129. doi:10.1108/LHT-10-2021-0342

Zhang, Q., Huang, B., & Chiu, D.K.W., & Ho, K.K.W. (2015). Learning Japanese Through Social Network Sites: A Case Study of Chinese Learners' Perceptions. *Micronesian Educators*, *21*, 55–71.

Zhang, X., Lo, P., So, S., Chiu, D. K. W., Leung, T. N., Ho, K. K. W., & Stark, A. (2021). Medical students' attitudes and perceptions towards the effectiveness of mobile learning: A comparative information-need perspective. *Journal of Librarianship and Information Science*, *53*(1), 116–129. doi:10.1177/0961000620925547

Zhuang, Y., Li, Q., Chiu, D. K., Wu, Z., & Hu, H. (2014). Efficient Personalized Probabilistic Retrieval of Chinese Calligraphic Manuscript Images in Mobile Cloud Environment. *ACM Transactions on Asian Language Information Processing*, *13*(4), 18. doi:10.1145/2629575

Zuo, Y., Lam, A. H. C., & Chiu, D. K. W. (2023). Digital protection of traditional villages for sustainable heritage tourism: A case study on Qiqiao Ancient Village, China. In A. Masouras, C. Papademetriou, D. Belias, & S. Anastasiadou (Eds.), *Sustainable Growth Strategies for Entrepreneurial Venture Tourism and Regional Development* (pp. 121–195). IGI Global. doi:10.4018/978-1-6684-6055-9.ch009

Chapter 8

Parents' View of Graphic Novels in Hong Kong Under the 21st Century Mobile Digital Environment

Samuel Kin Fung Chan
The University of Hong Kong, Hong Kong

Mimi Mei Wa Chan
The University of Hong Kong, Hong Kong

Apple Hiu Ching Lam
 https://orcid.org/0000-0002-2587-6979
The University of Hong Kong, Hong Kong

Dickson K. W. Chiu
 https://orcid.org/0000-0002-7926-9568
The University of Hong Kong, Hong Kong

ABSTRACT

This study reviews the current parents' perspectives and habits of using graphic novels, including electronic equipment, in Hong Kong to support children's reading and learning. Parents were invited to participate in semi-structured group interviews based on our five research questions. The authors employ a thematic analysis approach to analyze the data and determine the similarities and differences between the literature review and the current mobile digital environment. The findings indicated that respondents were not keen on using graphic novels to support their children's reading and learning. The two main reasons were inadequate understanding or bias on graphic novels and the wide availability of online digital materials for reading and learning support. Scant studies focus on graphic novels in parent-child reading from parents' perspectives, especially in the Asian context, though graphic novels are becoming a worldwide trend. Parents' opinions about this issue are invaluable for educators and librarians in curriculum design and collections development.

DOI: 10.4018/978-1-6684-8671-9.ch008

INTRODUCTION

Reading has recently regained attention (Chan & Chiu, 2023; Chiu & Ho, 2022; Kong et al., 2018; Lo et al., 2021; Tse et al., 2022) because of the massive lockdown of cultural organizations during the COVID pandemic (Huang et al., 2021, 2022, 2023; Meng et al., 2023; Yu et al., 2023). As for graphics novels, parents often avoid selecting these reading materials for their children due to the stereotypes among parents. Text and illustrations are the two major elements in graphic novels to narrate stories; comics and picture books are common examples (Library of Congress, 2014). Parents seldom recognize graphic novels as formal literature and resist their children's reading because of the widespread use of illustrations in books (Fletcher-Spear et al., 2005). Moreover, the popularity of superhero comics provides some negative and misleading impressions of graphic novels, such as graphic novels are all about superheroes and contain inappropriate content for children (Scholastic, 2018).

Over the last decade, America's academia and education sectors have paid more attention to graphic novels and affirmed their advantages in distinct aspects (Morrison, 2017), such as encouraging reluctant readers (Bunn, 2012), improving information literacy (Crowley, 2015), and aiding children's learning in different subjects (Cooper et al., 2011; Griffith, 2010; Oz & Efecioglu, 2015).

Although researchers have discovered the contribution of graphic novels to children's reading and learning, scant studies investigate its current usage and acceptance in parents' perspectives, especially in the Eastern context. It is necessary to bring new insights to parents and related stakeholders (such as librarians and educators) and help them regard graphic novels as valuable literature and learning resources.

This research examined the current usage of graphic novels in Hong Kong families and explored parents' opinions on using them to help children's reading and learning. Thus, group interviews were arranged for several local parents and focused on the following research questions and objectives.

RQ1. What is the current awareness of graphic novels among parents in Hong Kong?

RQ2. What are the pros and cons from the parents' perspectives in using graphic novels to cultivate children's reading habits and support their learning?

RQ3. What is the current usage among these parents of graphic novels to cultivate children's reading behavior and support their learning?

RQ4. How can online platforms and resources be effectively used to read graphic novels in parent-child reading?

RQ5. What can society and government do to promote graphic novel reading to parents and children?

LITERATURE REVIEW

Graphic Novels and Biases

Will Eisner published a series of stories called 'A Contract with God' in 1978 about a poor and crowded Jewish Bronx community and invented the term 'graphic novel' to describe such complex novels that combined texts and illustrations (Bucher & Manning, 2004).

The texts and illustrations have equal status and importance for presenting characters' moods, tones, actions, and plots in graphic novels (Griffith, 2010). Usually, graphic novels have longer stories and more complex story structures than traditional comics. Furthermore, the topics of graphic novels are very diverse, including science-fiction, superhero tales, historical stories, manga tales, etc. Actually, fiction is not the only genre in the graphic novel world: nonfiction, biography, and autobiography (Bucher & Manning, 2004).

Owing to the heavy use of illustrations, some deep-rooted biases of graphic novels, such as 'too easy' to read and only suitable for children, existed in many people's minds. They believe that children should 'grow out of it' and read 'real books' as they grow up, as over-focusing on illustrations may cause neglect to understand the whole story and misinterpretation of the author's original intention. In addition, the biases can wrongly strengthen the linkage of graphic novels and sensitive topics, such as violence, drugs, sex, antiauthoritarianism, etc. (Downey, 2011).

Graphic Novels In Parent-Child Reading

Graphic novels are suitable reading materials for assisting children's personal growth. In social development, imitation is an effective learning method. Parents can use the illustrations in graphic novels to teach their children about facial and body expressions, symbolism, and posture meaning vividly (Bucher & Manning, 2004). Besides, graphic novels provide relaxed platforms and opportunities for parent-child discussions of sensitive issues. It is because graphic novels always stand at the forefront of many social problems, e.g., the 911 terrorist attack in '*Heroes*' by Marvel Comics; AIDS in '*Pedro and me*' by Judd Winick; rape and pregnancy in '*Amazing "True" Story of a Teenage Single Mom*' by Katherine Arnoldi; and sexual abuse in '*The Tale of One Bad Rat*' by Bryan Talbot (Bucher & Manning, 2004). Graphic novels are not only another easy channel for children to understand serious social issues but also cultivate their empathy and positive values. Parent-child reading can generally enhance children's listening, reading, spelling, and literacy skills (Duursma et al., 2008; Mol & Bus, 2011). In particular, reading graphic novels can enhance parent-child relationships, value education, and emotional communication (Sung & Chiu, 2022).

How Graphic Novels Affect Children's Reading Behavior

It is hard to promote reading among children as many think it is boring, different to finish, and requires too much time and energy. These children are called 'reluctant readers,' who can but are unwilling to read (Snowball, 2005). Children's growth environment is a potential reason that brings them negative images of reading. In the 21st Century, there is an inseparable relationship between life and electronic equipment besides televisions, such as computers and smartphones. These devices provide fast visual stimulation anywhere and anytime. Many children misuse such enjoyment and expect similar excitement in other activities, including reading (Downey, 2011).

Interest is a crucial reason for reading desire (Snowball, 2005). Graphic novels have three advantages that can satisfy children's reading expectations, interests, and sensory needs. Firstly, the obstacles to reading graphic novels are low, requiring advanced reading skills (Crawford, 2004). Secondly, text and illustration connect the media people watch and read. The two elements assist each other in narrating stories in a comprehensive and multisensory way (Downey, 2011; Yang, 2008). Thirdly, the illustrations help children understand stories visually and without challenging vocabulary.

Graphic novels lower reading efforts among children and bring them joy, making it easier to build a reading habit. Teacher librarians have provided evidence that adding graphic novels to their collections can increase the check-out rate (Bucher & Manning, 2004; Crawford, 2004; Downey, 2011). Further, Downey (2011) believed that graphic novels are stepstones to help children discover the joy of reading, and the key is linking casual reading and traditional text reading.

Graphic Novels for Children's Learning

Language and movement in movies and animations are time-bound, as people cannot choose how fast to watch (until recently with playback control on recorded media), especially live shows. However, the graphic novel is a static media that allows readers to review and fast-forward or back the content whenever needed (Yang, 2008). This characteristic can benefit children in learning and absorb new knowledge and thematic information, spark learning interest, and facilitate various kinds of literacy in relatively adequate time and personalized learning progress (Downey, 2011). Besides, no extra equipment and technical support are needed for reading graphic novels compared with motion pictures.

In the US, graphic novels are getting more popular in supporting children's teaching and learning. For example, the Department of Education in Maryland State and the Diamond Comics Distributors co-developed graphic-novel-based learning materials for K-12 teachers (Richardson, 2017). Besides, the Department of Education in New York City has been promoting graphic novels in schools since 2008 (Downey, 2011). These cases illustrated that Western countries believed graphic novels fit education purposes.

Besides, language development is one of the aspects that graphic novels can contribute. Snowball (2005) discovered that graphic novels are suitable reading materials in linguistics, improving children's reading, writing, spelling, and grammar abilities. For instance, graphic novels contained 53.5 rare and complex vocabularies per 1,000 words, while text-based children and adult books only contained 30.9 and 52.7 per 1,000 words, respectively (James, 2014). Children can acquire more new vocabulary from graphic novels. Besides, graphic novels are suitable for learning second languages as children can learn the practical use of slang and colloquial phrases (Oz & Efecioglu, 2015).

After that, children can practice and strengthen complex, comprehensive, cognitive, and interpretation skills when they are reading graphic novels (Jennings, Rule, & Vander Zanden, 2014), as they are required to comprehend the relationship and meaning of texts and illustrations at the same time (Bucher & Manning, 2004). The illustrations also assist children in understanding story structures, improving literary skills (such as characters, plots, scenes, rhetoric, etc.), and developing deductive reasoning abilities (Downey, 2011). Moreover, illustrations can reinforce children's memory and better recall (Short et al., 2013). For children with learning disabilities, like attention deficit hyperactivity disorder (ADHD), the positive effects of illustrations are even more significant. Illustrations give them extra clues to process information as they struggle to read words and sentences (Cummings, 2020).

Besides language and literacy development, graphic novels also benefit media literacy development. For example, children can learn to use different colors to represent emotions, illustrate stereotypes, angles, and positions of drawing affect thinking, and balance realism and virtualism (Downey, 2011).

Many cases demonstrate how to use graphic novels to support children in learning different subjects in real life. For example, Christensen (2006) applied '*Palestine*' by Joe Sacco to help children learn Middle East history by discussing the economic, emotional, and psychological conflicts between America and the Middle East, and they were more willing to learn from graphic novels. On the other hand, some studies discovered that graphic novels help learn confusing concepts in science and graphics design in arts (Cooper et al., 2011).

Research Gap

According to the literature reviews, three research gaps have been identified. First, the studies and examples were conducted in Western countries, not the East. As reading cultures and education systems in the East (such as Hong Kong) are different, graphic novels' application in the literature review may not be directly applied. Besides, scant studies review the latest situation and difficulties using graphic novels in-depth, both physical and electronic formats. Further, many related studies focusing on reading promotion, including graphic novels, are in the context of school libraries but not from parents' perspectives. Therefore, the research aimed to investigate the current usage of graphic novels in Hong Kong families and explore parents' opinions on using them to help children's reading and learning.

METHODOLOGY

A qualitative interview research method was applied to this study, which has several advantages. First, it allows researchers to collect direct opinions from individuals and explore why they have these ideas without overlooking information. Second, individuals can provide deeper and more meaningful opinions through participant interactions. Especially the semi-structured interview method permits researchers to ask follow-up and impromptu questions according to participants' responses (Leedy & Ormord, 2015).

The target participants of this study were parents with children from kindergarten to junior secondary school because they had more recent experiences with parent-child reading. Purposive sampling was applied to select suitable participants, including diverse genders, children with different genders and in different grades, and recent experience with parent-child reading. Table 1 lists the six parents' demographic backgrounds (of 9 children). Due to the COVID-19 epidemic in Hong Kong during the data collection, two 1-hour group Cantonese interviews were conducted via online video calls according to the participants' requests. There were eight main open-ended questions for all participants in the interviews to maintain the consistency of the data. The questions asked for participants' opinions on their awareness of graphic novels (RQ1), graphic novels for cultivating children's reading behavior (RQ2), the relationship between graphic novels and children's learning (RQ2), their acceptance of using graphic novels for reading and learning cultivation (RQ3), their selection criteria for high-quality graphic novels (RQ3), their graphic novel usage through online platforms (RQ4), and social support and promotion for graphic novel reading among children and related facilities (RQ5). Participants reviewed the transcriptions in English to avoid any misunderstanding and mistranslation.

Table 1. Demographic background of participants

Participants	Gender	No. of Children	Grade of Children	Parent-Child Reading
P1	F	2 (daughters)	S.2 and P.3	No for now
P2	M	1 (son)	P.1	15-30 mins / day
P3	M	1 (son)	P.1	15-20 mins / day
P4	F	2 (daughters)	P.1 and P.3	No for now
P5	F	2 (sons)	P.2 and P.5	No for now
P6	F	1 (son)	S.1	No for now

RESEARCH ANALYSIS AND DISCUSSION

The thematic analysis approach was used for analyzing the data in this study to classify data into themes and illustrate the relationships among the data (Alhojailan, 2012). Participants' opinions were summarized into various themes to provide answers to the research questions. The opinions were summarized and analyzed with different coding. Most of the opinions were coded according to the interview questions and, thus, the research questions. The relationships between graphic novels and children's learning (see Table 4) under RQ2 used the 5E Instructional Model to code the opinions from Question 4 (see Appendix). The following sub-sections were analyzed for the results based on our five research questions.

Awareness of Graphic Novels Among Parents in Hong Kong (RQ1)

As summarized in Table 2, most participants expressed that they were unfamiliar with graphic novels, and some said they did not hear this phrase, whether in English or Cantonese. However, they got clues from the wordings *graphic* and *novel* and inferred it should be fiction with illustrations. They thought the main elements of graphic novels were pictures and colors. These ideas matched existing research results (Fletcher-Spear et al., 2005). Besides, some participants anticipated connections or relationships between graphic novels, comics, and picture books. Although participants could not provide a clear definition of graphic novels, their speculation still matched the explanation from professionals (Library of Congress, 2014).

On the other hand, most participants said they did not use graphic novels in parent-child reading and children's individual reading. Marvel's comics were the most approximative graphic novels used in home reading. Their impressions were in line with the stereotypes of graphic novels, including 'superhero comics and 'too easy to read' (Downey, 2011). Interestingly, two participants used Marvel's comics

Table 2. Coding for participants' awareness of graphic novel

Themes	Example Quotations	Coding
Basic cognition	*P2: Not very familiar with this phraseP4: I am not sure what it isP6: I don't know about this phrase*	Unfamiliar
Core elements	*P1: Widely used pictures and imagesP2: Many pictures and colorsP3: Major elements ... should be graphics and pictures*	Widely use illustrations and colors to represent the story
Relation with other genres	*P1: Some kind of comics or picture booksP2: Similar to comics and picture booksP5: A genre similar to comics*	Great connection between graphic novels, comics, and picture book
Using for parent-child reading	*P2,5: Marvel's comicsP1,3,4,6: Did not use it*	Low usage in readingReading superhero comics, in some cases

before having sons in their families. As Moeller (2011) pointed out, the gender of their children may affect their willingness to use graphic novels.

Pros and Cons for Parents to Use Graphic Novels to Support Children's Reading and Learning (RQ2)

All participants agreed that illustrations in graphic novels play an essential role in cultivating children's reading habits. They opined that illustrations were more attractive for children, and text was often an obstacle to motivating children to read more books, rendering them reluctant readers, which matched previous research. For example, Wilhelm (1995) showed that illustrated books help highly reluctant readers to enjoy reading and evoke story worlds in their minds. However, some participants thought graphic novels should be the very introduction for children to the reading world and thus should balance the proportion of reading graphic novels and text-based materials to avoid a wrong impression about reading. It would be too late to deal with their fear of reading text-based materials if they over-relied on illustrations. Their opinion was in line with why graphic novels provide only a stepstone in children reading, as Downey (2011) pointed out. Since participants already understand the functions and limitations of using graphic novels to cultivate children's reading habits, they and other parents should be able to maximize the advantages of graphic novels on this issue.

For using graphic novels to support children's learning, participants provided opinions based on the 5E Instructional Model. This instructional model has been widely used for design curricula and teaching materials since the late 1980s, with 5 phases to sum up the whole learning process (Bybee et al., 2006; Cheung et al.,

Table 3. Coding for graphic novels for cultivating children's reading behavior

Themes	Example Quotations	Coding
Illustrations	*P1: Pictures are more attractive for children compared with text ... willing to absorb knowledge from picturesP2: Images and pictures in graphic novels and comics are very attractive to children ... text is very boringP6: Illustrations are very attractive*	The selling point of graphic novels is colorful illustrationsIllustrations are more interesting and attractive than texts
Transition to text-based materials	*P3: Do not think ... can eliminate their fear of reading text-based materials ... guide to reading more text is another important issueP4: If you want your children to read classic literature, I do not think reading graphic novels can cultivate their reading habits ... over-reliance on reading pictures but not textP5: Entry for children to experience or enjoy what is reading*	Graphic novels cannot eliminate children's fear of reading text-based reading materials like classic literatureChildren will over-reliance on illustrationsText-based material is the only destination in children's reading

2022; Jiang, Lam, Chiu, & Ho; 2023; Lam et al., 2023; Tsang & Chiu, 2022; Tse et al., 2022; Xie, Chiu, & Ho, 2023):

- Engagement – Helping learners engage in new concepts
- Exploration – Providing platforms or activities for learners to explore new knowledge
- Explanation – Explaining the knowledge with learners' understanding and experiences
- Elaboration – Offering challenges for learners to apply their new knowledge
- Evaluation – Testing the abilities of learners

Participants were requested to consider the pros and cons of graphic novels in the above phases. Table 4 summarizes their responses and illustrates the suitability of graphic novels to support children's learning.

First, all participants agreed that graphic novels as good motivational tools for raising children's interest in new knowledge. Botzakis et al. (2017) stated similar opinions and believed graphic novels could motivate children to keep reading. However, a participant mentioned graphic novels as only a medium to deliver information and not change the nature of knowledge. Second, participants thought graphic novels could provide an easy and interesting way to explore and explain new knowledge, particularly suitable for demonstrating essential skills, like science experiments and sports, as children could follow illustrations step by step. However, some participants queried why not use educational videos or traditional in-person demonstrations in this mobile digital age. Third, participants believed drawing graphic novels to present what children would learn might complicate the learning

Table 4. Coding for relationships between graphic novels and children's learning

Themes	Example Quotations	Coding
Effectiveness	P1: Lower than text-based materials ... waste time and inefficiency (Con)P5: Not sure these books bring positive effects on academic results (Con) P6: Don't think graphic novels can contain enough knowledge or information in limited space (Con)	InefficiencyWaste timeUnable to contain much information within limited pagesUncertain to improve academic results
Engagement	P1: Raise their interest and motivation (Pro)P3: Engage them to learn from books ... similar to attract them to read more books (Pro)P4: They like GN because of the beautiful colors and illustrations, not because they want not to learn (Con)P5: A channel to learn but still cannot change the nature of knowledge (Con)	Good motivational toolProvides a new channel for children to learn, but similar knowledge Attracting children to reading does not mean they like to learn
Exploration	P1: Easily provide new concepts and inspire them to match experiences and explore in real life (Pro)P2: Interested to learn from graphics instead of text and words (Pro)P3: Willing to learn new knowledge in an interesting way and format (Pro)P4: Educational video is more useful to help them explore new knowledge ... video is moveable, but illustrations are immovable (Con)P5: Exploring in real life and using different equipment, like a microscope, must be more helpful (Con)P6: Picture is only one of the clues for exploring the in-depth meaning of stories (Con)	Illustrations allow children to explore new knowledge easily and interestinglyEducation videos and demonstrations in real life should more effectively for promoting children to explore the worldIllustrations are one of the clues for exploring the in-depth meaning of stories
Explanation	P1: Learn easily step by step, especially to demonstrate science experiments (Pro)P2: Explain complicated science concepts (Pro)P3: Useful to represent some concepts (actions or postures) in physical education (Pro)P4: Video may be more suitable to explain complicated ideas and concepts (Con)P5: Demonstration in real is more practical (Con)	Explaining science experiments in step by step wayExplaining movements and postures in sportsEducation videos and demonstrations in real life should more effectively explain complicated concepts
Elaboration and Evaluation	P1: Do not think drawing graphic novels are suitable for secondary students (Con)P2: Draw to represent the correct steps of science experiments (Pro)P6: Make learning processes more complex and annoying (Con)	Useful for representing the correct steps of science experimentsDrawing makes elaboration complicated
Application	P1: Tell historical storiesP2: Abstract concepts and the steps of experimentsP3: Art skills, like breakups, ratios, and colorsP6: Phrases and informal sentences for daily use	HistoryScienceVisual artLanguage learning

processes. In sum, most participants reputed that using graphic novels in learning would be inefficient and waste time under the current technological environment.

The above responses differed from the studies discussed in the literature review. Although many examples (Bischell, 2018) exist in other countries to explain how graphic novels effectively improve learning effectiveness, Hong Kong respondents opined different ideas. They believed other suitable teaching materials could assist or substitute traditional learning. As e-learning and mobile learning become very common, multimedia materials are widely available online and readily accessible with mobile devices (Clark & Mayer, 2016; Cheng et al., 2022; Ding et al., 2021; Wai et al., 2018; Wang et al., 2016; S.W.S. Wong & Chiu, 2023; Yip et al., 2021; Zhang et al., 2021; Zheng et al., 2023). Thus, the graphic novel is only a small and perhaps outdated part of multimedia learning. Therefore, some participants opined that online videos were more helpful in engaging, exploring, and explaining new knowledge than graphic novels, especially concerning efficiency.

Acceptance of Using Graphic Novels to Support Children's Reading and Learning (RQ3)

Based on the participant responses in the previous sub-section, their acceptance of using graphic novels to support children's reading and learning should not be high. As shown in Table 4, participants considered graphic novels inappropriate and outdated for learning purposes because graphic novels cannot deliver considerable knowledge within limited pages and time. However, they considered graphic novels suitable for non-core subjects, like visual art or physical education. In Hong Kong, academic stress is remarkably high among children, and academic attainment is a key cause of this unhealthy learning environment (Leung, 2007). Further, participants focused on the inconsistencies of graphic novels to academic results, tight syllabi, tight school calendars, and a huge amount of homework and examinations instead of enhancing learning motivations and interest. Thus, there is a high barrier to applying graphic novels to children's learning in Hong Kong.

Participants' negative attitudes toward using graphic novels to support children's learning are similar to the findings of Nesmith et al. (2016) about elementary-age students and their parents using graphic novels in mathematics and science education. They also believed that graphics are inappropriate, distracting, and inhibiting the reader's imagination, and the plots and content are often inappropriate and poorly developed. They only recommended graphic novels to some students, like struggling readers and English language learners. However, they agreed that graphic novels were good tools for introducing abstract concepts, less intimidating than traditional textbooks, and enhancing recall and accessibility. These ideas matched our participants' responses in the previous section's discussion.

For leisure reading, participants' attitudes were relatively positive. They admitted graphic novels are useful for leisure or entertainment. It would be best if the Education Bureau, public libraries, or schools could recommend booklists for selecting high-quality graphic novels (Kong et al., 2018; Leung et al., 2020; Lo et al., 2015; Lo &

Table 5. Coding for acceptance of using graphic novels for cultivating children's reading behavior and learning

Themes	Example Quotations	Coding
For leisure reading	*P1: Read graphic novels for leisureP2: Recommended booklists provided by the Education Bureau or school; I will tryP3: Used for leisure reading ... too many pictures and not suitable for learning / do not think comics are suitable for parent-child readingP4: Something for leisure reading*	Graphic novels only for leisure readingStill want some recommended booklists from the authorities
For academic purpose	*P1: Many public exams ... not suitable for them to accept and apply knowledge in a limited timeP2: Inefficient ... in the tense school calendar ... not suitable to deliver a huge amount of knowledgeP3: Non-core subjectsP4: Syllabi in Hong Kong schools is too tightP5: Textbooks are more suitableP6: Not suitable for traditional learning*	InappropriateTight syllabus, tense school calendar, and a huge amount of homework and exam are not allowed to have graphic-novel-based learning

Chiu, 2015). It reflected that respondents still distrusted graphic novels for serious learning for children.

Regarding selecting high-quality graphic novels, participants were concerned about whether graphic novels fit their needs and reduce their defense mentality. Their diversified criteria included author, publisher, content, additional features, etc. Middle school libraries in America also suggested similar criteria for graphic novel selection (Moeller & Becnel, 2021), reflecting our participants' opinions as reasonable, realistic, and in line with library recommendations. Notably, participants preferred graphic novels that were educational and endorsed by professionals (Kong et al., 2018).

Table 6. Coding for selection criteria for high-quality graphic novels

Themes	Example Quotations	Coding
Background	*P5: Published by companies with a good reputation*	AuthorPublisher
Content	*P1: Educational meaningP4: Scared there are inappropriate wordings*	Educational purposeLanguage
Additional features	*P3: Illustrations are the most important part*	Illustration
Price	*P6: Price and discount*	Price
Professional suggestion	*P2: Recommended booklists provided by teachers or librarians*	TeacherLibrarian

Table 7. Coding for graphic novel usage via online platforms

Themes	Example Quotations	Coding
Current situation	*P1,3,4,6: Have not used beforeP2: Using physical books*	Did not use online platforms
Difficulties	*P2: Not easy to deal with account problems*	Account problems
Advantages	*P1: Narration may be useful for learning vocabularyP2: Dictionary, note-taking, and enlarge the textP3: Read the books in cross-platformP6: Do not occupy limited home space*	Enrichment features not available in physical booksAnytime, anywhere
Disadvantages	*P4: Health problems, e.g., myopia, cervical spine problemsP5: Easily be distracted as digital devices are multi-functions and full of entertainment, messaging functions*	Health problems upon prolonged useDistracted by other functions
Conditions of e-book selection	*P1: Interactive elements, like narrationP3: Able to access via mobile phone and computerP4: User interfaces should be easy to use and followP5: Available for free access via HKPL*	Interaction Compatibility User-friendly Price

Parents' Opinions of Reading Graphic Novels on Online Platforms (RQ4)

Regarding the current mobile digital environment, we investigate parents' current situation, cognition, and willingness to use graphic novels via online platforms to support children's reading and learning. As summarized in Table 7, participants rarely used online platforms for parent-child reading. Similar to the findings of Sung and Chiu (2022) and Dai and Chiu (2023), participants cared about health issues like myopia and cervical spine problems upon prolonged usage and distraction from other apps (like entertainment and messaging), while they agreed to advantages like extra functions available (like dictionary and note-taking) and anywhere, anytime usability. If the government or publishers want to promote graphic novel reading in Hong Kong, maximizing the strong points of e-reading may induce parents to try these reading materials.

The popularity of smart devices is changing human lifestyle swiftly, including reading behavior (Ding et al., 2021; Sung & Chiu, 2022; Yu et al., 2022). Notably, as the multiformat of reading materials make reading vivid and interactive, more people are willing to read online instead of physical books (Shimray et al., 2015). Due to the suspension of face-to-face classes and social distancing during the COVID-19 pandemic, the dependence on online reading and learning is growing (Cheng et al., 2022; Renaissance Learning, 2023; Sung & Chiu, 2021; Yao et al., 2023). The above participants' opinions provided precious hints to libraries and e-book publishers to develop user-friendly online platforms and e-books. Besides, as mentioned before, the characteristics of graphic novels stimulate students' self-learning and motivation positively (Bucher & Manning, 2004). During the epidemic, exploiting and promoting electronic graphic novels are necessary and useful.

Parents' Opinions of Society and Government Involvement in Reading Graphic Novels (RQ5)

Last but not least, this study figured out how to increase parents' awareness of graphic novels and support using graphic novels properly and conveniently. Participants desired more library workshops to guide them in using and selecting suitable graphic novels for parent-child reading and well-planned and adequate support (Lu et al., 2023). They also persisted in the balanced use of text-based and graphics-based materials for supporting children's learning. Further, participants pointed out that the poor reading atmosphere in Hong Kong is a more severe problem than the indifference to reading graphic novels (Kong et al., 2018; Lo et al., 2021).

Many strategies can achieve participants' suggestions. McClanahan & Nottingham (2019) studied a systematic approach to reading graphic novels for schools. This approach focuses on visual literacy, key graphic novel vocabulary, and text and image synthesis. They believe teachers can only design suitable graphic-novel-based learning materials for students if they have systematically learned to appreciate graphic novels and discover the advantages. Therefore, the first step in school involvement is to provide training programs about graphic novels for teachers, librarians, and related professionals (Williams & Peterson, 2009).

For libraries, using credible booklists, such as the Young Adult Library Services Association's Great Graphic Novels for Teens List, to develop graphic novel collections

Table 8. Coding for social support and promotion for graphic novel reading among children and related difficulties

Themes	Example Quotations	Coding
In library	*P1: Provide some workshops for parents and teach them how to use graphic novels in parent-child readingP3: Recommend some books for parents … for selecting high-quality graphic novels*	Workshops about how to use graphic novels in parent-child readingProviding high-quality booklists
In school	*P2: Create professional graphic-novel-based learning materials for teachersP5: Should have good planning … balance the use of text-based and graphic-novel-based learning materials*	Teachers should plan to balance the use of graphic-novel-based and text-based learning materials
In the City	*P4: The reading atmosphere is poor in Hong Kong; promoting 'reading' is an urgent priority compared with promoting 'graphic novel reading.'*	The poor reading atmosphere in Hong Kong is an obstacle to promoting graphic novels.

and related activities is very useful and common (Kim & Myers, 2012). These aids provide detailed guidance and recommendation for libraries to select updated and suitable graphic novels to meet users' learning, psychological, and social needs, which helps promote graphic novels and support parent-child reading.

CONCLUSION

Graphic novels and related variants were born over 40 years, offering a more relaxed method for reading and leisure. For example, '*No Fear Shakespeare Graphic Novel*' by SparkNotes is a graphic novel adaptation of classical Shakespeare literature. Before reading the difficult originals, it is an excellent introduction to Shakespeare's works. Researchers discovered that graphic novels play a significant role in reading and learning in many studies. First, this low-pressure reading helps reduce children's resistance to texts and fosters a love of reading. Second, children can increase their language and visual literacy skills simultaneously. Third, illustrations in graphic novels allow children with dyslexia to explore and understand the content in another way (Li, Wong, & Chiu, 2023).

Undoubtedly, graphic novels can support children's reading and learning. However, in the mobile digital age in Hong Kong, this study revealed some different opinions and perceptions from parents. Our respondents seldom use graphic novels in parent-child reading and did not agree on the effectiveness or appropriateness of graphic novels for enhancing children's reading and learning. They believed the connection between graphic novels and reading behaviors was weak and limited. Though their opinions might result from their limited understanding and bias on graphic novels, the convenience and ubiquitousness of current online materials for reading and learning might be a more pragmatic reason. Besides, they mainly focus on reading skills and academic results, not the love of reading and learning. It will be a great challenge to convince parents to recognize graphic novels as formal literature or valuable support for children's reading and learning.

As for the limitations of this study, the limited number of participants might not represent the mainstream opinion of society. Moreover, the interactions between participants were restricted as the interviews took place via the Internet due to the COVID-19 pandemic. Additionally, this study found that children's gender may affect their willingness to use graphic novels in parent-child reading. Therefore, we plan to apply quantitative methods, like a survey, to collect more data to investigate a more accurate situation and various influencing factors of using graphic novels in Hong Kong families. We are particularly interested in the effect of using electronic versions of graphic novels (Dai & Chiu, 2023; Sung & Chiu, 2022) during the COVID-19 pandemic to confirm whether parents' attitudes toward graphic novels or the availability of competing media and channels matter. We are also interested in the promotion of reading and other information services on social media (Chan et al., 2020; Chung et al., 2020; Fong et al., 2020; Lam et al., 2019; 2023; Jiang, Chiu, & Chan, 2023) as well as learning aids (Ho et al., 2018; Leung et al., 2023; Mak et al., 2022; Yang et al., 2022). We further use social media analytics to elicit reading preference and evaluate reading promotion (Deng & Chiu, 2022; He et al., 2022; Li, Chiu, & Ho, 2023; Li, Xie, Chiu, & Ho, 2023; Liu et al., 2023; Wang et al., 2022; Xie, Wong, Chiu, & Lei, 2023). We are also interested in digitalizing reading and reference materials (Chan & Chiu, 2023; Sun et al., 2022; Tse et al., 2022; Zuo et al., 2023).

REFERENCES

Alhojailan, M. I. (2012). Thematic analysis: A critical review of its process and evaluation. *West East Journal of Social Sciences*, *1*(1), 39–47.

Bischell, J. C. (2018). *Examining parents' perceptions of and preferences toward the use of comics in the classroom* [Doctoral dissertation, The University of Tennessee]. UTC Scholar. https://scholar.utc.edu/cgi/viewcontent.cgi?article=1695&context= theses

Botzakis, S., Savitz, R., & Low, D. E. (2017). Adolescents reading graphic novels and comics: What we know from research. In K. A. Hinchman & D. A. Appleman (Eds.), *Adolescent literacies: A handbook of practice-based research* (pp. 310–322). The Guilford Press.

Bucher, K. T., & Manning, M. L. (2004). Bringing graphic novels into a school's curriculum. *The Clearing House: A Journal of Educational Strategies, Issues and Ideas*, *78*(2), 67–72. doi:10.3200/TCHS.78.2.67-72

Bunn, V. (2012). Researching the Tintin effect: How can the active promotion of graphic novels support and enhance boys' enthusiasm for leisure reading? *School Librarian*, *60*(2), 74–76.

Bybee, R. W., Taylor, J. A., Gardner, A., Van Scotter, P., Powell, J. C., Westbrook, A., & Landes, N. (2006). *The BSCS 5E instructional model: Origins and effectiveness*. BSCS.

Chan, T. T. W., Lam, A. H. C., & Chiu, D. K. W. (2020). From Facebook to Instagram: Exploring user engagement in an academic library. *Journal of Academic Librarianship*, *46*(6), 102229. Advance online publication. doi:10.1016/j.acalib.2020.102229 PMID:34173399

Chan, V. H. Y., & Chiu, D. K. W. (2023). Integrating the 6C's motivation into reading promotion curriculum for disadvantaged communities with technology tools: A Case Study of reading dreams foundation in rural China. In A. S. Etim (Ed.), *Adoption and use of technology tools and services by economically disadvantaged communities: Implications for growth and sustainability*. IGI Global.

Cheng, J., Yuen, A. H., & Chiu, D. K. (2022). Systematic review of MOOC research in mainland China. *Library Hi Tech*. doi:10.1108/LHT-02-2022-0099

Cheng, W. W. H., Lam, E. T. H., & Chiu, D. K. W. (2020). Social media as a platform in academic library marketing: A comparative study. *The Journal of Academic Librarianship, 46*(5). doi:10.1016/j.acalib.2020.102188

Cheung, L. S. N., Chiu, D. K. W., & Ho, K. K. W. (2022). A quantitative study on utilizing electronic resources to engage children's reading and learning: Parents' perspectives through the 5E instructional model. *The Electronic Library*, *40*(6), 662–679. doi:10.1108/EL-09-2021-0179

Chiu, D. K. W., & Ho, K. K. W. (2022). Special selection on contemporary digital culture and reading. *Library Hi Tech*, *40*(5), 1204–1209. doi:10.1108/LHT-10-2022-516

Christensen, L. L. (2006). Graphic global conflict: Graphic novels in the high school social studies classroom. *Social Studies*, *97*(6), 227–230. doi:10.3200/TSSS.97.6.227-230

Chung, C., Chiu, D. K. W., Ho, K. K. W., & Au, C. H. (2020). Applying social media to environmental education: Is it more impactful than traditional media? *Information Discovery and Delivery*, *48*(4), 255–266. doi:10.1108/IDD-04-2020-0047

Clark, R. C., & Mayer, R. E. (2016). *E-learning and the science of instruction: Proven guidelines for consumers and designers of multimedia learning*. John Wiley & Sons. doi:10.1002/9781119239086

Cooper, S., Nesmith, S., & Schwarz, G. (2011). Exploring graphic novels for elementary science and mathematics. *School Library Media Research*, *14*, 1–17.

Crawford, P. (2004). Using graphic novels to attract reluctant readers. *Library Media Connection*, *27*, 26–28.

Crowley, J. (2015). Graphic novels in the school library: Using graphic novels to encourage reluctant readers and improve literacy. *School Librarian*, *63*(3), 140–142.

Cummings, W. R. (2020, March 6). Why are graphic novels so popular amongst students who have ADHA? *Psych Central*. https://blogs.psychcentral.com/childhood-behavioral/2020/03/why-are-graphic-novels-so-popular-amongst-students-who-have-adhd/

Dai, C., & Chiu, D.K.W. (2023). Impact of COVID-19 on reading behaviors and preferences: Investigating high school students and parents with the 5E instructional model. *Library Hi Tech*. doi:10.1108/LHT-10-2022-0472

Deng, S., & Chiu, D. K. W. (2022). Analyzing Hong Kong Philharmonic Orchestra's Facebook community engagement with the Honeycomb Model. In M. Dennis & J. Halbert (Eds.), *Community Engagement in the Online Space* (pp. 31–47). IGI Global.

Ding, S. J., Lam, E. T. H., Chiu, D. K. W., Lung, M. M. W., & Ho, K. K. W. (2021). Changes in reading behaviour of periodicals on mobile devices: A comparative study. *Journal of Librarianship and Information Science*, *53*(2), 233–244. doi:10.1177/0961000620938119

Downey, E. M. (2011). Graphic novels in curriculum and instruction collections. *Reference and User Services Quarterly*, *49*(2), 181–188. doi:10.5860/rusq.49n2.181

Duursma, E., Augustyn, M., & Zuckerman, B. (2008). Reading aloud to children: The evidence. *Archives of Disease in Childhood*, *93*(7), 554–557. doi:10.1136/adc.2006.106336 PMID:18477693

Fletcher-Spear, K., Jenson-Benjamin, M., & Copeland, T. (2005). The truth about graphic novels: A format, not a genre. *The ALAN Review*, *32*(2), 37–44. doi:10.21061/alan.v32i2.a.7

Fong, K. C. H., Au, C. H., Lam, E. T. H., & Chiu, D. K. W. (2020). Social network services for academic libraries: A study based on social capital and social proof. *Journal of Academic Librarianship*, *46*(1), 102091. Advance online publication. doi:10.1016/j.acalib.2019.102091

Griffith, P. E. (2010). Graphic novels in the secondary classroom and school libraries. *Journal of Adolescent & Adult Literacy*, *54*(3), 181–189. doi:10.1598/JAAL.54.3.3

He, Z., Chiu, D. K. W., & Ho, K. K. W. (2022). Weibo analysis on Chinese cultural knowledge for gaming. In Z. Sun & Z. Wu (Eds.), *Handbook of Research on Foundations and Applications of Intelligent Business Analytics* (pp. 320–349). IGI Global. doi:10.4018/978-1-7998-9016-4.ch015

Ho, K. K. W., Takagi, T., Ye, S., Au, C. K., & Chiu, D. K. W. (2018). *The Use of Social Media for Engaging People with Environmentally Friendly Lifestyle – A Conceptual Model.* SIG Green Pre ICIS Workshop. https://aisel.aisnet.org/sprouts_proceedings_siggreen_2018/2/

Huang, P. S., Paulino, Y., So, S., Chiu, D. K. W., & Ho, K. K. W. (2021). Special issue editorial - COVID-19 pandemic and health informatics (part 1). *Library Hi Tech*, *39*(3), 693–695. doi:10.1108/LHT-09-2021-324

Huang, P.-S., Paulino, Y. C., So, S., Chiu, D. K. W., & Ho, K. K. W. (2022). Guest editorial: COVID-19 pandemic and health informatics part 2. *Library Hi Tech*, *40*(2), 281–285. doi:10.1108/LHT-04-2022-447

Huang, P.-S., Paulino, Y. C., So, S., Chiu, D. K. W., & Ho, K. K. W. (2023). Guest editorial: COVID-19 pandemic and health informatics part 3. *Library Hi Tech*, *41*(1), 1–6. doi:10.1108/LHT-02-2023-585

James, B. (2014, March 11). Redefining literacy with graphic novels. *Minnesota English Journal*. Retrieved from https://minnesotaenglishjournalonline.org/2014/03/11/praxis-strategies-for-teaching-literature-3-graphic-novels/

Jennings, K. A., Rule, A. C., & Vander Zanden, S. M. (2014). Fifth graders' enjoyment, interest, and comprehension of graphic novels compared to heavily-illustrated and traditional novels. *International Electronic Journal of Elementary Education*, *6*(2), 257–274.

Jiang, M., Lam, A.H.C., Chiu, D.K.W., & Ho, K.K.W. (2023). Social media aids for business learning: A quantitative evaluation with the 5E instructional model. *Education and Information Technology*. doi:10.1007/s10639-023-11690-z

Jiang, X., Chiu, D. K. W., & Chan, C. T. (2022). Application of the AIDA model in social media promotion and community engagement for small cultural organizations: A case study of the Choi Chang Sau Qin Society. In M. Dennis & J. Halbert (Eds.), *Community Engagement in the Online Space* (pp. 48–70). IGI Global.

Kim, J., & Myers, R. (2012). Discovering greatness: YALSA's great graphic novels for teens list. *Young Adult Library Services*, *10*(3), 39–41.

Kong, E. W. S., Lo, P., Yu, P. J., Ip, K. K. L., & Chiu, D. K. W. (2018). 閱讀推手: 學校圖書館專業 [Reading Promoter: School Library Professions]. Hong Kong: Commercial Press.

Lam, A. H. C., Ho, K. K. W., & Chiu, D. K. W. (2023). Instagram for student learning and library promotions? A quantitative study using the 5E Instructional Model. *Aslib Journal of Information Management*, *75*(1), 112–130. doi:10.1108/AJIM-12-2021-0389

Lam, E. T. H., Au, C. H., & Chiu, D. K. W. (2019). Analyzing the use of Facebook among university libraries in Hong Kong. *Journal of Academic Librarianship*, *45*(3), 175–183. doi:10.1016/j.acalib.2019.02.007

Leedy, P., & Ormrod, J. (2015). *Practical research planning and design* (11th ed.). Pearson Harlow.

Leung, G. S. M. (2007). A study of the effects of parental support and children's resourcefulness on the academic stress of senior primary school students in Hong Kong (China). *Dissertation Abstracts International. A, The Humanities and Social Sciences*, *67*(8-A), 3180.

Leung, L. M., Chiu, D. K. W., & Lo, P. (2020) School librarians' view of cooperation with public libraries: A win-win situation in Hong Kong. *School Library Research, 23*. www.ala.org/aasl/slr/volume23/leung-chiu-lo

Li, Q., Wong, J., & Chiu, D. K. W. (2023). (in press). School library reading support for students with dyslexia: A qualitative study in the digital age. *Library Hi Tech*. Advance online publication. doi:10.1108/LHT-03-2023-0086

Li, S., Chiu, D. K. W., Kafeza, E., & Ho, K. K. W. (2023). Social media analytics for non-governmental organizations: A case study of Hong Kong Next Generation Arts. In Z. Sun (Ed.), *Driving socioeconomic development with big data: theories, technologies, and applications* (pp. 277–295). IGI Global. doi:10.4018/978-1-6684-5959-1.ch013

Li, S., Xie, Z., Chiu, D. K. W., & Ho, K. K. W. (2023). Sentiment analysis and topic modeling regarding online classes on the Reddit Platform: Educators versus learners. *Applied Sciences (Basel, Switzerland)*, *13*(4), 2250. Advance online publication. doi:10.3390/app13042250

Library of Congress. (2014, June 19). *Cataloging Graphic Novels*. https://www.loc.gov/aba/cyac/graphicnovels.html

Liu, Y., Chiu, D. K. W., & Ho, K. K. W. (2023). Short-form videos for public library marketing: Performance analytics of Douyin in China. *Applied Sciences (Basel, Switzerland)*, *13*(6), 3386. Advance online publication. doi:10.3390/app13063386

Lo, P., Cheuk, M. K., Leedy, S. K., & Chiu, D. K. W. (2021). 文武之道【上冊】:從城市森林中尋訪對閱讀有想法的武者與文化人 [The Tao of Arts and Warriorship Vol. 1: Looking for martial artists and cultural people with reading ideas in a metropolis]. Hong Kong: Systech Publications.

Lo, P., & Chiu, D. K. W. (2015). Enhanced and changing roles of school librarians under the digital age. *New Library World*, *116*(11/12), 696–710. doi:10.1108/NLW-05-2015-0037

Lo, P., Yu, K., & Chiu, D. (2015). A research agenda for enhancing teacher librarians' roles and practice in Hong Kong's 21st century learning environments. *School Libraries Worldwide*, *21*(1), 19–37. doi:10.29173lw6881

Lu, S.S., Tian, R., & Chiu, D.K.W. (2023). Why do people not attend public library programs in the current digital age? *Library Hi Tech*. doi:10.1108/LHT-04-2022-0217

Mak, M. Y. C., Poon, A. Y. M., & Chiu, D. K. W. (2022). Using Social Media as Learning Aids and Preservation: Chinese Martial Arts in Hong Kong. In S. Papadakis & A. Kapaniaris (Eds.), *The Digital Folklore of Cyberculture and Digital Humanities* (pp. 171–185). IGI Global. doi:10.4018/978-1-6684-4461-0.ch010

McClanahan, B. J., & Nottingham, M. (2019). A suite of strategies for navigating graphic novels: A dual coding approach. *The Reading Teacher*, *73*(1), 39–50. doi:10.1002/trtr.1797

Meng, Y., Chu, M. Y., & Chiu, D. K. W. (2023). The impact of COVID-19 on museums in the digital era: Practices and challenges in Hong Kong. *Library Hi Tech*, *41*(1), 130–151. doi:10.1108/LHT-05-2022-0273

Moeller, R. A. (2011). "Aren't these boy books?": High school students' readings of gender in graphic novels. *Journal of Adolescent & Adult Literacy*, *54*(7), 476–484. doi:10.1598/JAAL.54.7.1

Moeller, R. A., & Becnel, K. E. (2021). Recommended reading: Comparing elementary/middle school graphic novel collections to recommended reading lists. *Children & Libraries*, *19*(2), 6–13. doi:10.5860/cal.19.2.6

Mol, S. E., & Bus, A. (2011). To read or not to read: A meta-analysis of print exposure from infancy to early adulthood. *Psychological Bulletin*, *137*(2), 267–296. doi:10.1037/a0021890 PMID:21219054

Morrison, L. (2017, April 14). *The research behind graphic novels and young learners* [Blog post]. https://www.ctd.northwestern.edu/blog/research-behind-graphic-novels-and-young-learners

Nesmith, S., Cooper, S., Schwarz, G., & Walker, A. (2016). Are graphic novels always "cool"? Parent and student perspectives on elementary mathematics and science graphic novels: The need for action research by school leaders. *Planning and Changing*, *47*(3-4), 228–245.

Oz, H., & Efecioglu, E. (2015). Graphic novels: An alternative approach to teach English as a foreign language. *Journal of Language and Linguistic Studies*, *11*(1), 75–90.

Renaissance Learning. (2021). *What kids are reading 2021*. https://www.renaissance. com/wkar-report/?int=https://www.renaissance.com/wkar/

Richardson, E. M. (2017). "Graphic novels are real books": Comparing graphic novels to traditional text novels. *Delta Kappa Gamma Bulletin*, *83*(5), 24–31.

Scholastic. (2018). *A guide to using graphic novels with children and teens*. https:// www.scholastic.com/content/dam/teachers/lesson-plans/18-19/Graphic-Novel-Discussion-Guide-2018.pdf

Shimray, S. R., Keerti, C., & Ramaiah, C. K. (2015). An overview of mobile reading habits. *DESIDOC Journal of Library and Information Technology*, *35*(5), 343–354. doi:10.14429/djlit.35.5.8901

Short, J. C., Randolph-Seng, B., & McKenny, A. F. (2013). Graphic presentation: An empirical examination of the graphic novel approach to communicate business concepts. *Business Communication Quarterly*, *76*(3), 273–303. doi:10.1177/1080569913482574

Snowball, C. (2005). Teenage reluctant readers and graphic novels. *Young Adult Library Services*, *3*(4), 43–45.

Sun, X., Chiu, D. K. W., & Chan, C. T. (2022). Recent digitalization development of buddhist libraries: A comparative case study. In S. Papadakis & A. Kapaniaris (Eds.), *The Digital Folklore of Cyberculture and Digital Humanities* (pp. 251–266). IGI Global. doi:10.4018/978-1-6684-4461-0.ch014

Sung, Y. Y. C., & Chiu, D. K. W. (2022). E-book or print book: Parents' current view in Hong Kong. *Library Hi Tech*, *40*(5), 1289–1304. doi:10.1108/LHT-09-2020-0230

Tsang, A. L. Y., & Chiu, D. K. W. (2022). Effectiveness of virtual reference services in academic libraries: A qualitative study based on the 5E Learning Model. *Journal of Academic Librarianship*, *48*(4), 102533. Advance online publication. doi:10.1016/j. acalib.2022.102533

Tse, H. L., Chiu, D. K. W., & Lam, A. H. C. (2022). From reading promotion to digital literacy: An analysis of digitalizing mobile library services with the 5E Instructional Model. In A. Almeida & S. Esteves (Eds.), *Modern reading practices and collaboration between schools, family, and community* (pp. 239–256). IGI Global. doi:10.4018/978-1-7998-9750-7.ch011

Wai, I. S. H., Ng, S. S. Y., Chiu, D. K. W., Ho, K. K., & Lo, P. (2018). Exploring undergraduate students' usage pattern of mobile apps for education. *Journal of Librarianship and Information Science*, *50*(1), 34–47. doi:10.1177/0961000616662699

Wang, J., Deng, S., Chiu, D. K. W., & Chan, C. T. (2022). Social network customer relationship management for orchestras: A case study on Hong Kong Philharmonic Orchestra. In N. B. Ammari (Ed.), *Social Customer Relationship Management (Social-CRM) in the Era of Web 4.0* (pp. 250–268). IGI Global. doi:10.4018/978-1-7998-9553-4.ch012

Wang, P., Chiu, D. K. W., Ho, K. K., & Lo, P. (2016). Why read it on your mobile device? Change in reading habit of electronic magazines for university students. *Journal of Academic Librarianship*, *42*(6), 664–669. doi:10.1016/j.acalib.2016.08.007

Wilhelm, J. D. (1995). Reading is seeing: Using visual response to improve the literary reading of reluctant readers. *Journal of Reading Behavior*, *27*(4), 467–503. doi:10.1080/10862969509547896

Williams, V. K., & Peterson, D. V. (2009). Graphic novels in libraries supporting teacher education and librarianship programs. *Library Resources & Technical Services*, *53*(3), 166–173. doi:10.5860/lrts.53n3.166

Wong, S. W. S., & Chiu, D. K. W. (2023). Re-examining the value of remote academic library storage in the mobile digital age: A comparative study. *Portal (Baltimore, Md.)*, *23*(1), 89–109.

Xie, Z., Chiu, D.K.W., & Ho, K.K.W. (2023). The role of social media as aids for accounting education and knowledge sharing: Learning effectiveness and knowledge management perspectives in Mainland China. *Journal of Knowledge Economy*. doi:10.1007/s13132-023-01262-4

Xie, Z., Wong, G. K. W., Chiu, D. K. W., & Lei, J. (2023). Bridging K-12 Mathematics and computational thinking in the Scratch community: A Big Data Analysis. *IT Professional*, *25*(2), 64–70. doi:10.1109/MITP.2023.3243393

Yang, G. (2008). Graphic novels in the classroom. *Language Arts*, *85*(3), 185.

Yang, Z., Zhou, Q., Chiu, D. K. W., & Wang, Y. (2022). Exploring the factors influencing continuance usage intention of academic social network sites. *Online Information Review*, *46*(7), 1225–1241. doi:10.1108/OIR-01-2021-0015

Yao, L., Lei, J., Chiu, D. K. W., & Xie, Z. (2023). Adult learners' perception of online language English learning platforms in China. In A. Garcés-Manzanera & M. E. C. García (Eds.), *New approaches to the investigation of language teaching and literature* (pp. 123–149). IGI Global.

Yip, K. H. T., Chiu, D. K. W., Ho, K. K. W., & Lo, P. (2021). Adoption of Mobile Library Apps as Learning Tools in Higher Education: A Tale between Hong Kong and Japan. *Online Information Review*, *45*(2), 389–405. doi:10.1108/OIR-07-2020-0287

Yu, H. Y., Tsoi, Y. Y., Rhim, A. H. R., Chiu, D. K. W., & Lung, M. M. W. (2022). Changes in habits of electronic news usage on mobile devices in university students: A comparative survey. *Library Hi Tech*, *40*(5), 1322–1336. doi:10.1108/LHT-03-2021-0085

Yu, P. Y., Lam, E. T. H., & Chiu, D. K. W. (2023). Operation management of academic libraries in Hong Kong under COVID-19. *Library Hi Tech*, *41*(1), 108–129. doi:10.1108/LHT-10-2021-0342

Zhang, X., Lo, P., So, S., Chiu, D. K. W., Leung, T. N., Ho, K. K. W., & Stark, A. (2021). Medical students' attitudes and perceptions towards the effectiveness of mobile learning: A comparative information-need perspective. *Journal of Librarianship and Information Science*, *53*(1), 116–129. doi:10.1177/0961000620925547

Zheng, J., Lam, A. H. C., & Chiu, D. K. W. (2023). Evaluating the Effectiveness of Learning Commons as Third Space with the 5E Usability Model: The Case of Hong Kong University of Science and Technology Library. In C. Kaye, J. H. Writer, & J. Batsaikhan (Eds.), *Third-Space Exploration in Education*. IGI Global.

Zuo, Y., Lam, A. H. C., & Chiu, D. K. W. (2023). Digital protection of traditional villages for sustainable heritage tourism: A case study on Qiqiao Ancient Village, China. In A. Masouras, C. Papademetriou, D. Belias, & S. Anastasiadou (Eds.), *Sustainable Growth Strategies for Entrepreneurial Venture Tourism and Regional Development*. IGI Global. doi:10.4018/978-1-6684-6055-9.ch009

KEY TERMS AND DEFINITIONS

5E Instructional Model: A constructivist instructional model developed by Bybee et al. (2006) for science education but later employed in diverse disciplines and education levels. It consists of five phases, i.e., engagement, exploration, explanation, elaboration, and evaluation.

Biases on Graphic Novels: Some stereotyping biases on the values of graphic novels, for example, easy content only for young children that is not conducive to children's growth because of the heavy use of illustrations and misunderstanding of graphic novels relating to sensitive and parental-guidance-required topics.

Children's Learning: Learning new knowledge and skills through a wide variety of resources, for instance, learning through reading literature such as graphic novels to improve literacy skills, second language skills, and cognitive skills. Therefore, reading literature is commonly used to support children's learning.

Children's Reading Behavior: A behavior significantly influenced by children's reading desire or motivation, including but not limited to their reading expectations, interest in reading the literature, and sensory needs.

Graphic Novels: A type of literature majorly comprising text and illustrations for presenting characters' moods, actions, tones, and plots in the form of narration. Examples can include comics and picture books with varied themes.

Parent-Child Reading: A type of parent-child activity that encourages parents to read literature with children, potentially improving children's literacy skills and parent-child relationships, value education, and emotional communication.

Reluctant Readers: A type of reader with an unwillingness to read due to consideration of reading as boring, challenging to achieve, and time-consuming.

APPENDIX: INTERVIEW QUESTIONS

1. May you simply introduce the current situation of storytelling time with your children and children's individual reading in your family? Including but not limited to…
    a. How often of the parent-child reading session?
    b. When, Where, Who?
    c. How long for each session?
    d. What type of reading materials do you choose? Fiction? Nonfiction?
    e. What platform have you used for storytelling? Borrow from the library or retrieve online database?
2. Have you heard the phrase "graphic novel"?
    a. What is your first impression of this kind of reading material?
    b. Can you give me some examples?
    c. Have you used it for your parent-child reading?
3. From your understanding and / or experience, do you agree that graphic novels benefit from cultivating children's reading habits? Why or why not?
    a. Are there any features you believe can attract the children to read?
    b. Compared with traditional text-based reading materials, do you think graphic novels are more attractive to children? Why or why not?
    c. In order to attract children to read more books, will you use (more) graphic novels in parent-child reading? Why or why not?
4. From your understanding and / or experience, do you agree that graphic novels are benefiting to children's learning? Especially in academic results? Why or why not?
    a. Are there any features you think are good for children's learning?
    b. Have you seen or known children using graphic-novel-based learning materials in school? If so, what are they?
    c. Have you heard using graphic novels in the classroom is the new trend in the West? If so, can you briefly describe what you heard?
    d. Do you agree to increase the proportion of using graphic novels in school? Why or why not?
    e. Do you think there is a relationship between graphic novels, children learning **engage**ment, motivation, and interest? How?
    f. Do you think graphic novels are useful for children to **explore** new knowledge and phenomenon? For example, exploring the in-depth meaning of the story.
    g. Do you think graphic novels are useful for **explaining** complicated ideas and concepts and helping children learn easily? Or can children

use graphic novels to explain, **elaborate,** and **evaluate** what they have discovered and learned?

h.   What subjects do you think are the easiest to apply graphic-novel-based learning materials? Why and how?

i.   Will you use graphic novels in parent-child reading for learning (both general school curricular and extracurricular)? Why or why not?

5.   No matter for cultivating children's reading habits or supporting children's learning, are there any conditions you think may be very important for selecting high-quality graphic novels?

a.   Author? Publisher? Theme? Language? Pictures? Interaction?

6.   Have you used any online platform for parent-child reading? Such as OverDrive, Kindle? Why or why not?

a.   From your understanding and / or experience, any advantage and disadvantage to using online platforms / e-books?

b.   Have you used the online platform for reading graphic novels with children? If yes, is there any difference between electronic and physical versions? If not, will you try to use e-versions instead of a printed version?

c.   What are your requirements if you need to choose a suitable e-version of graphic novels for children? Price? Easy to access and use? Cross-platform? Or what else?

7.   Do you think there is enough support from society or the government for promoting graphic novel reading among children?

a.   Any difficulties you have faced when you choose or read graphic novels?

b.   Anything other support or inducement society or government should be provided to attract parents to use graphic novels, both physical and e-version, in parent-child reading?

8.   Finally, after this interview and sharing (with each other), do you change your mind about graphic novels or have any valuable ideas you have not shared? Please feel free to raise and discuss this with us.

a.   For example, do you always disagree with using any graphic novel in parent-child reading for any reason?

b.   Will you share good graphic novels with other parents / friends? What channels do you use for sharing?

Chapter 9

Accessing User Satisfaction of Makerspace in Academic Libraries:
A Comparative Study Based on the 5E Instructional Model

Cho Yiu Cheung
The University of Hong Kong, Hong Kong

Apple Hiu Ching Lam
iD https://orcid.org/0000-0002-2587-6979
The University of Hong Kong, Hong Kong

Dickson K. W. Chiu
iD https://orcid.org/0000-0002-7926-9568
The University of Hong Kong, Hong Kong

ABSTRACT

Makerspaces have developed from a trend to a core service in higher education libraries. Many academic libraries have been actively expanding the "makerspace" within the physical library and revitalizing the library as a center of learning and innovation. This case study investigates the application of makerspace technologies in a major comprehensive library in Hong Kong, which has designed a specific makerspace to encourage innovation and creativity. Few studies have focused on in-depth studies of makerspaces in East Asian academic libraries and how patrons perceive makerspace services and innovative spaces. A survey instrument was developed using the 5E instructional model (engage, explore, explain, elaborate, and evaluate) to evaluate makers' experiences systematically. The finding revealed that respondents applauded the importance of innovative spaces and the perceived outcomes from makerspaces, including nurturing creativity and critical thinking. However, they did not have sufficient skills to use emerging technologies, resulting in low usage.

DOI: 10.4018/978-1-6684-8671-9.ch009

INTRODUCTION

Traditionally, a library is a place to collect, access, and preserve print collections, and information services, circulation, and rich repositories have historically been most libraries' core functions (T. T. W. Chan et al., 2020; Chiu & Wong, 2023; Stover et al., 2019; T. Yip et al., 2019). Several trends highlight the value and demand for library learning spaces, including changing students' learning habits (Chiu & Ho, 2022a, 2022b; Q. Deng et al., 2019; Ding et al., 2021; Dong et al., 2021; Fan et al., 2020; Hui et al., 2023; P. Wang et al., 2016; Webb, 2018; H. Y. Yu et al., 2022; Zhou et al., 2022) and cultivating a maker culture (Melo & Nichols, 2020; Moorefield-lang, 2019). Students have expected to use emerging technologies to express ideas or present data visually (Suen et al., 2020; Webb, 2018). The evolving information and learning environment have led to a transformation of physical libraries (T. Jiang et al., 2019; Kee & Chiu, 2023; Y. Ni et al., 2023; Tse et al., 2022; K. C. Wong & Chiu, 2023; J. Zhang et al., 2023), and the academic libraries that play a vital part in supporting campus learning (Lo et al., 2017; Radniecki, 2017; Wu et al., 2023; Xue et al., 2023; Y. Zhang et al., 2020).

To fully support multiple forms of learning, academic libraries in Hong Kong have implemented renovation works, aiming at transforming the library spaces into collaborative, interactive, informal learning spaces, learning commons, or information commons (D. L. H. Chan & Spodick, 2014; Leung, Chiu, Ho, & Luk, 2022; Ho et al., 2023) and shifting from studying and learning into creative, innovative, inspiring spaces (Bieraugel & Neill, 2017). The rise of makerspaces as a concept in recent years, and the service is widely available internationally in academic and school libraries as a new learning space (Burke, 2015; Curry, 2017; Willingham & De Boer, 2015). This learning environment grows in popularity in academic libraries (Lee, 2017) and offers students a wide range of learning opportunities. Makerspaces is a term used in concurrence with diverse laboratories and hackerspaces (Curry, 2017; Kroski, 2017; Willingham & De Boer, 2015). Kroski (2017) defines makerspaces as "a place where people gather to make things, collaborate and share knowledge" (p.3). The value of makerspace is to facilitate knowledge exchange, inspire students to become a maker, and foster creativity and innovative ideas (Bieraugel & Neill, 2017; Gstalder, 2017).

Few studies have focused on the deployment of makerspaces in East Asia, especially Hong Kong, and even less on the experience of academic libraries. As makerspaces are relatively new for academic libraries, there is very little analysis about how library patrons perceive makerspace services and innovative spaces and the satisfaction of makerspace services provided by academic libraries in Hong Kong. This study seeks to fill the gap in the existing literature by focusing on makerspaces in Hong Kong academic libraries and finding a fully operational and newly developing makerspace for analysis. This study gathers and analyzes user survey data to investigate the perspective of library patrons toward innovative services in a major academic library in Hong Kong according to the 5E Instructional Model. This research investigates why academic libraries in Hong Kong implement library renovation projects and how makerspace services support creativity and innovation to propose improvements to the current makerspace services to meet users' needs.

This case study provides valuable information for understanding the university students' needs and their reflection on engagement in makerspaces and learning outcomes. As few studies focus on makerspace services offered in Asia, librarians planning for makerspace deployment and running library space projects can better understand the application of innovative and informal learning spaces in higher education, such as the range of makerspace services offered and user perception of ideal makerspaces. Besides practical suggestions, this study allows other scholars to conduct scholarly literature analyzing innovative and technology-based community learning spaces.

Some background information concerning the makerspace services offered by the case is briefly introduced as follows. The case adopts technology-based makerspaces to support the library's vision and mission by providing emerging technologies such as 3D printers, 3D scanners, laser engraver and cutter, vinyl cutter, editing rooms with equipment facilitating audio recording and audiovisual editing, VR Salon, immersive technology space equipped with virtual reality (VR), augmented reality (AR), and mixed reality (MR). Some spaces and rooms are also designated for brainstorming and collaboration in the making activities. Specialized library guides are developed with detailed text, graphics, and video explanations for library patrons. Consultation sections on how to use the technologies and equipment are provided on the library's online booking system.

LITERATURE REVIEW

The Origin of Makerspaces

Burke (2015) introduced the origin of makerspaces in 2005 with the beginning of "Make:" magazine. The makerspace concept was born out of the maker movement (Melo & Nichols, 2020), and they are interchangeably referred to as hackerspaces and fabrication labs (Curry, 2017; Kroski, 2017; Willingham & De Boer, 2015) and are even closely associated with Galleries, Libraries, Archives, and Museums (GLAM) lab sector (Mahey et al., 2019). Unlike hackerspaces, makerspace emphasizes making something tangible with tools or equipment. Specifically, libraries serve as a "third place," forming a community space of makers (Choy & Goh, 2016; Willingham &

De Boer, 2015; J. Zhang et al., 2023). The term "third place" is a concept developed by Ray Oldenburg that characterized the library as a public place outside the home or workplace (Bailin, 2011; Choy & Goh, 2016; Willingham & De Boer, 2015). Additionally, Bagley (2014) defines a makerspace as a community-oriented space where people can create, build and learn. Johnson (2017) considers makerspace a creative library space that promotes creativity and provides a collaborative learning environment where faculty and students can use shared digital resources and develop new skills. Users can learn through engagement, exploration, and experimentation individually or in a group within a controlled setting (D. L. H. Chan & Spodick, 2014; Osawaru et al., 2020). Current makerspaces typically provide the hardware and software support for 3D modeling, VR, AR (Suen et al., 2020; Lo et al., 2019), and Internet of Things (IoTs) (Chang et al., 2018).

The Emergence of Makerspaces in Academic Libraries

Learning spaces in academic libraries have evolved over the years from traditional style to informal and collaborative learning commons (Leung, Chiu, Ho, & Luk, 2022; Zhou et al., 2022). Collaborative and technology-enriched spaces are developed throughout the transformation journey of library spaces (Choy & Goh, 2016; Q. Liu et al., 2022; Lo et al., 2018; 2020). Indeed, the engagement with book collection as a primary use of library space has been shifted to serve library patrons' information and technology needs (Chiu & Wong, 2023). Makerspaces have been widely available in schools, public libraires, and academic libraries. In an era of rapidly changing technology, makerspace services are an emerging phenomenon in Hong Kong's academic libraries. Recently, the literature has introduced the value of makerspace in academic librarianship and discussed the application of innovation labs in the United States and the United Kingdom (Burke, 2015; Curry, 2017), but few cover academic libraries in East Asia.

Libraries have been places of learning, discovery, and idea sharing. Kyle Bowen, Director of Innovation for Teaching and Learning with Technology at Penn State University, says libraries are the ideal destination to deploy makerspaces on university campuses (Grush, 2015). Adams Becker et al. (2017) and Moorefield-lang (2019) concurred with the argument as libraries have traditionally been collaborative and interdisciplinary and have long been the center of information and knowledge (F. Wang et al., 2016). Based on the historical role of the library, many academic libraries adopted the trend of makerspaces and emerging technologies to provide a platform for generating new information and research projects.

In recent years, many universities have incorporated makerspace to facilitate the creation, innovation, experiential education, and interdisciplinary collaboration (Farritor, 2017; Michalak & Rysavy, 2019; A. Wong & Partridge, 2016). University libraries offer a shared community space for makers and create opportunities for people interested in making. The value of the makerspaces is to support teaching and learning across disciplines, especially university programs that integrate the arts with STEM disciplines (Burke, 2015; Koh & Abbas, 2015). Various makerspaces share common elements in different regions. Burke (2015) shared the library profile with existing makerspace services offered in the United States and pointed out that the most common technologies used in academic library makerspaces are workstations, 3D printers, and image/video editing software.

Meanwhile, Moorefield-Lang (2015) and Kroski (2017) identified that some learning laboratories focus on crafting and painting elements while some makerspaces emphasize cross-disciplinary applications and are technology-driven, such as 3D modeling, robotics, and digital cameras equipped with 4K video recording (Burke, 2015; Moorefield-Lang, 2015). In particular, the younger generation is more attracted to videos than images and images than text (W. Cheng et al., 2023; V. S. Y. Cheung et al., 2023). To support the technological-based makerspace services, academic libraries offer regular workshops and instruction sessions to teach students how to use the equipment or the software (Burke, 2015). Equally importantly, libraries provide on-the-job training and self-learning resources to the staff to gain the competencies required in this changing environment (Koh & Abbas, 2015; Willingham & De Boer, 2015). Curry (2017) highlighted the concerns of developing makerspaces and learning labs, such as intellectual property rights and safety issues, and suggested establishing makerspaces policy and safety guidelines.

To ensure the quality of makerspace services meeting user needs, academic libraries should better conduct user satisfaction surveys or the like to receive user feedback for service improvement. Radniecki (2017) evaluated the instructional services regarding 3D modeling through usage statistics, attendance, and user satisfaction surveys with questions including need fulfillment, the likelihood of recommendation, expertise satisfaction, and helpfulness to the future. Slightly differently, Radniecki and Winterman (2020) investigated the use of student employees in supporting student's makerspace activities from both user and student employee's perspectives through booking data, consultation data, and user surveys with questions including but not limited to user satisfaction and user confidence. Although a few studies investigated the makerspace services from the user satisfaction perspective, they only asked simple and single questions for each aspect and rarely examined them with inferential statistics (Radniecki, 2017; Radniecki & Winterman, 2020).

Perceived Benefits of Makerspaces

Ito et al. (2009) stated that makerspaces and innovative labs represented a principle of connected learning, providing educational opportunities through social connections.

A rich literature has affirmed the benefit and outcomes of makerspace services and how the library spaces help foster creative and innovative thinking. For example, Bieraugel and Neill (2017) claimed that library spaces influence students' learning preferences and behaviors. These innovation labs can stimulate innovative and creative thinking, encourage invention, and boost physical library usage (Bagley, 2014; Bieraugel & Neill, 2017; Curry, 2017; Efe, 2021; Gstalder, 2017; Moorefield-Lang, 2015). Curry (2017) proposes a similar argument that people engaged in the makerspace can develop a sense of self-efficacy and a higher level of confidence. Burke (2015) identified that the impact of makerspaces may be derived from the learning theory of constructionism by Seymour Papert and the concept of participatory culture by Henry Jenkins. As makerspaces allow students to define problems and design solutions during creation, they may stimulate students to develop teamwork, communication, problem-solving, critical thinking, and decision-making skills (Burke, 2015; Lee, 2017). Meanwhile, students can share their experiences and become teachers in this informal learning environment. Overall, makerspaces can bring great benefits in terms of learning.

Case Studies in Hong Kong Academic Libraries

As a nascent library technology, most Hong Kong universities have adopted technology-based makerspaces to support the library's vision and mission, promote lifelong learning and add value to current library services (Willingham & De Boer, 2015). The major factors for reforming the physical library spaces: (i) overcrowding and green development (Ho et al., 2023), (ii) changes in learning preferences from individual study to collaborative learning (Dong et al., 2021; J. Zhang et al., 2023), (iii) the increasing demand for learning spaces (Burke, 2015; Leung, Chiu, Ho, & Luk, 2022; Zhou et al., 2022), and (iv) demand for 24/7 open hours (M. K. Y. Chan et al., 2020; T. Yip et al., 2019). Besides, the changing pattern of collections and the trend of deploying learning commons are major reasons for reshaping library spaces (D. L. H. Chan & Spodickm, 2014; Chiu & Wong, 2023). They see makerspaces as an emerging trend in academic libraries for innovative learning experiences (D. L. H. Chan & Spodick, 2014). Sidorko and Lee (2014) also expressed that Joint University Librarians Advisory Committee (JULAC) in Hong Kong predicted the development of learning spaces and established a collaborative storage facility named the Joint University Research Archive to free up space for creating innovative learning spaces. To expand library services and support learning, libraries created makerspaces by repurposing unused rooms or constructing new areas (Burke, 2015; Xue et al., 2023).

From the observation, the first university that integrated the idea of makerspaces within librarianship was the Hong Kong University of Science and Technology (HKUST) Library. The objective of the renovation projects operated by HKUST is to enrich students' learning experiences by repurposing the library spaces (D. L. H. Chan & Spodick, 2014). D. L. H. Chan and Spodick (2014) outlined the makerspaces as Creative Media Zone, including a media production studio and workstations for editing. A series of strategic plans were illustrated to highlight the success of learning commons, such as deep cooperation with faculties and departments; and the promotion of training and seminars by giving credit to respondents. Insights on innovative spaces have been emphasized by providing 3D printers and makerspace to foster creativity. Makerspaces in libraries are a trending topic in discussion and literature, but a few in-depth studies of makerspace services have contributed significantly to the accumulated knowledge on this topic. Suen et al. (2020) investigated the virtual reality services offered in Hong Kong academic libraries and compared the deployment of Virtual Reality (VR) services between two case studies of the City University of Hong Kong (CityU) and The Hong Kong Polytechnic University (PolyU). Major challenges of using VR are listed, such as technical capability, space, and budget constraints (Suen et al., 2020).

Research Gap and Questions

A literature review on makerspaces in Hong Kong academic libraries revealed a preliminary but growing body of research. The research focused on related topics of makerspace development was only found published by HKUST Library (D. L. H. Chan & Spodick, 2014). The research focused on the transformation of learning commons, which makerspace is only part of the implementation plan. The study did not explain new technologies and equipment adopted in recent years. Besides, much literature has introduced various VR services provided by different Hong Kong academic libraries, including the case study of CityU and PolyU in Hong Kong by Suen et al. (2020). However, that case study focused on the implementation and limitation of virtual reality services in academic libraries but not the students' views on makerspaces, which is significant for the future development of makerspace services.

In other words, relatively few studies of innovation labs and makerspaces have contributed significantly to the knowledge on this topic. Thus, this research studied the implementation of makerspace (i.e., facilities/services offered), attempted to fill the knowledge gap in students' perception and satisfaction with makerspace services in university libraries, and made recommendations to improve the service model and users' experience in makerspaces. The following key research questions (RQ) guide this research:

RQ1. What are the kinds of makerspace services students intend to use in the library?
RQ2. What is the respondents' satisfaction with the makerspace services?
RQ3. How can the library improve its current makerspace services?

METHODOLOGY

Research Design and Rationale

This research adopted a quantitative online survey that focused on the attitude of library patrons and their interests in using makerspaces, targeting the major library user group, university students. It also investigated what makerspace services the regular makers use to gain initial insights into users' preferences of makerspaces. Meanwhile, the research relied on the primary data collected through library websites and site visits.

The survey was based on the questionnaire used in previous studies (Efe, 2021; Hunt et al., 2020) and included 24 questions. It was designed to take approximately 15 minutes to complete and aimed at exploring users' perspectives towards makerspaces. It comprises five sections: consent, demographic, interests in makerspaces, use of makerspace, and satisfaction level. Closed-ended questions were used with Likert scales and checklist choices, while open-ended questions were used for respondents expressing their opinions. Part II included the basic demographic questions, for example, age, gender, and majors. Part III included the interests of maker communities. Parts IV and V included perspectives from students towards the library's makerspace and satisfaction level of the library's makerspace. A survey instrument was developed utilizing the 5E Instructional Model (Engage, Explore, Explain, Elaborate, and Evaluate) (Bybee et al., 2006) as a framework to describe distinctive stages of the learning experience with emerging technologies, specifically in this digital age (L. S. N. Cheung et al., 2022; M. Jiang et al., 2023). Moersch (2014) stated that the 5E Instructional Model includes:

- Engage - Begin with the active engagement of students through stimulation and curiosity.
- Explore - Apply understandings/skills and develop ideas through observation and investigation.
- Explain - Justify the ideas through discussion and sharing.
- Elaborate -Apply newly formed understandings/skills and new skills reinforced.
- Evaluate - Self-reflection and review have changed beliefs, understanding, and behaviors.

Additionally, respondents were asked to use a 9-point Likert scale representing the level of satisfaction with makerspace services (1 - Very Dissatisfied to 9 - Very Satisfied) and the degree of agreeing or disagreeing with the statement (1- Strongly disagree to 9 - Strongly agree), the degree of importance (1 – Not at all Important to 9 – Extremely important) and the degree of likelihood/interest (1- Extremely unlikely to 9 – Extremely likely).

Overall, it allows the researcher to collect comprehensive quantitative data to know students' perspectives toward makerspace services, how well the makerspace services suit library users' needs, and how the library can improve and expand the current services offered. Respondents were current undergraduate and postgraduate students at a major university in Hong Kong. Microsoft Excel and IBM SPSS were applied for data analysis. Results were analyzed using Mann-Whitney U Test (Mann & Whitney, 1947) for comparing means of different genders across gender, skills level, user preference, and interests of maker communities demonstrated by figures and tables. The researcher conducted a pilot study (n=5) to clarify the understanding of questions and further revised the survey instrument. Afterward, an online anonymous survey on Google Forms was distributed via social networking sites, learning commons, and university libraries. Supermarket coupons were available to the respondents as lucky draw prizes to encourage responses.

RESULTS

Demographics

Table 1 shows the demographic information of 50 respondents, with 26 male and 24 female responses collected. Most respondents are pursuing bachelor's degrees (72%), followed by master's degrees (26%) and doctorate degrees (2%). Over 64% of the students are between the age range of 20 and 23, rendering the age diversity of these spaces low. Faculty are fairly evenly divided by Science (20%), Engineering (18%), Arts (18%), and Education (16%), respectively.

General User Interests

As shown in Table 2, respondents generally loved to be a maker to create things (mean=6.10) and liked the idea of innovative spaces (mean=7.00). In addition, Table 3 shows the similarities and differences in interests between genders. Both genders were interested in 3D modeling, photography, production studio, video editing, and virtual reality (VR), rated over 6 points. The most popular interests were VR, followed by video editing and production. Notably, female users were

Table 1. Demographics of respondents (n=50)

Demographics		n	%
Gender	Male	26	52%
	Female	24	48%
Age	20-21	15	30%
	22-23	17	34%
	24-25	5	10%
	26-27	9	18%
	28-29	4	8%
Educational level	Undergraduate	36	72%
	Postgraduate – Master	13	26%
	Postgraduate – Doctoral	1	2%
Faculty	Architecture	3	6%
	Arts	9	18%
	Business and Economics	1	2%
	Dentistry	1	2%
	Education	8	16%
	Engineering	9	18%
	Medicine	4	8%
	Science	10	20%
	Social Sciences	5	10%
STEAM-related curriculum	Yes	27	54%
	No	23	46%

less interested in science, computing, and engineering-related fields. Maric (2018) further explained that it may be caused by "gender stereotyping, male-dominated environments" (p.10).

Table 2. Interests in making and makerspaces (by gender)

Interests (9-Point Likert Scale)	Mean	SD	P-Value
I like to build, make or create things (including models, handcrafting, video production, etc.)	6.10	1.705	.850
Male	6.15	1.759	
Female	6.04	1.681	
I like the idea of innovative spaces and makerspaces.	7.00	1.512	.826
Male	7.00	1.697	
Female	7.00	1.319	

Table 3. Means of users' interests (by gender)

Area (9-Point Likert Scale)	Female	Male	P-Value
3D Modeling	6.08	6.15	.527
Architecture and Drafting	4.96	4.46	.225
Art, Painting, and Graphic Design	6.17	5.15	.046
Audio Recording or Editing	5.63	5.15	.407
Digital Photography and Photo Editing	6.42	6.15	.373
Coding, Software, and Programming	4.13	5.23	.026
Electronics	4.21	5.27	.034
Game Creation	5.63	6.12	.025
Handcraft	5.71	4.42	.001
Hardware and Machining	4.21	4.58	.292
Internet of Things (IoT)	4.63	5.00	.172
Metalworking	3.88	4.15	.385
Production Studio and Green Screen Video Backdrops	6.25	6.27	.747
Robotics	4.46	5.38	.091
Video Editing and Production	6.38	6.62	.545
Virtual Reality (VR) Equipment	6.63	7.08	.149

User Perceptions on the Library's Makerspace

Table 4 illustrates the level of agreement with the statements based on the 5E Instructional Model. Cronbach's Alpha was employed to check if the scale is reliable and estimate the consistency reliability (Cronbach, 1951), with results between 0.864 and 0.940, indicating a reliable and satisfactory level of high internal consistency for the question responses (Cronbach, 1951).

We present the highest means or interesting results at each stage for further investigation. At the engagement stage, respondents generally agreed that the library's makerspace could stimulate interest in making things (mean=6.76), interacting with friends, and working collaboratively (mean=6.80). Respondents were more likely to agree that the library's makerspace allows users to learn new things and expand their knowledge (mean=7.08) in the exploration stage. At the explanation stage, respondents could form maker communities to share ideas and explore solutions (mean=6.7). At the elaboration stage, respondents believed the library's makerspace could stimulate critical thinking skills (mean=6.96) and creativity and design thinking (mean=7.12). Respondents believed the library's makerspace could promote multidisciplinary thinking and learning (mean=7.18) at the evaluation

Table 4. Perceptions towards makerspaces by the 5E instructional model

Statement (9-point Likert scale, p < 0.009)	Mean (SD)	Cronbach's Alpha
Engagement		**0.918**
a. Makerspace stimulates my interest in making things.	6.76 (1.135)	
b. Makerspace stimulates my interest in learning in general.	6.62 (1.176)	
c. Makerspace generates curiosity and raises my attention to learn new skills through maker activities/projects.	6.66 (1.287)	
d. Makerspace stimulates my interest in interacting with my peers and working collaboratively.	6.80 (1.325)	
Exploration		**0.877**
e. Makerspace supports exploring the topics in my curriculum.	6.12 (1.534)	
f. Makerspace provides the opportunity to learn new things and expand my knowledge.	7.08 (1.226)	
g. Makerspace supports hands-on learning, inquiring, and discovery.	6.48 (1.165)	
Explanation		**0.864**
h. Makerspace helps me understand ideas/concepts for solving problems.	6.10 (1.055)	
i. Makerspace offers various resources/tools for me to analyze the findings and reach a solution.	6.28 (1.179)	
j. Makerspace facilitates the connection of maker communities for sharing ideas and solutions.	6.70 (1.111)	
Elaboration		**0.940**
k. Makerspace facilitates developing a deep understanding of my projects.	5.96 (1.245)	
l. Makespace helps me apply knowledge to new situations or other disciplines.	6.70 (1.446)	
m. Makerspace offers a space to organize and reinforce what I have learned.	5.86 (1.325)	
n. Makerspace facilitates interactions and resource sharing with other learners and/or library staff.	6.52 (1.282)	
o. Makerspace stimulates critical thinking skills.	6.96 (1.340)	
p. Makerspace stimulates creativity and design thinking.	7.12 (1.380)	
q. Makerspace helps me generate new ideas for prototyping.	5.82 (1.207)	
Evaluation		**0.885**
r. Makerspace allows for self-directed learning and self-reflection.	6.32 (1.151)	
s. Makespace promotes multidisciplinary thinking and learning in general.	7.18 (1.190)	
t. Makespace helps improve my skillsets in general.	6.74 (1.175)	
u. Makerspace helps improve my academic results in general.	5.00 (1.355)	

stage. Results indicated much lower values of "makerspace helps improve academic results in general" (mean=5.00) than other statements.

User Experiences and Skills

Table 5 indicates the frequent hands-on learning experience that respondents participated in during their study. Respondents have built, made, or created in the following areas: video editing and production (24%), 3D modeling and printing (14%), and webpage design (12%). In addition to user experiences, Table 6 shows respondents do not have adequate skills using the makerspace (mean=4.22).

Table 5. Type of hands-on learning experience during study

Category of Hands-On Learning Experience	n	%
3D modeling and printing	7	14
Animation and Cinematography	3	6
Architecture and Drafting	5	10
Art, Painting, and Graphic Design	5	10
Audio Recording or Editing	1	2
Coding, Software, and Programming	4	8
Electronics	2	4
Game Creation	1	2
Handcraft	4	8
Hardware and Machining	3	6
Product Design	1	2
Robotics	2	4
Video Editing and Production	12	24
Webpage Design	6	12

User Satisfaction

Overall, the user satisfaction level with the library's makerspace was somewhat satisfactory (mean=5.04), mainly due to the experienced high-quality services (see Table 7). However, according to Table 8, 44 out of 50 respondents (88%) mainly reported that it lacks promotion (mean=6.66) and training (mean=6.52).

Table 6. Skills and competence of users (by gender)

Statement (9-Point Likert Scale)	Mean	SD	P-Value
I have the skills needed to use a makerspace.	4.22	1.489	.767
Male	4.15	1.567	
Female	4.29	1.429	

Table 7. Satisfaction level of the makerspaces

(9-Point Likert Scale, p <0.001)	Mean	SD	Min	Max
Satisfaction level	5.04	1.228	3	8

Table 8. Reason for not using makerspaces (n=44)

(9-Point Likert Scale, p <0.005)	Mean	SD
Lack of motivation	4.91	1.476
Lack of interests	4.52	1.592
Lack of awareness or promotion	6.66	1.584
Lack of training and workshop	6.52	1.577
Current makerspace services do not suit my needs	5.61	1.466

Proposed Improvements for Makerspaces

To address the satisfaction level of makerspaces, respondents were also asked to rate the importance of the proposed improvements (see Table 9). Results indicated that respondents have various services and facilities (mean=6.70) to support interdisciplinary education. Also, they rated higher points on training (mean=7.56) and cross-collaboration activities between faculties and libraries (mean=7.02) than an upgrade of maker tools (mean=5.66) and sufficient staffing (mean=6.60). Meanwhile, one respondent commented about makerspace services and pointed out that more introductory level training is desirable, as she was not skillful in handling the digital equipment.

Table 10 indicates that individual study space (mean=7.70), collaboration and discussion space (mean=7.74), relaxation space (mean=7.52), self-check-in/out equipment (mean=6.98), and social space (mean=6.96) were the top five elements of ideal makerspaces among respondents. Table 11 shows that respondents preferred

Table 9. Proposed improvements for makerspaces

Proposed Improvements (9-Point Likert Scale, p < 0.001)	Mean	SD
Offer a wider range of services and facilities	6.70	1.607
Upgrade and enhance current hardware equipment	5.66	1.649
Provide more training and workshops to users	7.56	1.128
Provide sufficient staff to support the use of equipment	6.60	1.161
Cross-promotion is required between faculties and the library	7.02	1.097

Table 10. Makerspaces facilities preferences

Preference (9-Point Likert Scale, p < 0.003)	Mean	SD
Computer workstation	6.88	1.304
Display Screen	6.50	1.182
Self-service check-in/out Equipment	6.98	1.000
Whiteboard	6.60	1.125
Individual Study Space	7.70	.955
Collaboration and Discussion Space	7.74	.803
Lockers and Storage Space	5.42	1.401
Relaxation Space	7.52	1.074
Social Space	6.96	1.068

Table 11. User preferences on training offered

Types of Training (9-Point Likert Scale, p < 0.001)	Mean	SD
Instruction Sessions and Workshops	7.40	1.195
Orientation	6.88	1.100
One-on-One Instructions	6.76	1.506

to be trained during instructional classes and workshops (mean=7.40), followed by orientation (mean=6.88). Table 12 illustrates that they were more likely to join hands-on workshops (mean=6.76) to learn new technologies or skills, followed by guided tours (mean=6.32) and showcase events (mean=5.48).

Table 12. User preferences on maker activities

Types of maker activities (9-Point Likert Scale, p < 0.134)	Mean	SD
Attend hands-on workshops to learn new skills/machines/technologies	6.76	1.611
Attend seminars or presentations by makers	5.22	1.461
Attend makerspace gallery showings of projects	5.48	1.607
Join an orientation or guided tours	6.32	1.463
Join the competition organized by the university library	4.52	1.798

DISCUSSION

This research has investigated the deployment of makerspace services in a major comprehensive university in Hong Kong and explored respondents' perceptions toward makerspaces. This section will respond to the research questions by discussing the students' intentions in using the Makerspace services, user perceptions of makerspace, usage of library makerspace, satisfaction level of makerspace services, and further recommendations.

Students' Interests and Perceptions of General Makerspace Services (RQ1)

User Interests

According to our findings, respondents generally love to be a maker to create things and like the idea of innovative spaces. In addition, students, regardless of gender, seem to be interested in 3D modeling, photography, production studio, video editing, and VR, with the most interest in VR. Notably, female students appear to be less interested in science, computing, and engineering-related fields. Maric (2018) further explained that it may be caused by "gender stereotyping, male-dominated environments" (p. 10).

User Perceptions

Our findings suggest students' perceptions of the makerspace services according to the 5E Instructional Model. At the engagement stage, students tend to agree that makerspaces can stimulate interest in making things, interacting with friends, and working collaboratively. This result aligns with Hatch (2013), who claimed that making is at the core of human beings. Stimulation in learning can boost students' motivation to make and learn new technologies. Makerspaces provide various ways of informal social interaction, including peer learning (Vuorikari et al., 2019). Delgado et al. (2020) highlighted that innovation spaces could promote collaboration and interaction between makers, and the relationship of the library community can be strengthened in makerspaces through collaborative learning.

The next stage is the exploration stage. Makerspaces are creative and innovative spaces designed to create knowledge and expand knowledge based on different expertise (F. Wang et al., 2016). Students are more likely to agree that the makerspace allows users to learn new things and expand their knowledge. At the explanation stage, students might form maker communities to share ideas and explore solutions. New emerging technologies, making topics, and related skill sets can be discussed and refined from any misconceptions.

At the elaboration stage, students believed the makerspace could stimulate critical thinking skills, creativity, and design thinking. This finding is consistent with other studies' findings (Curry, 2017; Moorefield-Lang, 2019; A. Wong & Partridge, 2016). Such a creative learning process can help develop critical thinking and problem-solving.

Finally, students believed the makerspace helped promote multidisciplinary thinking and learning at the evaluation stage. This view is also supported by Hynes (2017), who recognized that the makerspace could promote interdisciplinary collaboration and provide an opportunity for rich cross-disciplinary interaction among students from multiple disciplines. Interestingly, results indicated that students appear to be less likely to perceive makerspace as a support to improve their academic results in general. It seems to be no significant correlation between academic results and makerspaces. However, no doubt enhanced skills may positively impact academic performance, but the low values of this statement bring up the need for future studies on the potential impact of makerspaces on different majors' students.

Students' Usage and Satisfaction on the Library's Makerspace Services (RQ2)

User Experiences and Skills

Our findings indicated that respondents have built, made, or created in the following areas: video editing and production, 3D modeling and printing, and webpage design. These results align with Burke (2015)'s investigation of the trend of makerspace activities in higher education settings, which indicated the high popularity of 3D printing and multimedia creation and editing. Regarding user skills, most respondents appeared to lack skills in utilizing makerspace services, especially emerging technologies, aligning with users' preference for using instructional services and attending hands-on workshops and presentations. Therefore, the library is recommended to organize more hands-on workshops, seminars, and other instructional programs to enhance users' skills.

User Satisfaction

According to our findings, the overall satisfaction level is somewhat satisfactory, mainly due to the experienced high-quality services. However, students might consider that the makerspaces lack promotion and training, which is quite a common problem in academic libraries in East Asia (W. W. H. Cheng et al., 2020; Fong et al., 2020; A. H. C. Lam et al., 2023; E. T. H. Lam et al., 2019; Tsang & Chiu, 2022). Most of them might not be aware of the makerspaces in the library.

Suggestions for Improving Current Makerspace Services (RQ3)

Ideal Makerspaces From the Students' Perspectives

Our findings indicate that makerspaces should have various services and facilities to support interdisciplinary education. Also, user training and cross-collaboration activities between faculties and libraries may be important for developing an ideal makerspace. Most students seem to prefer to attend instructional classes, workshops, and orientation. They might be more likely to join hands-on workshops to learn new technologies or skills, guided tours, and showcase events. User training in this section echoes the students' previous needs expressed in the user skills section. Radniecki (2017) emphasized the importance of instructions and training in university libraries. Bad user experience will limit the overall utilization of makerspaces and hinder potential users from learning new technologies. These results indicate several problems that should be addressed: lack of training, workshops, and learning aids on makerspaces and lack of maker activities and events to encourage participation and engagement.

Besides, our findings show that individual study space, collaboration and discussion space, relaxation space, self-check-in/out equipment, and social space are essential elements of an ideal makerspace. It shows an intense desire among respondents for more dynamic spaces in makerspaces. It is undeniable that offering individual study spaces and collaborative spaces is equally important to the students. Somewhat surprisingly, the researcher found that users have a demand for self-services and relaxation space in makerspaces. Collaboration and discussion spaces promote brainstorming, interaction, and knowledge sharing (Lee, 2017), allow students to collaborate on different projects, and support social activities. Individual study spaces and computer workstations are still valuable for students to quiet study, research, and self-reflection. Significantly, libraries should provide more relaxation space and self-service (Lo et al., 2013; Zhou et al., 2022). Bieraugel (2019) highlighted the importance of addressing the diverse needs of the students, including relaxation and restoration. A casual and relaxing environment can promote information sharing and build connections between library communities. Additionally, self-service kiosks to borrow and return digital equipment can be added.

Accessibility, Inclusion, and Diversity

Yi and Baumann (2020) identified three core values when building a makerspace. This design thinking framework includes "accessibility, inclusion, and diversity." An ideal makerspace should be accessible to all users (AccessEngineering, 2015; Yi & Baumann, 2020) and meet diverse training and instruction styles (Yi & Baumann, 2020). Yi and Baumann (2020) highlighted that libraries should treat all makers equally regardless of abilities and majors and make the makerspace accessible to people with disabilities (V. H. Y. Chan et al., 2023; Cheung et al., 2021; Li, Xie, Chiu, & Ho, 2023). Flexible and adjustable furniture is highly recommended to accommodate different learning styles and usages (AccessEngineering, 2015; Yi & Baumann, 2020), for example, seated and standing places, collaborative workspaces, and classrooms.

Concerning access, the makerspaces of this case study run on a full-service model rather than self-service. For example, users must check in and out equipment at the circulation counter, and library staff must go through the printing process of 3D products. Although it provides first-time users immediate mentoring and technical support, it also discourages autonomy and accessibility. In the long run, the overall learning and innovative environment should be more open, and all makers can easily access the resources.

Concerning flexible seating (e.g., chairs and study carrels with wheels) and a dynamic layout, the Digital Interactive Lab and Multipurpose Area are designed for individual study, collaborative spaces, teaching, and conference purposes. These spaces can be turned into unique venues that increase the seating capacity, allow flexibility, and encourage active engagement. Nevertheless, adjustable-height workbenches should be considered to increase accessibility (Steele et al., 2018).

Yi and Baumann (2020) emphasized that offering diversity in training and instruction is essential. However, according to the webpages and Facebook, no formal makerspace workshops have been conducted, and it mainly provides orientation, consultation sessions, and online tutorials. The instructional information on the library website is static, such as operation guidelines and YouTube videos provided by the vendor. In response to meeting diverse needs and instruction styles, the library should seek various instruction styles, for example, task-specific instructions and skill-based guidance (Yi & Baumann, 2020).

Recommendations for Improvements

This section proposes recommendations to improve the overall user experience and awareness of library's makerspace services.

Active role in organizing outreach activities. There was low makerspace usage in this case, and most respondents were unaware of the makerspace services. The library serves as a bridge to connect the maker community to learn and develop a broader community. To raise awareness of makerspace services and grow more engaged library patrons, the library should actively organize outreach programs and promote makerspace. It can reach the target users to know more about the existing services and tools. Hosting and co-organizing events could promote engagement and innovation (Lu et al., 2023; Stover et al., 2019). To drive new makers and encourage participation, the library can regularly organize outreach activities that partner with faculties to develop partnerships for this creative space, such as small-scale guided tours and first-year orientations even during COVID-19 (Huang et al., 2021; 2022; 2023; P. Y. Yu et al., 2023).

Adopt different promotion strategies. The case primarily uses word of mouth and eye-catching posters to promote makerspaces. Herron and Kaneshiro (2017) shared some effective strategies for promoting the services. For example, the libraries can prepare promotional items for the respondents attending library workshops and orientation, like 3D printing products or 3D printing coupons, as small welcoming gifts. In addition, Lee (2017) claimed that promoting special events can increase library traffic. A. Wong and Partridge (2016) stated that library websites are generally used to promote makerspaces and upcoming maker activities and workshops. The library should introduce new services and facilities on the library's social media channels and websites and send bulk emails advertising discipline-related activities to the students, especially STEAM-related curricula. Besides, posting printed and digital posters inside the libraries and hard copies on bulletin boards within the campus help spread the news. These measures align with Moorefield-lang (2019)'s advocacy in word-of-mouth (WOM) marketing so that library users can spread the word about makerspaces within their community. Thus, this implies that the library should spread positive feedback on VR services to stimulate students to learn new technology. More specifically, libraries should continue to concentrate on these makerspace services and maintain quality services to achieve high user satisfaction.

Conduct regular user surveys and focus group discussions. Lee (2017) emphasized that feedback from students, staff, and faculty is crucial to support the growth of makerspaces. Stover et al. (2019) explained that focus group interviews and surveys help libraries assess the effectiveness of the makerspace. This view is supported by F. Yi and Baumann (2020), who expressed that qualitative analysis is appropriate for understanding both makers' physical and emotional needs. The library should continue to analyze usage and feedback collected from interviews and user surveys to expand the learning opportunities and eventually increase the usage of the innovative space and improve the user experience. QR codes linked to user surveys can be placed at the service spot, library website, and social media to reach more library patrons. In addition, surveys can be distributed after the consultation sessions and library workshops/orientation. Users can share their views on specific services, and the library can learn from patrons' behaviors and actual needs.

Offer more training workshops to support teaching and learning practices. From the findings, most respondents did not have adequate equipment skills, emphasizing the need for training. F. and Baumann (2020) highlighted that libraries should aim at serving both experienced and unknowledgeable users. The library can offer more beginner to advanced digital equipment workshops. Specifically, introductory workshops for unknowledgeable users help them to get started and gain more hands-on experience. F. Yi and Baumann (2020) emphasized that offering diversity in training and instruction is essential. However, a few training and instructional classes have been provided in this case. To stimulate a culture of creativity and innovation, the library should offer training in different forms to establish close ties with the makers and stimulate learning, such as demonstrations, "try & build workshops," and short library courses with credits. Besides, the library can take different learning styles, such as organizing seminars by hiring instructors beyond the library or collaborating with staff familiar with the emerging technology and STEAM-related curriculum (Okpala, 2016). It also facilitates collaboration by sharing expertise (Julian & Parrott, 2017).

Staff and training. Besides the training offered to the users, the library should provide professional development opportunities for library staff to expand their knowledge and skill sets (Lin et al., 2022). Overall, the above suggestions involve many extra human resources. The library may consider employing more volunteers and interns to assist with these tasks (K. K. Li & Chiu, 2022; Ng et al., 2022; Yew et al., 2022). Besides, virtual reference services should also be provided as makerspaces are often open 24/7 so that volunteers and staff may support users facing problems (Guo et al., 2022; Tsang & Chiu, 2022) and other general inquiries anytime, anywhere (J. Cheng et al., 2022; Dukic et al., 2015; Ezeamuzie et al., 2023; Fung et al., 2016; K. P. Lau et al., 2017; K. S. N. Lau et al., 2020; Law et al., 2019; J. Ni et al., 2022; Wai et al., 2018; K. H. T. Yip et al., 2021; X. Zhang et al., 2021). Librarians, volunteers, and users can also form various communities of practices to share their experiences (Chung et al., 2021; Lei et al., 2021; Leung, Hui, Luk, Chiu, & Ho, 2022; Mak et al., 2022; W. Wang et al., 2021; Xie, Chiu, & Ho, 2023; Xie, Wong, Chiu, & Lei, 2023; Yang et al., 2023)

CONCLUSION

Makerspace is ideal for students learning, collaborating, and communicating across disciplines, emphasizing the importance of makerspaces and the huge perceived benefits to both individual and maker communities. It can encourage students to experience new emerging technologies and stimulates creativity, critical thinking, and collaboration. This case study covers an overview of the makerspace services in a comprehensive university in Hong Kong and students' perspectives toward these services. Respondents reflected their interests in makerspace activities and use of makerspaces. They were interested in VR, 3D modeling, video editing, and production studio services. However, they might lack skills in dealing with digital equipment, and the library offered a few training workshops. This study attempts to extend the existing literature on library makerspace in several ways, including understanding students' perspectives, frequent maker activities conducted, and preference for joining maker activities. The findings support previous findings and confirm that makerspaces can encourage students' engagement from multiple disciplines and improve students' skillset, including creativity, communication, collaboration, design thinking, and critical thinking.

Students' ideas on ideal makerspaces and improvements on current library's makerspace have also been investigated. Respondents believed that individual and collaboration spaces and PC workstations are significant amenities and spaces in the makerspace and that the libraries generally fulfilled the expectation of an ideal makerspace and academic needs from users. However, it seems the library should still improve the current makerspace services and consider the proposed suggestions for further developing the makerspace. The library should lead in organizing outreach activities and offering more workshops to bring external parties into the makerspace (i.e., unknowledgeable and potential makers) to encourage active engagement and involvement of the students. Also, the library should conduct more user surveys to review whether the deployment of the makerspace can meet user needs and rethink any new services worth developing.

The study specifically focuses on a single case study of the deployment of makerspace services in a single university and highlights its emerging technologies. The findings of this study provide specific recommendations and best practices for the future application of makerspace services or second makerspaces.

Long and Hicks (2020) highlight makerspace resources offered by faculties and departments designed for integration into existing academic curricula. We see makerspace services' potential to continue growing in popularity and expand in academic libraries. Moreover, future studies should compare diverse makerspaces in multiple universities in different countries.

Additionally, this case study was limited by a small sample size of 50. This small-scale quantitative study can serve as a pilot study for future research. Additional research can include interviews and focus-group qualitative studies to triangulate the results. Further, social media analytics can investigate users' general comments and sentiments regarding different kinds of makerspaces (S. Deng & Chiu, 2013; Fang et al., 2021; He et al., 2022; S. Li, Chiu, Kafeza, & Ho, 2023; Y. Liu et al., 2023; J. Wang et al., 2022). We are also interested in the changes in usage, habits, and preferences of different learning (S. Li, Xie, Chiu, & Ho, 2023; Yau et al., 2023), information (Dai & Chiu, 2023; Sung & Chiu, 2022; Y. Yi & Chiu, 2023), library (V. H. Y. Chan et al., 2022; Sun et al., 2022; P. Y. Yu et al., 2023), and cultural (Chen et al., 2018; Chin & Chiu, 2023; W. Deng et al., 2022; Gao et al., 2021; X. Jiang et al., 2023; S. Li, Lam, & Chiu, 2023; Meng et al., 2023; A. K. -k. Wong & Chiu, 2023; H. H. K. Yu et al., 2023; Xu et al., 2023) services after COVID-19.

REFERENCES

AccessEngineering. (2015). *Making a makerspace? Guidelines for accessibility and universal design.* https://www.washington.edu/doit/sites/default/files/atoms/files/Making_a_Makerspace_8_03_15.pdf

Adams Becker, S., Cummins, M., Davis, A., Freeman, A., Giesinger Hall, C., Ananthanarayanan, V., Langley, K., & Wolfson, N. (2017). *NMC horizon report: 2017 library edition.* The New Media Consortium.

Bagley, C. A. (2014). *Makerspaces top trailblazing projects: a LITA guide.* American Library Association.

Bailin, K. (2011). Changes in academic library space: A case study at the University of New South Wales. *Australian Academic and Research Libraries, 42*(4), 342–359. doi:10.1080/00048623.2011.10722245

Bieraugel, M. (2019). Do your library spaces help entrepreneurs? Space planning for boosting creative thinking. *Supporting Entrepreneurship and Innovation, 40,* 21–32. doi:10.1108/S0732-067120190000040001

Bieraugel, M., & Neill, S. (2017). Ascending bloom's pyramid:fostering student creativity and innovation in academic library spaces. *College & Research Libraries, 78*(1), 35–52. doi:10.5860/crl.78.1.35

Burke, J. (2015). *Making sense: Can makerspaces work in academic libraries?* https://www.ala.org/acrl/sites/ala.org.acrl/files/content/conferences/confsandpreconfs/2015/Burke.pdf

Bybee, R. W., Taylor, J. A., Gardner, A., Van Scotter, P., Powell, J. C., Westbrook, A., & Landes, N. (2006). *The BSCS 5E instructional model: Origins and effectiveness*. BSCS. https://media.bscs.org/bscsmw/5es/bscs_5e_full_report.pdf

Chan, D. L. H., & Spodick, E. (2014). Space development: A case study of HKUST Library. *New Library World*, *115*(5/6), 250–262. doi:10.1108/NLW-04-2014-0042

Chan, M. K. Y., Chiu, D. K. W., & Lam, E. T. H. (2020). Effectiveness of overnight learning commons: A comparative study. *Journal of Academic Librarianship*, *46*(6), 102253. Advance online publication. doi:10.1016/j.acalib.2020.102253 PMID:34173399

Chan, T. T. W., Lam, A. H. C., & Chiu, D. K. W. (2020). From Facebook to Instagram: Exploring user engagement in an academic library. *Journal of Academic Librarianship*, *46*(6), 102229. Advance online publication. doi:10.1016/j.acalib.2020.102229 PMID:34173399

Chan, V. H. Y., & Chiu, D. K. W. (2023). Integrating the 6C's motivation into reading promotion curriculum for disadvantaged communities with technology tools: A case study of Reading Dreams Foundation in rural China. In A. S. Etim (Ed.), *Adoption and use of technology tools and services by economically disadvantaged communities: Implications for growth and sustainability*. IGI Global.

Chan, V. H. Y., Chiu, D. K. W., & Ho, K. K. W. (2022). Mediating effects on the relationship between perceived service quality and public library app loyalty during the COVID-19 era. *Journal of Retailing and Consumer Services*, *67*, 102960. Advance online publication. doi:10.1016/j.jretconser.2022.102960

Chang, V., Chiu, D. K. W., Ramachandran, M., & Li, C.-S. (2018). Internet of Things, Big Data and Complex Information Systems: Challenges, solutions and outputs from IoTBD 2016, COMPLEXIS 2016 and CLOSER 2016 selected papers and CLOSER 2015 keynote. *Future Generation Computer Systems*, *79*(3), 973–974. doi:10.1016/j.future.2017.09.013

Chen, Y., Chiu, D. K. W., & Ho, K. K. W. (2018). Facilitating the learning of the art of Chinese painting and calligraphy at Chao Shao-an Gallery in Hong Kong. *Micronesian Educators*, *26*, 45–58.

Cheng, J., Yuen, A. H. K., & Chiu, D. K. W. (2022). (in press). Systematic review of MOOC research in mainland China. *Library Hi Tech*. Advance online publication. doi:10.1108/LHT-02-2022-0099

Cheng, W., Tian, R., & Chiu, D. K. W. (2023). Travel vlogs influencing tourist decisions: Information preferences and gender differences. *Aslib Journal of Information Management*. Advance online publication. doi:10.1108/AJIM-05-2022-0261

Cheng, W. W. H., Lam, E. T. H., & Chiu, D. K. W. (2020). Social media as a platform in academic library marketing: A comparative study. *Journal of Academic Librarianship*, *46*(5), 102188. Advance online publication. doi:10.1016/j.acalib.2020.102188

Cheung, L. S. N., Chiu, D. K. W., & Ho, K. K. W. (2022). A quantitative study on utilizing electronic resources to engage children's reading and learning: Parents' perspectives through the 5E instructional model. *The Electronic Library*, *40*(6), 662–679. doi:10.1108/EL-09-2021-0179

Cheung, T. Y., Ye, Z., & Chiu, D. K. W. (2021). Value chain analysis of information services for the visually impaired: A case study of contemporary technological solutions. *Library Hi Tech*, *39*(2), 625–642. doi:10.1108/LHT-08-2020-0185

Cheung, V. S. Y., Lo, J. C. Y., Chiu, D. K. W., & Ho, K. K. W. (2023). Evaluating social media's communication effectiveness on travel product promotion: Facebook for college students in Hong Kong. *Information Discovery and Delivery*, *51*(1), 66–73. doi:10.1108/IDD-10-2021-0117

Chin, G. Y. L., & Chiu, D. K. W. (2023). RFID-based robotic process automation for smart museums with an alert-driven approach. In R. K. Tailor (Ed.), *Application and adoption of robotic process automation for smart cities*. IGI Global.

Chiu, D. K. W., & Ho, K. K. W. (2022a). Editorial: Special selection on contemporary digital culture and reading. *Library Hi Tech*, *40*(5), 1204–1209. doi:10.1108/LHT-10-2022-516

Chiu, D. K. W., & Ho, K. K. W. (2022b). Editorial: 40th anniversary: Contemporary library research. *Library Hi Tech*, *40*(6), 1525–1531. doi:10.1108/LHT-12-2022-517

Chiu, D. K. W., & Wong, S. W. S. (2023). Reevaluating remote library storage in the digital age: A comparative study. *portal. Portal (Baltimore, Md.)*, *23*(1), 89–109. doi:10.1353/pla.2023.0009

Choy, F. C., & Goh, S. N. (2016). A framework for planning academic library spaces. *Library Management*, *37*(1/2), 13–28. doi:10.1108/LM-01-2016-0001

Chung, C., Chiu, D. K. W., Ho, K. K. W., & Au, C. H. (2020). Applying social media to environmental education: Is it more impactful than traditional media? *Information Discovery and Delivery*, *48*(4), 255–266. doi:10.1108/IDD-04-2020-0047

Cronbach, L. J. (1951). Coefficient alpha and the internal structure of tests. *Psychometrika*, *16*(3), 297–334. doi:10.1007/BF02310555

Curry, R. (2017). Makerspaces: A beneficial new service for academic libraries? *Library Review*, *66*(4/5), 201–212. doi:10.1108/LR-09-2016-0081

Dai, C., & Chiu, D. K. W. (2023). Impact of COVID-19 on reading behaviors and preferences: Investigating high school students and parents with the 5E instructional model. *Library Hi Tech*. Advance online publication. doi:10.1108/LHT-10-2022-0472

Delgado, L., Galvez, D., Hassan, A., Palominos, P., & Morel, L. (2020). Innovation spaces in universities: Support for collaborative learning. *Journal of Innovation Economics and Management*, *31*(1), 123–153. doi:10.3917/jie.pr1.0064

Deng, Q., Allard, B., Lo, P., Chiu, D. K. W., See-To, E. W. K., & Bao, A. Z. R. (2019). The role of the library café as a learning space: A comparative analysis of three universities. *Journal of Librarianship and Information Science*, *51*(3), 823–842. doi:10.1177/0961000617742469

Deng, S., & Chiu, D. K. W. (2023). Analyzing the Hong Kong Philharmonic Orchestra's Facebook community engagement with the Honeycomb Model. In M. Dennis & J. Halbert (Eds.), *Community engagement in the online space* (pp. 31–47). IGI Global. doi:10.4018/978-1-6684-5190-8.ch003

Deng, W., Chin, G. Y.-l., Chiu, D. K. W., & Ho, K. K. W. (2022). Contribution of literature thematic exhibition to cultural education: A case study of Jin Yong's Gallery. *Micronesian Educators*, *32*, 14–26.

Ding, S. J., Lam, E. T. H., Chiu, D. K. W., Lung, M. M., & Ho, K. K. W. (2021). Changes in reading behavior of periodicals on mobile devices: A comparative study. *Journal of Librarianship and Information Science*, *53*(2), 233–244. doi:10.1177/0961000620938119

Dong, G., Chiu, D. K. W., Huang, P.-S., Ho, K. K. W., Lung, M. M., & Geng, Y. (2021). Relationships between research supervisors and students from coursework-based Master's degrees: Information usage under social media. *Information Discovery and Delivery*, *49*(4), 319–327. doi:10.1108/IDD-08-2020-0100

Dukic, Z., Chiu, D. K. W., & Lo, P. (2015). How useful are smartphones for learning? Perceptions and practices of Library and Information Science students from Hong Kong and Japan. *Library Hi Tech*, *33*(4), 545–561. doi:10.1108/LHT-02-2015-0015

Efe, R. T. (2021). Awareness of the concept of makerspace: the scenario of university libraries in Nigeria. *Library Philosophy and Practice*, 1-17. https://digitalcommons.unl.edu/libphilprac/4952/

Ezeamuzie, N. M., Rhim, A. H. R., Chiu, D. K. W., & Lung, M. M. (2022). (in press). Exploring gender differences in foreign domestic helpers' mobile information usage. *Library Hi Tech*. Advance online publication. doi:10.1108/LHT-07-2022-0350

Fan, K. Y. K., Lo, P., Ho, K. K. W., So, S., Chiu, D. K. W., & Ko, K. H. T. (2020). Exploring the mobile learning needs amongst performing arts students. *Information Discovery and Delivery*, *48*(2), 103–112. doi:10.1108/IDD-12-2019-0085

Fang, J., Chiu, D. K. W., & Ho, K. K. W. (2021). Exploring cryptocurrency sentiments with clustering text mining on social media. In Z. Sun (Ed.), *Intelligent analytics with advanced multi-industry applications* (pp. 157–171). IGI Global. doi:10.4018/978-1-7998-4963-6.ch007

Farritor, S. (2017). University-based makerspaces: A source of innovation. *Technology and Innovation*, *19*(1), 389–395. doi:10.21300/19.1.2017.389

Fong, K. C. H., Au, C. H., Lam, E. T. H., & Chiu, D. K. W. (2020). Social network services for academic libraries: A study based on social capital and social proof. *Journal of Academic Librarianship*, *46*(1), 102091. Advance online publication. doi:10.1016/j.acalib.2019.102091

Fung, R. H. Y., Chiu, D. K. W., Ko, E. H. T., Ho, K. K. W., & Lo, P. (2016). Heuristic usability evaluation of University of Hong Kong Libraries' mobile website. *Journal of Academic Librarianship*, *42*(5), 581–594. doi:10.1016/j.acalib.2016.06.004

Gao, W., Lam, K. M., Chiu, D. K. W., & Ho, K. K. W. (2021). A big data analysis of the factors influencing Movie Box Office in China. In Z. Sun (Ed.), *Intelligent analytics with advanced multi-industry applications* (pp. 232–249). IGI Global. doi:10.4018/978-1-7998-4963-6.ch011

Grush, M. (2015, December 1). *Rethinking the makerspace*. Campus Technology. https://campustechnology.com/Articles/2015/12/01/Rethinking-the-Makerspace.aspx?p=1

Gstalder, S. H. (2017). *Understanding library space planning* (Order No. 10289537). Available from ProQuest Dissertations & Theses A&I. (1954088754). https://repository.upenn.edu/dissertations/AAI10289537

Guo, Y., Lam, A. H. C., Chiu, D. K. W., & Ho, K. K. W. (2022). Perceived quality of reference service with WhatsApp: A quantitative study from user perspectives. *Information Technology and Libraries*, *41*(3). Advance online publication. doi:10.6017/ital.v41i3.14325

Hatch, M. (2013). *The maker movement manifesto: rules for innovation in the new world of crafters, hackers, and tinkerers.* McGraw Hill Professional.

He, Z., Chiu, D. K. W., & Ho, K. K. W. (2022). Weibo analysis on Chinese cultural knowledge for gaming. In Z. Sun & Z. Wu (Eds.), *Handbook of research on foundations and applications of intelligent business analytics* (pp. 320–349). IGI Global. doi:10.4018/978-1-7998-9016-4.ch015

Herron, J., & Kaneshiro, K. (2017). A university-wide collaborative effort to designing a makerspace at an academic health sciences library. *Medical Reference Services Quarterly*, *36*(1), 1–8. doi:10.1080/02763869.2017.1259878 PMID:28112641

Ho, C. Y., Chiu, D. K. W., & Ho, K. K. W. (2023). Green space development in academic libraries: A case study in Hong Kong. In V. Okojie & M. O. Igbinovia (Eds.), *Global perspectives on sustainable library practices* (pp. 142–156). IGI Global. doi:10.4018/978-1-6684-5964-5.ch010

Huang, P.-S., Paulino, Y. C., So, S., Chiu, D. K. W., & Ho, K. K. W. (2021). Editorial. *Library Hi Tech*, *39*(3), 693–695. doi:10.1108/LHT-09-2021-324

Huang, P.-S., Paulino, Y. C., So, S., Chiu, D. K. W., & Ho, K. K. W. (2022). Guest editorial: COVID-19 pandemic and health informatics part 2. *Library Hi Tech*, *40*(2), 281–285. doi:10.1108/LHT-04-2022-447

Huang, P.-S., Paulino, Y. C., So, S., Chiu, D. K. W., & Ho, K. K. W. (2023). Guest editorial: COVID-19 pandemic and health informatics part 3. *Library Hi Tech*, *41*(1), 1–6. doi:10.1108/LHT-02-2023-585

Hui, S. C., Kwok, M. Y., Kong, E. W. S., & Chiu, D. K. W. (2023). (in press). Information security and technical issues of cloud storage services: A qualitative study on university students in Hong Kong. *Library Hi Tech*. Advance online publication. doi:10.1108/LHT-11-2022-0533

Hunt, J. M., Goodner, R. E., & Jay, A. (2020). *Comparing male and female student responses on MIT maker survey: understanding the implications and strategies for more inclusive spaces.* https://ijamm.pubpub.org/pub/m71isalo/release/1?reading Collection=ee0f734d

Jiang, M., Lam, A. H. C., Chiu, D. K. W., & Ho, K. K. W. (2023). (in press). Social media aids for business learning: A quantitative evaluation with the 5E instructional model. *Education and Information Technologies*. Advance online publication. doi:10.100710639-023-11690-z PMID:37361768

Jiang, T., Lo, P., Cheuk, M. K., Chiu, D. K. W., Chu, M. Y., Zhang, X., Zhou, Q., Liu, Q., Tang, J., Zhang, X., Sun, X., Ye, Z., Yang, M., & Lam, S. K. (2019). 文化新語:兩岸四地傑出圖書館、檔案館及博物館傑出工作者訪談 [New cultural dialog: Interviews with outstanding librarians, archivists, and curators in Greater China]. Systech Publications.

Jiang, X., Chiu, D. K. W., & Chan, C. T. (2023). Application of the AIDA model in social media promotion and community engagement for small cultural organizations: A case study of the Choi Chang Sau Qin Society. In M. Dennis & J. Halbert (Eds.), *Community engagement in the online space* (pp. 48–70). IGI Global. doi:10.4018/978-1-6684-5190-8.ch004

Johnson, E. D. M. (2017). The right place at the right time: creative spaces in libraries. *The Future of Library Space, 36*, 1-35. doi:10.1108/S0732-067120160000036001

Julian, K. D., & Parrott, D. J. (2017). Makerspaces in the library: Science in a student's hands. *Journal of Learning Spaces*, 6(2), 13–21. https://eric.ed.gov/?id=EJ1152687

Kee, H. C. L., & Chiu, D. K. W. (2023). Building social capital in major U.S. public libraries. In D. K. W. Chiu & K. K. W. Ho (Eds.), *Emerging technology-based services and systems in libraries, educational institutions, and non-profit organizations*. IGI Global.

Koh, K., & Abbas, J. (2015). Competencies for information professionals in learning labs and makerspaces. *Journal of Education for Library and Information Science*, 56(2), 114–129. Advance online publication. doi:10.12783/issn.2328-2967/56/2/3

Kroski, E. (Ed.). (2017). *The makerspace librarian's sourcebook*. Facet Publishing.

Lam, A. H. C., Ho, K. K. W., & Chiu, D. K. W. (2023). Instagram for student learning and library promotions? A quantitative study using the 5E Instructional Model. *Aslib Journal of Information Management*, 75(1), 112–130. doi:10.1108/AJIM-12-2021-0389

Lam, E. T. H., Au, C. H., & Chiu, D. K. W. (2019). Analyzing the use of Facebook among university libraries in Hong Kong. *Journal of Academic Librarianship*, 45(3), 175–183. doi:10.1016/j.acalib.2019.02.007

Lau, K. P., Chiu, D. K. W., Ho, K. K. W., Lo, P., & See-To, E. W. K. (2017). Educational usage of mobile devices: Differences between postgraduate and undergraduate students. *Journal of Academic Librarianship*, *43*(3), 201–208. doi:10.1016/j.acalib.2017.03.004

Lau, K. S. N., Lo, P., Chiu, D. K. W., Ho, K. K. W., Jiang, T., Zhou, Q., Percy, P., & Allard, B. (2020). Library and learning experiences turned mobile: A comparative study between LIS and non-LIS students. *Journal of Academic Librarianship*, *46*(2), 102103. Advance online publication. doi:10.1016/j.acalib.2019.102103

Law, T. Y., Leung, F. C. W., Chiu, D. K. W., Lo, P., Lung, M. M.-W., Zhou, Q., Xu, Y., Lu, Y., & Ho, K. K. W. (2019). Mobile learning usage of LIS students in Mainland China. *International Journal of Systems and Service-Oriented Engineering*, *9*(2), 12–34. doi:10.4018/IJSSOE.2019040102

Lee, R. J. (2017). Campus-library collaboration with makerspaces. *Public Services Quarterly*, *13*(2), 108–116. doi:10.1080/15228959.2017.1303421

Lei, S. Y., Chiu, D. K. W., Lung, M. M., & Chan, C. T. (2021). Exploring the aids of social media for musical instrument education. *International Journal of Music Education*, *39*(2), 187–201. doi:10.1177/0255761420986217

Leung, T. N., Chiu, D. K. W., Ho, K. K. W., & Luk, C. K. L. (2022). User perceptions, academic library usage, and social capital: A correlation analysis under COVID-19 after library renovation. *Library Hi Tech*, *40*(2), 304–322. doi:10.1108/LHT-04-2021-0122

Leung, T. N., Hui, Y. M., Luk, C. K. L., Chiu, D. K. W., & Ho, K. K. W. (2022). Evaluating Facebook as aids for learning Japanese: Learners' perspectives. *Library Hi Tech*. Advance online publication. doi:10.1108/LHT-11-2021-0400

Li, K. K., & Chiu, D. K. W. (2022). A worldwide quantitative review of the iSchools' archival education. *Library Hi Tech*, *40*(5), 1497–1518. doi:10.1108/LHT-09-2021-0311

Li, S., Chiu, D. K. W., Kafeza, E., & Ho, K. K. W. (2023). Social media analytics for non-governmental organizations: A case study of Hong Kong Next Generation Arts. In Z. Sun (Ed.), *Handbook of research on driving socioeconomic development with big data* (pp. 277–295). IGI Global. doi:10.4018/978-1-6684-5959-1.ch013

Li, S., Xie, Z., Chiu, D. K. W., & Ho, K. K. W. (2023). Sentiment analysis and topic modeling regarding online classes on the Reddit Platform: Educators versus learners. *Applied Sciences (Basel, Switzerland)*, *13*(4), 2250. Advance online publication. doi:10.3390/app13042250

Li, S. M., Lam, A. H. C., & Chiu, D. K. W. (2023). Digital transformation of ticketing services: A value chain analysis of POPTICKET in Hong Kong. In J. D. Santos, I. V. Pereira, & P. B. Pires (Eds.), *Management and marketing for improved retail competitiveness and performance*. IGI. Global.

Lin, C.-H., Chiu, D. K. W., & Lam, K. T. (2022). Hong Kong academic librarians' attitudes towards robotic process automation. *Library Hi Tech*. Advance online publication. doi:10.1108/LHT-03-2022-0141

Liu, Q., Lo, P., Zhou, Q., Chiu, D. K. W., & Cheuk, M. K. (2022). 走進大專院校圖書館: 圖書館員視角下的大中華區高等教育 [Why the Library? The Role of Librarians in the Higher Education Systems of Greater China]. City University Press.

Liu, Y., Chiu, D. K. W., & Ho, K. K. W. (2023). Short-form videos for public library marketing: Performance analytics of Douyin in China. *Applied Sciences (Basel, Switzerland)*, *13*(6), 3386. Advance online publication. doi:10.3390/app13063386

Lo, P., Allard, B., Anghelescu, H. G. B., Xin, Y., Chiu, D. K. W., & Stark, A. J. (2020). Transformational leadership practice in the world's leading academic libraries. *Journal of Librarianship and Information Science*, *52*(4), 972–999. doi:10.1177/0961000619897991

Lo, P., Chan, H. H. Y., Tang, A. W. M., Chiu, D. K. W., Cho, A., See-To, E., Ho, K. K. W., He, M., Kenderdine, S., & Shaw, J. (2019). Visualising and revitalising traditional Chinese martial arts: Visitors' engagement and learning experience at the 300 years of Hakka KungFu. *Library Hi Tech*, *37*(2), 269–288. doi:10.1108/LHT-05-2018-0071

Lo, P., Chiu, D. K. W., Cho, A., & Allard, B. (2018). *Conversations with leading academic and research library directors: International perspectives on library management*. Chandos Publishing.

Lo, P., Chiu, D. K. W., & Chu, W. (2013). Modeling your college library after a commercial bookstore? The Hong Kong Design Institute Library experience. *Community & Junior College Libraries*, *19*(3-4), 59–76. doi:10.1080/02763915.2014.915186

Lo, P., Cho, A., Law, B. K.-K., Chiu, D. K. W., & Allard, B. (2017). Progressive trends in electronic resources management among academic libraries in Hong Kong. *Library Collections, Acquisitions & Technical Services*, *40*(1-2), 28–37. doi:10.1080/14649055.2017.1291243

Long, J., & Hicks, J. (2020). *Makerspaces for adults: Best practices and great projects*. Rowman & Littlefield.

Lu, S. S., Tian, R., & Chiu, D. K. W. (2023). (in press). Why do people not attend public library programs in the current digital age? A mix method study in Hong Kong. *Library Hi Tech*. Advance online publication. doi:10.1108/LHT-04-2022-0217

Mahey, M., Al-Abdulla, A., Ames, S., Bray, P., Candela, G., Chambers, S., Derven, C., Dobreva-McPherson, M., Gasser, K., Karner, S., Kokegei, K., Laursen, D., Potter, A., Straube, A., Wagner, S., & Wilms, L. (2019). *Open a GLAM Lab*. Qatar University Press.

Mak, M. Y. C., Poon, A. Y. M., & Chiu, D. K. W. (2022). Using social media as learning aids and preservation: Chinese martial arts in Hong Kong. In S. Papadakis & A. Kapaniaris (Eds.), *The digital folklore of cyberculture and digital humanities* (pp. 171–185). IGI Global., doi:10.4018/978-1-6684-4461-0.ch010

Mann, H. B., & Whitney, D. R. (1947). On a test of whether one of two random variables is stochastically larger than the other. *Annals of Mathematical Statistics*, *18*(1), 50–60. https://www.jstor.org/stable/2236101. doi:10.1214/aoms/1177730491

Maric, J. (2018). The gender-based digital divide in maker culture: Features, challenges and possible solutions. *Journal of Innovation Economics and Management*, *3*(27), 147–168. doi:10.3917/jie.027.0147

Melo, M., & Nichols, J. T. (Eds.). (2020). *Re-making the library makerspace: Critical theories, reflections, and practices*. Library Juice Press.

Meng, Y., Chu, M. Y., & Chiu, D. K. W. (2023). The impact of COVID-19 on museums in the digital era: Practices and challenges in Hong Kong. *Library Hi Tech*, *41*(1), 130–151. doi:10.1108/LHT-05-2022-0273

Michalak, R., & Rysavy, M. D. T. (2019). Academic libraries in 2018: A comparison of makerspaces within academic research libraries. *Supporting Entrepreneurship and Innovation*, *40*, 67–88. doi:10.1108/S0732-067120190000040008

Moersch, C. M. (2014). *Improving achievement with digital age best practices*. Corwin. doi:10.4135/9781506374482

Moorefield-Lang, H. (2015). Change in the making: makerspaces and the ever-changing landscape of libraries. *TechTrends, 59*(3), 107-112. doi:10.1007/s11528-015-0860-z

Moorefield-lang, H. (2019). Lesson learned: Intentional implementation of second makerspaces. *RSR. Reference Services Review*, *47*(1), 37–47. Advance online publication. doi:10.1108/RSR-07-2018-0058

Ng, T. C. W., Chiu, D. K. W., & Li, K. K. (2022). Motivations of choosing archival studies as major in the i-Schools: Viewpoint between two universities across the Pacific Ocean. *Library Hi Tech*, *40*(5), 1483–1496. doi:10.1108/LHT-07-2021-0230

Ni, J., Rhim, A. H. R., Chiu, D. K. W., & Ho, K. K. W. (2022). Information search behavior among Chinese self-drive tourists in the smartphone era. *Information Discovery and Delivery*, *50*(3), 285–296. doi:10.1108/IDD-05-2020-0054

Ni, Y., Lam, A. H. C., & Chiu, D. K. W. (2023). Leveraging online communities for building social capital in university libraries: A case study of Fudan University Medical Library. In S. H. Jarvie, & C. Metz (Eds.), Balance and boundaries in creating meaningful relationships in online higher education. IGI Global.

Okpala, H. N. (2016). Making a makerspace case for academic libraries in Nigeria. *New Library World*, *117*(9/10), 568–586. doi:10.1108/NLW-05-2016-0038

Osawaru, K. E., Dime, A. I., & Okonjo, E. H. (2020). The right time for makerspaces in Nigerian academic libraries: Perceived benefits and challenges. *International Journal on Integrated Education*, *3*(10), 103–115. doi:10.31149/ijie.v3i10.694

Radniecki, T. (2017). Supporting 3D modeling in the academic library. *Library Hi Tech*, *35*(2), 240–250. doi:10.1108/LHT-11-2016-0121

Radniecki, T., & Winterman, M. (2020). Leveraging student expertise for niche services. *RSR. Reference Services Review*, *48*(2), 287–306. doi:10.1108/RSR-11-2019-0083

Sidorko, P., & Lee, L. (2014). JURA: A collaborative solution to hong kong academic libraries storage challenge. *Library Management*, *35*(1), 46–68. doi:10.1108/LM-03-2013-0025

Steele, K. M., Cakmak, M., & Blaser, B. (2018). Accessible making: Designing a makerspace for accessibility. *International Journal of Designs for Learning*, *9*(1), 114–121. doi:10.14434/ijdl.v9i1.22648

Stover, M., Jefferson, C., & Santos, I. (2019). Innovation and creativity: A new facet of the traditional mission for university libraries. *Supporting Entrepreneurship and Innovation*, *40*, 135–151. doi:10.1108/S0732-067120190000040006

Suen, R. L. T., Chiu, D. K. W., & Tang, J. K. T. (2020). Virtual reality services in academic libraries: Deployment experience in Hong Kong. *The Electronic Library*, *38*(4), 843–858. doi:10.1108/EL-05-2020-0116

Sun, X., Chiu, D. K. W., & Chan, C. T. (2022). Recent digitalization development of buddhist libraries: A comparative case study. In S. Papadakis & A. Kapaniaris (Eds.), *The digital folklore of cyberculture and digital humanities* (pp. 251–266). IGI Global. doi:10.4018/978-1-6684-4461-0.ch014

Sung, Y. Y. C., & Chiu, D. K. W. (2022). E-book or print book: Parents' current view in Hong Kong. *Library Hi Tech*, *40*(5), 1289–1304. doi:10.1108/LHT-09-2020-0230

Tsang, A. L. Y., & Chiu, D. K. W. (2022). Effectiveness of virtual reference services in academic libraries: A qualitative study based on the 5E Learning Model. *Journal of Academic Librarianship*, *48*(4), 102533. Advance online publication. doi:10.1016/j.acalib.2022.102533

Tse, H. L. T., Chiu, D. K. W., & Lam, A. H. C. (2022). From reading promotion to digital literacy: An analysis of digitalizing mobile library services with the 5E Instructional Model. In A. P. Almeida & S. Esteves (Eds.), *Modern reading practices and collaboration between schools, family, and community* (pp. 239–256). IGI Global. doi:10.4018/978-1-7998-9750-7.ch011

Vuorikari, R., Ferrari, A., & Punie, Y. (2019). *Makerspaces for education and training: Exploring future implications for Europe.* https://publications.jrc.ec.europa.eu/repository/handle/JRC117481

Wai, I. S. H., Ng, S. S. Y., Chiu, D. K. W., Ho, K. K. W., & Lo, P. (2018). Exploring undergraduate students' usage pattern of mobile apps for education. *Journal of Librarianship and Information Science*, *50*(1), 34–47. doi:10.1177/0961000616662699

Wang, F., Wang, W., Wilson, S., & Ahmed, N. (2016). The state of library makerspaces. *International Journal of Librarianship*, *1*(1), 2–16. doi:10.23974/ijol.2016.vol1.1.12

Wang, J., Deng, S., Chiu, D. K. W., & Chan, C. T. (2022). Social network customer relationship management for orchestras: A case study on Hong Kong Philharmonic Orchestra. In N. B. Ammari (Ed.), *Social customer relationship management (Social-CRM) in the era of Web 4.0* (pp. 250–268). IGI Global. doi:10.4018/978-1-7998-9553-4.ch012

Wang, P., Chiu, D. K. W., Ho, K. K. W., & Lo, P. (2016). Why read it on your mobile device? Change in reading habit of electronic magazines for university students. *Journal of Academic Librarianship*, *42*(6), 664–669. doi:10.1016/j.acalib.2016.08.007

Wang, W., Lam, E. T. H., Chiu, D. K. W., Lung, M. M., & Ho, K. K. W. (2021). Supporting higher education with social networks: Trust and privacy vs perceived effectiveness. *Online Information Review*, *45*(1), 207–219. doi:10.1108/OIR-02-2020-0042

Webb, K. K. (2018). *Development of creative spaces in academic libraries: a decision maker's guide*. Chandos Publishing.

Willingham, T., & De Boer, J. (2015). *Makerspaces in libraries*. Rowman & Littlefield.

Wong, A., & Partridge, H. (2016). Making as learning: Makerspaces in universities. *Australian Academic and Research Libraries*, *47*(3), 143–159. doi:10.1080/00048 623.2016.1228163

Wong, A. K.-k., & Chiu, D. K. W. (2023). Digital transformation of museum conservation practices: A value chain analysis of public museums in Hong Kong. In R. Pettinger, B. B. Gupta, A. Roja, & D. Cozmiuc (Eds.), *Handbook of research on the digital transformation digitalization solutions for social and economic needs* (pp. 226–242). IGI Global. doi:10.4018/978-1-6684-4102-2.ch010

Wong, K. C., & Chiu, D. K. W. (2023). Promoting the use of electronic resources of international schools: A case study of ESF King George V School in Hong Kong. In E. Meletiadou (Ed.), *Handbook of research on redesigning teaching, learning, and assessment in the digital era* (pp. 123–143). IGI Global. doi:10.4018/978-1-6684-8292-6.ch007

Wu, M., Lam, A. H. C., & Chiu, D. K. W. (2023). Transforming and promoting reference services with digital technologies: A case study on Hong Kong Baptist University Library. In B. Holland (Ed.), Handbook of research on advancements of contactless technology and service innovation in library and information science (pp. 128-145). IGI Global. doi:10.4018/978-1-6684-7693-2.ch007

Xie, Z., Chiu, D. K. W., & Ho, K. K. W. (2023). The role of social media as aids for accounting education and knowledge sharing: Learning effectiveness and knowledge management perspectives in Mainland China. *Journal of the Knowledge Economy*. Advance online publication. doi:10.100713132-023-01262-4

Xie, Z., Wong, G. K. W., Chiu, D. K. W., & Lei, J. (2023). Bridging K-12 Mathematics and computational thinking in the Scratch community: Implications drawn from a creative learning context. *IT Professional*, *25*(2), 64–70. doi:10.1109/MITP.2023.3243393

Xu, C., Lam, A. H. C., Gao, X., & Chiu, D. K. W. (2023). Antique bookstores marketing strategies as urban cultural landmark: A case analysis for Suzhou Antique Bookstore. In M. A. Rodrigues & M. A. M. Carvalho (Eds.), *Exploring niche tourism business models, marketing, and consumer experience*. IGI Global.

Xue, Y., Lam, A. H. C., & Chiu, D. K. W. (2023). Redesigning library information literacy education with the BOPPPS model: A case study of the HKUST Library. In R. Taiwo, B. Idowu-Faith, & S. Ajiboye (Eds.), *Transformation of higher education through institutional online spaces* (pp. 279–296). IGI Global. doi:10.4018/978-1-6684-8122-6.ch017

Yang, Z., Zhou, Q., Chiu, D. K. W., & Wang, Y. (2022). Exploring the factors influencing continuance usage intention of academic social network sites. *Online Information Review, 46*(7), 1225–1241. doi:10.1108/OIR-01-2021-0015

Yew, A., Chiu, D. K. W., Nakamura, Y., & Li, K. K. (2022). A quantitative review of LIS programs accredited by ALA and CILIP under contemporary technology advancement. *Library Hi Tech, 40*(6), 1721–1745. doi:10.1108/LHT-12-2021-0442

Yi, F., & Baumann, M. (2020). Guiding principles for designing an accessible, inclusive, and diverse library makerspace. *International Journal of Academic Makerspaces and Making*. https://ijamm.pubpub.org/pub/op9ltdpf

Yi, Y., & Chiu, D. K. W. (2023). Public information needs during the COVID-19 outbreak: A qualitative study in mainland China. *Library Hi Tech, 41*(1), 248–274. doi:10.1108/LHT-08-2022-0398

Yip, K. H. T., Lo, P., Ho, K. K. W., & Chiu, D. K. W. (2021). Adoption of mobile library apps as learning tools in higher education: A tale between Hong Kong and Japan. *Online Information Review, 45*(2), 389–405. doi:10.1108/OIR-07-2020-0287

Yip, T., Chiu, D. K. W., Cho, A., & Lo, P. (2019). Behavior and informal learning at night in a 24-hour space: A case study of the Hong Kong Design Institute Library. *Journal of Librarianship and Information Science, 51*(1), 171–179. doi:10.1177/0961000617726120

Yu, H. H. K., Chiu, D. K. W., & Chan, C. T. (2023). Resilience of symphony orchestras to challenges in the COVID-19 era: Analyzing the Hong Kong Philharmonic Orchestra with Porter's five force model. In W. J. Aloulou (Ed.), *Handbook of research on entrepreneurship and organizational resilience during unprecedented times* (pp. 586–601). IGI Global. doi:10.4018/978-1-6684-4605-8.ch026

Yu, H. Y., Tsoi, Y. Y., Rhim, A. H. R., Chiu, D. K. W., & Lung, M. M.-W. (2022). Changes in habits of electronic news usage on mobile devices in university students: A comparative survey. *Library Hi Tech*, *40*(5), 1322–1336. doi:10.1108/LHT-03-2021-0085

Yu, P. Y., Lam, E. T. H., & Chiu, D. K. W. (2023). Operation management of academic libraries in Hong Kong under COVID-19. *Library Hi Tech*, *41*(1), 108–129. doi:10.1108/LHT-10-2021-0342

Zhang, J., Lam, A. H. C., & Chiu, D. K. W. (2023). Evaluating the effectiveness of learning commons as third space with the 5E usability model: The case of Hong Kong University of Science and Technology Library. In C. Kaye & J. H. Writer (Eds.), *Third-Space Exploration in Education*. IGI Global.

Zhang, X., Lo, P., So, S., Chiu, D. K. W., Leung, T. N., Ho, K. K. W., & Stark, A. (2021). Medical students' attitudes and perceptions towards the effectiveness of mobile learning: A comparative information-need perspective. *Journal of Librarianship and Information Science*, *53*(1), 116–129. doi:10.1177/0961000620925547

Zhang, Y., Lo, P., So, S., & Chiu, D. K. W. (2020). Relating library user education to business students' information needs and learning practices: A comparative study. *RSR. Reference Services Review*, *48*(4), 537–558. doi:10.1108/RSR-12-2019-0084

Zhou, J., Lam, E., Au, C. H., Lo, P., & Chiu, D. K. W. (2022). Library café or elsewhere: Usage of study space by different majors under contemporary technological environment. *Library Hi Tech*, *40*(6), 1567–1581. doi:10.1108/LHT-03-2021-0103

Chapter 10
Building Social Capital in Contemporary Major U.S. Public Libraries:
Leading Information Services and Beyond

Helen Chui Ling Kee
The University of Hong Kong, Hong Kong

Mimi Mei Wa Chan
The University of Hong Kong, Hong Kong

Dickson K. W. Chiu
ⓘ https://orcid.org/0000-0002-7926-9568
The University of Hong Kong, Hong Kong

ABSTRACT

As the US is a highly ethnically diverse country and the origin of public libraries, this chapter explores how US public libraries construct social capital for the public via various services and activities. This chapter selected nine cases of US public libraries with interviews of their management for analysis of library services and activities related to empowering library users and building their social capital. The findings indicate that libraries are community meeting places connecting community members through library programs, activities, and information services as community educational institutions that empower underprivileged people and new immigrants by satisfying their information needs. The process contributes to the social development of library users and their communities, building social capital. Scant studies summarize the good practice of renowned public libraries in building social capital. This chapter contributes to understanding the good practice of US public libraries as a creator of social capital, serving as a reference for public libraries worldwide.

DOI: 10.4018/978-1-6684-8671-9.ch010

INTRODUCTION

With more than a century of history, many public libraries have been transformed from places providing physical books for reading to community builders and developers of social capital (American Library Association, 2017). During the past decade, the discussion of public libraries as a creator of social capital for library users has been ubiquitous in Library and Information Services (LIS) literature (Ferguson, 2012; Leung et al., 2022; Lu et al., 2023). Technology has become an integral part of contemporary society, transforming the way individuals interact and engage with one another. In the United States, public libraries have harnessed the power of technology to create and sustain social capital within communities.

This chapter explores how US public libraries construct social capital for library users via various services and activities through nine cases of US public libraries in the contemporary globalized information environment. While there are case studies for some individual libraries, scant research summarizes good practices of renowned public libraries to build social capital. This research can provide a reference for public libraries in other countries to build social capital for library users to fill the gap.

This research chooses US libraries as the study cases as public libraries originated in the US. The Peterborough Town Library, established in 1833, is the first tax-supported public library globally (Peterborough Town Library). Another example is the Boston Public Library. Established in 1848, it is considered the US's earliest free major municipal library (Boston Public Library, 2019).

Furthermore, it is worthwhile to explore how public libraries in the US, a highly ethnically diverse country and the origin of public libraries, offer library services and activities to empower library users of different ethnicities and build their social capital. As such, this study aims to answer the following research questions (RQ):

RQ1. What are the library services and activities provided for building the social capital of library users?

RQ2. How do library managers lead and manage these library services and activities to empower library users and build social capital?

LITERATURE REVIEW

Social Capital Theory

Social capital is a concept from sociology. Its earliest reported use can be traced to 1916 when Lyda J. Hanifan first wrote about rural school community centers (Ferguson, 2012). It refers to "those tangible assets [that] count for most in the daily lives of people: namely goodwill, fellowship, sympathy, and social intercourse among the individuals and families who make up a social unit" (Hanifan, 1920, p.78). However, the concept did not draw much attention in the academic field until the late 1980s (Halpern, 2005). At that time, sociologists such as Bourdieu (1992) pointed out that the concepts of social capital and cultural capital were ignored in economic orthodoxy, which only "limited itself to the study of a narrow band of *practices* that were socially recognized as *economic*" (p. 119). He (1992) defined social capital as below:

Social capital is the sum of the resources, actual or virtual, that accrue to an individual or a group by virtue of possessing a durable network of more or less institutionalized relationships of mutual acquaintance and recognition. Acknowledging that capital can take a variety of forms is indispensable to explain the structure and dynamics of differentiated societies. (p. 119)

American sociologist Coleman (1988) had similar views to Bourdieu and proposed a broad definition of social capital:

Social capital is defined by its function. It is not a single entity but a variety of different entities with two elements in common: they all consist of some aspect of social structures, and they facilitate certain actions of actors - whether persons or corporate actors - within that structure. Like other forms of capital, social capital is productive, making possible the achievement of certain ends that, in its absence, would not be possible. (p. 96)

Others have also tried to explain what social capital is. Halpern (2005) argued that social capital refers to different social networks, including bonds like family and associations, that keep the networks together. Besides, Putnam (2002) suggested that social capital refers to "social networks and the norms of reciprocity associated with them" (p.3). The World Bank (2001) suggested that "the social capital of society includes the institutions, the relationships, the attitudes and values that govern interactions among people and contribute to economic and social development" (p.4). A widely quoted definition of social capital is from Putnam (1995), who proposed that "social capital, in short, refers to social connections and the attendant norms and trust" (p.664-5). Vårheim, a scholar in the LIS field, suggested that "social capital is a strong predictor of individual health and human well-being" (2011, p.1).

However, various definitions of social capital may not facilitate understanding the concept. Halpern (2005) thus proposed three main components of social capital for more natural understanding. Firstly, it consists of a *network* that its density and closure can further characterize. The remaining components include a cluster of *social norms and values* shared by group members and *sanctions* that refer to punishments and rewards that help to maintain the norms and network.

Social capital has been classified into different forms by scholars. Putnam (2002) classified social groupings as "bonding social capital," through which similar people come together, and "bridging social capital," which gathers different types of people. Putnam (2002) also distinguished "inward-looking" groups, such as chambers of commerce, from "outward-looking groups" (environmental movement is an example). Moreover, Hart (2013) differentiated horizontal social bonds from vertical social bonds, whereas the former refers to groupings that only work "for their narrow self-interest."

Social Capital and Public Libraries

Social capital has been a widely discussed concept not only in the field of Sociology but also in LIS literature. Public institutions, especially public libraries, are considered the "most efficient generators of social capital" (Khoir, Du, Davison, & Koronios, 2017, p.34). Johnson (2010) explained that public libraries' openness to community members of all ages, genders, and ethnicities makes them a prominent generator of social capital. Much research literature has discussed such roles of public libraries (Chen & Ke, 2017; Ferguson, 2012; Hart, 2013; Johnson, 2011, 2015; Johnson & Griffis, 2014; Khoir et al., 2017; Kranich, 2001; Vårheim, 2007, 2011).

To generate social capital, public libraries are often converted into spaces for social activities (Khoir et al., 2017; Lu et al., 2023; Jiang et al., 2019; Lo et al., 2021; Chiu & Ho, 2022a; 2022b) while providing information services through various means and channels. For example, they offer training in digital literacy skills to facilitate communication between citizens and access to information, provide a place for citizens to meet and socialize, and facilitate community involvement to build social capital. Libraries also offer a social space that makes people feel safe to meet each other (Vårheim, Steinmo, & Ide, 2008; Yip et al., 2019). Furthermore, public libraries reach out to the community to promote their services to members of the public who do not enjoy the benefits of library services, encouraging more citizens to participate in social activities in libraries (Ferguson, 2012; Lu et al., 2023).

Vårheim (2007) discussed the strategies of how public libraries can build social capital. Firstly, libraries can cooperate with voluntary organizations to increase participation in these organizations, enhancing the public's involvement in community activities. Secondly, libraries can develop their function as informal meeting spaces for citizens to facilitate social gatherings. Thirdly, libraries can develop social capital by providing general services to the public.

For the reasons and evidence of whether public libraries can generate social capital, Stevens and Campbell (2006) believed that libraries' role in promoting lifelong learning makes them cultural agencies, which are the prerequisite for creating social capital. In this information society, there is a danger that people without adequate knowledge of Information Technology (IT) for Internet access are excluded from society. Therefore, by offering information literacy courses and Internet services, public libraries can connect different citizens and strengthen communities (Butcher & Street, 2009). A previous study by Johnson and Griffis (2010) concluded that there is "a close relationship between the use of the public library and high levels of community social capital" (p.188), which demonstrated the ability of public libraries to generate social capital.

Public libraries are often the first contact point in the community for new immigrants to get information about social services, schools, and language classes. In this connection, public libraries provide "an interim stock of social capital" (Johnson, 2010, p. 150).

Public libraries play a crucial and multifaceted role in fostering the development of social capital within local communities. As James (2018) noted, these libraries serve as inclusive community hubs that provide physical spaces for individuals to gather, interact, and engage in various social activities, fostering a sense of belonging and creating support networks. Through community events, book clubs, and workshops, libraries facilitate face-to-face interactions that enable community members to connect, share experiences, and cultivate social relationships (James, 2018).

In addition, public libraries actively promote information and digital literacy, which are fundamental skills for active participation in today's society (Sharma, 2019). By offering free access to books, educational resources, and digital technologies, libraries empower individuals to enhance their knowledge, skills, and capacity to engage with others. Literacy programs, computer training, and internet access provided by libraries bridge the digital divide, ensuring equitable opportunities for individuals to access information, connect with others, and participate in the digital world. This inclusive approach enhances individual empowerment and contributes to forming social capital by enabling informed, connected, and involved community members (Sharma, 2019).

Furthermore, public libraries serve as valuable repositories of local history, culture, and traditions (White, 2012). By preserving and sharing community heritage, libraries contribute to cultivating social capital by fostering a collective identity and a sense of pride among community members. These libraries provide spaces for exploring and celebrating shared cultural experiences, strengthening social connections and intergenerational bonds within the community.

In summary, public libraries are pivotal in developing social capital within local communities, serving as inclusive gathering places, promoting information and digital literacy, and preserving community heritage. These multifaceted efforts create a vibrant social fabric, fostering connections, empowerment, and a sense of belonging among community members.

Technology-based Services in Libraries

Technology plays a crucial role in enhancing the capacity of public libraries to build social capital within local communities. Numerous studies highlight the importance of technology-based services in fostering social connections, user empowerment, and information access. These services contribute to social capital formation by enabling individuals to interact online, access digital resources, and share knowledge.

One of the key functions of libraries is to provide access to information. Technology is essential in expanding access by enabling digital collections, online databases, and electronic resources. Public libraries leverage technologies such as online catalogs, digital repositories, and digital lending platforms to ensure that users can access a wide range of materials anywhere, anytime (Ding et al., 2021; Lo & Chiu, 2015; Lo et al., 2015, 2017; Suen et al., 2020; Sun et al., 2022; Wang et al., 2016; K.C. Wong & Chiu, 2023; Zuo et al., 2023). These digital platforms offer convenience and flexibility, allowing users to browse, borrow, and engage with resources remotely. Thus, people have changed their information and learning consumption habits to electronic and online resources, particularly led by the trend of the younger generation (Chan & Chiu, 2022; Dai & Chiu, 2023; Guo et al., 2022; Sung & Chiu, 2022; Yi et al., 2023; Wai et al., 2018; S.W.S. Wong & Chiu, 2023; Yu et al., 2022; Zhang et al., 2021).

Further, the COVID-19 pandemic has transformed many services into intelligent and online services (Cheng et al., 2022; Chin & Chiu, 2023; Q. Li et al., 2023; S.M. Li et al., 2023; Lin et al., 2022; Lo, Allard, Anghelescu, Xin, Chiu, 2020; Tsang & Chiu, 2022; A. K.-k. Wong & Chiu, 2023; Wu et al., 2023; Xue et al., 2023), especially supported by ubiquitous mobile Internet services, apps, and devices (Dukic et al., 2015; Ezeamuzie et al., 2023; Fan et al., 2020; Fung et al., 2016; Gong et al., 2017; Hui et al., 2023; Lau et al., 2017, 2020; Law et al., 2019; Ni et al., 2022; Yip et al., 2021).

According to ALA President Leslie Burger, libraries with modern technology are crucial for fostering thriving communities. Libraries serve as vital access points for millions of Americans, connecting them to e-government services, online homework assistance, and employment resources. However, without increased federal, state, and local funding, many libraries may struggle to sustain the technology-based services that patrons rely on (Davis & Bertot, n.d.).

The concept of "digital social capital" has emerged to describe the social connections, trust, and support built through online interactions facilitated by technology (Lightfoot, 2016). Digital platforms and tools provided by libraries, such as social media platforms, online discussion forums, and virtual communities, offer opportunities for individuals to connect, collaborate, and share knowledge beyond the physical library space (Cheng et al., 2020, 2023; Fong et al., 2020; Lam et al., 2019, 2023; S. Li, Xie, Chiu & Ho, 2023; Lo et al., 2014; Liu et al., 2023; Ni et al., 2023; Thompson et al., 2014; Yang et al., 2022), forming communities of practice (Chung et al., 2020; Ho et al., 2018; Lei et al., 2021; Leung et al., 2023; Jiang, Lam, Ho, Chiu, 2023; Mak et al., 2022; Xie, Chiu, & Ho, 2023; Xie, Wong, Chiu, & Lei, 2023; Wang et al., 2021; Zhang et al., 2020).

Studies have shown that library technology-based initiatives empower users by enhancing digital literacy skills and providing access to online resources (Claark & Perry, 2015; Tse et al., 2022). Access to technology and digital resources enables individuals to expand their information networks, acquire new skills, and actively participate in digital society, thus enhancing their social capital (Lybeck, Koiranen, & Koivula, 2023).

Furthermore, library technology-based services profoundly impact community development, fostering social inclusion and empowering individuals. By providing equal access to information and digital resources, libraries reduce the digital divide and promote digital equity within communities (Gerner Nielsen, 2014). Moreover, libraries serve as community technology hubs, offering free internet access, computer facilities, and training programs that enable users to apply for jobs, complete online forms, and participate in online learning initiatives (Khosrow-Pour, 2014; Lu et al., 2023). These services enhance users' digital skills and facilitate active engagement in the digital society.

Overall, technology-based services in libraries have the potential to enhance social capital formation by enabling individuals to connect, collaborate, and access information in digital spaces. These services empower users, facilitate community engagement, and expand social networks, fostering a sense of belonging, trust, and shared knowledge within local communities.

DATA ANALYSES

This study employed a qualitative method of a case study approach, as this is more appropriate to study why a research subject (public libraries in the US in this case) behave in a particular way and to gain insights and a deeper understanding of a problem (Creswell & Guetterman, 2019). Nine cases of US public libraries from nine interviews with leaders of US public libraries in the book *World´s Leading National, Public, Monastery and Royal Library Directors – Leadership, Management, Future of Libraries* by Lo, Cho, and Chiu (2017) were analyzed. The book only publishes the original conversion data of the interviews verbose in each of the chapters without any analysis of the details of the content, but just a brief concluding chapter. Thus, the book is just the raw data without any detailed analysis. As interviews of renowned librarians of renowned major public libraries are difficult to obtain, a systematic analysis according to an established conceptual model provides important insights into how systemically good library practices and strategies can be designed and implemented.

The content analysis method for the comparative case study was adopted because it helped gather detailed information from the interviews, which cannot be otherwise observed easily, especially for how the libraries generate social capital and why they have succeeded (Connaway, 2017). Further, as the programs and activities showcased in these interviews were probably those they were highly proud of and considered the best, it helped the researchers select good practices from their first-hand experience instead of their subjective judgment.

Research data was collected through document analysis from the book, as each chapter contains the interview conversations in verbose. Details of library services and activities related to empowering library users and building their social capital for the nine US public libraries were analyzed to support the assertion that public libraries develop social capital for library users. The Appendix lists the names of libraries, their locations, and their various programs and activities relevant to this research.

The library services and activities offered by the nine US public libraries that contribute to generating social capital for library users were highlighted. According to the interviews, various types of community members in different regions of the US, including new immigrants, underprivileged children and elderly, homeless people, and unemployed people, are identified as users of public libraries who can benefit from the generation of social capital in libraries (Lo, Cho, & Chiu, 2017).

Apart from library programs that are tailor-made for these community members, some activities and services target all users instead of specific groups. Some of these library programs for all users aim to build the community. In contrast, many are non-traditional library services, and some are related to learning increasingly important areas of Science, Technology, Engineering, and Mathematics (STEM). The following section explains how these library programs enable the creation of social capital.

Library Programs Targeting New Immigrants

The New York Public Library provides English language and literacy classes for new immigrants with a low education but without many learning resources. They rely so much on these classes that the library decided to expand its Adult Literacy Center to become the third-largest provider of English as a second language (ESL) instruction to immigrants in New York. These new immigrants may not get used to online learning (Yau et al., 2023; Cheng et al., 2022). Some US public libraries offer similar services for these immigrants. The Boston Public Library hosts ESL conversation groups to let them develop English oral skills. In contrast, the Seattle Public Library System and McAllen Public Library both provide ESL courses and preparation classes for related exams.

Los Angeles's population originates worldwide, including Korea, China, Japan, and Armenia. Thus, one of the major initiatives of the Los Angeles Public Library is immigrant integration, and the library launched a program named "Your Path to Citizenship Begins at the Los Angeles Public Library" to offer new immigrants with information that they need, such as citizenship and English language classes.

The Orange County Library System offers computer classes taught in English, Spanish, and Haitian Creole for new immigrants who speak these languages. Moreover, there are classes for immigrants who plan to attend the US Naturalization Test and interview and those who require ESL instruction and aid in essential language skills, such as pronunciation and writing.

Library Programs Targeting the Underprivileged, Homeless, and Unemployed

Cleveland has the second-highest level of childhood poverty in the US (Lo, Cho, & Chiu, 2017, p.229). During the summer holiday, many children in Cleveland do not have enough nutritious food because schools no longer provide them lunch as they need not go to school. Thus, the Cleveland Public Library (CPL) works with the food bank to launch a free lunch program to provide meals to school-age children. Besides, they offer after-school tutoring to these children so they can get help and concentrate on schoolwork after a meal. CPL also cooperates with external parties to provide free legal and financial aid services for community members who cannot afford these services. Other libraries launch similar programs. The Los Angeles Public Library offers free lunches to students, whereas the New York Public Library provides after-school programming to offer homework help to students.

The Orange County Library System designs and implements a "Right Service Right Time" database to cater to citizens' emergency economic needs, like job loss and unforeseen health risks. The database provides them with useful information on social services for dormitories, food, clothing, etc.

Among the nine libraries, some of them offer library programs relating to literacy to the underprivileged. The San Diego Public Library launched a "READ/San Diego" literacy program to empower residents who do not know reading and writing. In contrast, the Seattle Public Library System developed a "Books on Bikes" program to provide mobile library services to older and disabled people.

The poverty and literacy rates in the Rio Grande Valley of Texas are 26% and 50%, respectively. Therefore, McAllen Public Library (MAPL) takes the initiative to provide literacy education, such as launching a Summer Reading Program for children and babies and offering homework help and free lunches and snacks to children and teens (Cheung et al., 2022). Moreover, MAPL organizes the annual McAllen Book Festival for children and teens in Rio Grande Valley to allow them to meet and chat with some authors and illustrators.

The New York Public Library provides comprehensive job search resources for job seekers. The Orange County Library System hosts classes about job research, resume, and interview preparation to assist employment. In the McAllen Public Library, library users can search for jobs and learn to write resumes.

Library Programs for Building Community

To connect general community members, Anythink Libraries (AL) hosts block parties and dog parties to gather residents to watch a local football game in a park. To perform its role as a social connector, AL hosts community conversations with the Aspen Institute Citizenship and American Identity Program to let community members discuss what Americans should know to be "culturally and civically literate" (Lo, Cho, & Chiu, 2017, p.174).

The Los Angeles (LA) Public Library (LAPL) conducts outreach services with its LAPL Book Bike. To connect residents with the library, library staff on bike distributes a free book to residents and tells them about the free resources available in LAPL. They can also sign up for a library card on the spot. Besides, LAPL formed an LGBTQ library staff committee for staff training and outreach to the LGBT community. Notably, LAPL has a vast photo collection, and the digital Feathers Map Collection enables the community to access more information about LA history (Wong and Chiu, 2023; Deng et al., 2022; Chen et al., 2018).

Similarly, the Orange County Library System (OCLS) operates two projects to preserve local history: Orlando Memory and EPOCH (Electronically Preserving Obituaries as Cultural Heritage). The former is to store Orlando residents' favorite memories and stories digitally to preserve Orlando's history. In contrast, the latter is "for the community to save local personal histories, photos, documents, and recorded memory of loved ones" (Lo, Cho, & Chiu, 2017, p.241-242). Furthermore, OCLS sets up its e-book platform, ePulp, to host self-published e-books that the local community has reviewed (Chan et al., 2022b, Sung & Chiu, 2022).

The Seattle Public Library System creates mini-libraries at community events like farmers' markets and concerts to connect the library with the community. As mentioned previously, MAPL has its annual McAllen Book Festival. Apart from offering children and teens the opportunity to meet with authors and illustrators, it also includes outdoor rides and rescue animal shows to encourage community member connections through the festival activities (Lu et al., 2023).

Learning and Other Non-Traditional Services for the Public

To help people develop social capital, many libraries provide non-traditional services to assist them in acquiring knowledge, such as digital skills and STEM knowledge, which have become popular and essential for everyday life. For instance, Anythink Libraries accidentally realized to turn their act of hiring goats for the sustainability of the library environment into learning experiences about science knowledge for library users. Thus, they bring goats to the library every year to give people a chance to feed goats. Moreover, the library establishes unique "experience zones" to make information and learning more attractive within the library to foster self-learning (Lo, Cho, & Chiu, 2017).

Some libraries provide courses relating to technology and digital skills. For example, the Boston Public Library offers research and technology classes. In contrast, the Orange County Library System (OCLS) provides technology and software courses for adults and hands-on STEM programs for children. To fulfill its role in supporting life-learning, OCLS even provides non-traditional services like writer programs that help local authors enhance their writing skills and cooking demonstrations from local chefs (Chan et al., 2019; 2022a).

Apart from education courses, a few libraries build facilities in the library to address the technological needs of community members. For instance, the San Diego Public Library has an IDEA Lab to provide users with hands-on experience in contemporary technology to make them more competitive in this digital world. Moreover, the library has a Bio-Safety lab and an Information, Innovation & Incubation Lab ("I^3 Lab") to offer users the tools, space, and technology for invention. Besides, the Cleveland Public Library established its technology lab, "TechCentral Downtown," to let all residents access the latest technology (Lo, Cho, & Chiu, 2017).

Both the Seattle Public Library System and McAllen Public Library offer untraditional services to address the educational needs of library users. The former provides in-person homework help for students. The latter supplies space for various purposes to let the community gather; for example, users can book spaces in their Meeting Center for presentations or meetings and free study rooms (Zhou et al., 2022).

DISCUSSION

As Kranich (2001) proposed, there are two main themes in the literature about why public libraries can generate social capital: the library is a community meeting place and a *community educational institution*. We adopt such a framework to explore how public libraries construct social capital for different users. Library programs are tailor-made for specific user groups and discussed separately from library programs for all users. Notably, technology plays an integrative role in various information services and activities.

How Library Programs for Specific Users Build Social Capital

For New Immigrants

Khoir et al. (2017) suggested that new language skills, access to community information, and community bonding are indispensable for new immigrants. In a multicultural country like the US, as new immigrants from different parts of the world may not know much about English, ESL classes and conversation groups offered by NYPL, BPL, OCLS, SPLS, and MAPL enable these immigrants to grasp basic English language skills to communicate with others. LAPL's "Your Path to Citizenship Begins at the Los Angeles Public Library" provides information about the community, which is necessary for new immigrants to acquire knowledge about their community and nation.

These libraries act as community educational institutions by offering language classes and information services to empower new immigrants and contribute to social development. They generate opportunities for further education or find a job with a higher salary (Vårheim, 2011; Chan et al., 2022a; 2019). This echoes the World Bank's definition of social capital on contribution to social development (2001), and such community empowerment programs trigger the development of social capital (Vårheim, 2011).

For the Underprivileged, Homeless, and Unemployed

This part discusses the notion of the public library as a community meeting place to create social capital (Chan, Chiu, & Lam, 2020; Deng et al., 2019; Leung et al., 2022; Zhou et al., 2022). Social capital can be constructed via public libraries for library programs aiming at the underprivileged and homeless because they serve as community meeting places that create "social cohesion and trust in the community" (Johnson, 2010, p.147).

Facing poverty and lack of nutritious food for underprivileged children, free lunches are provided in CPL, LAPL, and MAPL. These libraries directly respond to the needs of the underprivileged to improve their living standard and simultaneously turn their library into community meeting places. The libraries provide social support and act as the hub of the social network trusted by the underprivileged, thus fostering social capital building (Vårheim, 2007).

With CPL's free legal and financial aid services and OCLS's Right Service Right Time database providing information about social assistance, their patrons have equal chances of receiving information services without monetary costs, irrespective of their social status and financial situation. In this case, the public library has become a place where diverse people meet, and all users are treated with respect, facilitating the development of social capital for the underprivileged (Vårheim, 2011; Cheung et al., 2021).

Although some people cannot go to libraries due to problems like illness, disability, and old age, SPLS's "Books on Bikes program" serves older and disabled people by offering mobile library services in the neighborhood instead of inside the library. This expands the boundary of the library from inside specific buildings outward to the community and strengthens the library's bonding with the public. Moreover, as the library is open to all community members regardless of one's health condition and social status, this avoids social exclusion or marginalizing these underprivileged people. Thus, the public library serves as a "significant generator of social capital" (Ferguson, 2012, p.25).

Vårheim (2011) argues that building a public library in an underprivileged community is a kind of community building that can construct social capital. MAPL's McAllen Book Festival is an excellent example. In response to the community's high poverty and low literacy rates, MAPL organizes such festivals to link residents in the neighborhood, especially teens and children, with famous authors, illustrators, and publishers. Thus, residents in this underprivileged community can improve their literacy skills, connect with people in the publishing industry, and get to know others in the community. This enhancement in human well-being and the community-building process facilitates the generation of social capital.

Next, we discuss the notion of the public library as a *community education institution* for building social capital. Among the nine US public libraries, many offer support for learning and literacy skills to school kids, such as the literacy program of SDPL and MAPL and the after-school programming and homework help services in CPL, NYPL, and MAPL. These programs help consolidate children's knowledge learned from school and equip them with the basic skills required in society (Lu et al., 2023). With such a contribution to social development for the next generation in the community (World Bank, 2001), these programs generate social capital.

Unemployed people have information needs for job search. Therefore, NYPL and MAPL provide job market information and resources in their library. They may also need to strengthen their interview skills, have advice on writing an appropriate resume, etc. Thus, MAPL and OCLS offer courses to teach job searching, writing resumes, and interview preparation.

To satisfy unemployed people's information needs and train them in job searching skills, these public libraries have become community education institutions so that community members can find jobs for better living. The assistance from public libraries to unemployed people forms a social network that benefits these people. Thus, social capital is generated according to its definition, "social networks and the norms of reciprocity associated with them" (Putnam 2002, p.3).

How Library Programs for All Users Build Social Capital

Public libraries can build social capital as the library is a *community meeting place* (Kranich, 2001; Lu et al., 2023). The US public libraries offer various programs for all users that can be classified into two types. The first type of program provides community members with chances to communicate and know about each other and connect the library with the neighborhood, such as neighborhood parties and community conversation events organized by AL, LAPL's book bike, and outreach to the LGBT community, SPLS's pop-up libraries in community events, MAPL's McAllen Book Festival to connect community members and library space offered for meeting and study (Zhou et al., 2022). Another example is ePulp, the e-book platform established by OCLS, which links community members with one another as resident reviews are necessary before a community member can publish an e-book on this platform. Therefore, social capital can be built due to the roles of the public library as a *community meeting place* and as an "important location for creating social cohesion and trust in the community" (Johnson, 2010, p.147; Lu et al., 2023).

Apart from gathering residents in the neighborhood, the second type of library program relates to the notion of "library as a *community meeting place*" and focuses on the history and memories of the local community. Examples include LAPL's digitalized map and photo collections, the Orlando Memory program, and the EPOCH project operated by OCLS. These programs collect and preserve the memories of the neighborhood, which helps unite community members. Through these programs and collections, public libraries serve as a platform for residents to be connected after a deeper understanding of the history and memories of the community (Lo et al., 2021; Jiang et al., 2019).

As found in the interviews, many US public libraries promote digital skills and STEM (Science, Technology, Engineering, and Mathematics) knowledge (Ma et al., 2016), such as AL's goat feeding activity and the establishment of "experience zones" for learning, technology, and software courses and STEM programs in BPL and OCLS. Some public libraries even have technical laboratories in the library for public use, such as SDPL's IDEA Lab, I³ Lab, and CPL's TechCentral downtown. These classes and facilities help library users acquire digital skills vital in this globalized economy, making them more competitive in society. In this case, the public library serves as a *community education institution* that educates community members. People with low information literacy may not know how to use the Internet properly. With library training courses related to the Internet and computer to teach them digital skills and free Internet access, their deficit in social capital can be filled (Johnson, 2010).

CONCLUSION

As discussed and analyzed in this paper, the public library contributes significantly to building social capital. Through the nine cases of US public libraries, our findings revealed that public libraries offer different types of library programs for various groups of users, which enable the generation of social capital for library users.

The concepts of the library as a community meeting place and a community education institution were adopted to explain how library programs from the US public libraries help construct social capital from the perspective of a community. For the notion of "library as a community meeting place," the meaning of "library" does not confine to space within the library building. Instead, the library now becomes a platform to connect community members through library programs inside or outside the library building, while various technologies enable libraries to provide their services through various online channels (Chan et al., 2022b; Ding et al., 2021; Ezeamuzie et al., 2022; Yu, Tsoi, et al., 2022; Zhang et al., 2021). Notably, technology plays an integrative role in various information services and activities.

By empowering underprivileged people and new immigrants through education, catering to their information needs, and providing a safe space and platform for interacting with other community members (Cheung et al., 2021), public libraries contribute to the social development of library users and their community, and, therefore, social capital can be generated. Further, we can see how these libraries follow technology trends closely in their educational programs to fulfill patrons' needs (Lo, Cho, Law, Chiu, & Allard, 2017).

In this study, only nine US public libraries were included, though these libraries are renowned for a long history. The sampling size of nine cases may be relatively small and thus cannot represent all US public libraries, especially smaller and less established ones, or reflect the situation of all public libraries worldwide. Yet, they have provided rich examples of excellent services to create social capital, which these public librarians considered to be their showcases. Besides, the portion of interview contents relating to the theme of social capital and the effectiveness of library programs in generating social capital might be limited in the book where data were collected for this study. Many questions with different focuses, such as the library leader's experience and leadership (Lo et al., 2020a; 2020b), were included in the interviews, though this also confirmed the library directors' determination to build social capital through their servant leadership philosophy. We plan to invite these library leaders to interviews to further discuss their good practice in building social capital.

On the other hand, we are investigating the development of social capital and promotions for libraries on social networks (Wong et al., 2023; Leung et al., 2022; Fong et al., 2020). We are concerned with the impact on social capital due to the massive lockdown of public and cultural facilities during the COVID-19 pandemic (Huang et al., 2021, 2022, 2023; Yu, Lam, & Chiu, 2023; Yu, Chiu, & Chan, 2023; Meng et al., 2023). We are also interested in the promotion of reading and other information services on social media (Chan et al., 2020; Fong et al., 2020; Cheng et al., 2020; Lam et al., 2022; Jiang et al., 2022) and further use of social media analytics to elicit reading preference and evaluate reading promotion (Wang et al., 2022; Deng & Chiu, 2023; He et al., 2022). We are also interested in the digitalization of reading and reference materials (Sun et al., 2022; Tse et al., 2022; Chan and Chiu, 2023)

REFERENCES

American Library Association. (2017). *The state of America's libraries 2017.* Retrieved from https://www.ala.org/news/sites/ala.org.news/files/content/State-of-Americas-Libraries-Report-2017.pdf

Boston Public Library. (2019). *BPL History*. Retrieved from https://www.bpl.org/bpl-history/

Bourdieu, P. (1992). *An invitation to reflexive sociology*. University of Chicago Press.

Butcher, W., & Street, P.-A. (2009). Lifelong learning with older adults. *Australasian Public Libraries and Information Services*, 22(2), 64–70.

Chan, A. W. Y., Chiu, D. K. W., & Ho, K. K. (2022). Workforce Information Needs for Vocational Guidance System Design. *International Journal of Systems and Service-Oriented Engineering*, 12(1), 1–16. doi:10.4018/IJSSOE.297134

Chan, A. W. Y., Chiu, D. K. W., Ho, K. K. W., & Wang, M. (2019). Information Needs of Vocational Training from Training Providers' Perspectives. *International Journal of Systems and Service-Oriented Engineering*, 8(4), 26–42. doi:10.4018/IJSSOE.2018100102

Chan, M. K. Y., Chiu, D. K. W., & Lam, E. T. H. (2020). Effectiveness of overnight learning commons: A comparative study. *Journal of Academic Librarianship*, 46(7), 102253. doi:10.1016/j.acalib.2020.102253 PMID:34173399

Chan, M. M. W., & Chiu, D. K. W. (2022). Alert Driven Customer Relationship Management in Online Travel Agencies: Event-Condition-Actions rules and Key Performance Indicators. In A. Naim & S. Kautish (Eds.), *Building a Brand Image Through Electronic Customer Relationship Management* (pp. 250–268). IGI Global. doi:10.4018/978-1-6684-5386-5.ch012

Chan, T. T. W., Lam, A. H. C., & Chiu, D. K. W. (2020). From Facebook to Instagram: Exploring user engagement in an academic library. *Journal of Academic Librarianship*, *46*(6), 102229. doi:10.1016/j.acalib.2020.102229 PMID:34173399

Chan, V. H. Y., & Chiu, D. K. W. (2023). Integrating the 6C's Motivation into Reading Promotion Curriculum for Disadvantaged Communities with Technology Tools: A Case Study of Reading Dreams Foundation in Rural China. In A. Etim & J. Etim (Eds.), *Adoption and Use of Technology Tools and Services by Economically Disadvantaged Communities: Implications for Growth and Sustainability*. IGI Global.

Chan, V. H. Y., Ho, K. K. W., & Chiu, D. K. W. (2022). Mediating effects on the relationship between perceived service quality and public library app loyalty during the COVID-19 era. *Journal of Retailing and Consumer Services*, *67*, 102960. doi:10.1016/j.jretconser.2022.102960

Chen, T.-T., & Ke, H.-R. (2017). Public library as a place and breeding ground of social capital: A case of Singang Library. *Malaysian Journal of Library and Information Science*, *22*(1), 45–58. doi:10.22452/mjlis.vol22no1.4

Chen, Y., Chiu, D. K. W., & Ho, K. K. W. (2018). Facilitating the learning of the art of Chinese painting and calligraphy at Chao Shao-an Gallery. *Micronesian Educators*, *26*, 45–58.

Cheng, J., Yuen, A. H., & Chiu, D. K. (2022). Systematic review of MOOC research in mainland China. *Library Hi Tech*. Advance online publication. doi:10.1108/LHT-02-2022-0099

Cheng, W., Tian, R., & Chiu, D. K. W. (2023). Travel vlogs influencing tourist decisions: Information preferences and gender differences. *Aslib Journal of Information Management*. Advance online publication. doi:10.1108/AJIM-05-2022-0261

Cheng, W. W. H., Lam, E. T. H., & Chiu, D. K. W. (2020). Social media as a platform in academic library marketing: A comparative study. *Journal of Academic Librarianship*, *46*(5), 102188. doi:10.1016/j.acalib.2020.102188

Cheung, L. S. N., Chiu, D. K. W., & Ho, K. K. W. (2022). A quantitative study on utilizing electronic resources to engage children's reading and learning: Parents' perspectives through the 5E instructional model, The Electronic Libraries. *The Electronic Library, 40*(6), 662–679. doi:10.1108/EL-09-2021-0179

Cheung, T. Y., Ye, Z., & Chiu, D. K. W. (2021). Value chain analysis of information services for the visually impaired: A case study of contemporary technological solutions. *Library Hi Tech, 39*(2), 625–642. doi:10.1108/LHT-08-2020-0185

Chin, G. Y. L., & Chiu, D. K. W. (2023). RFID-based Robotic Process Automation for Smart Museums with an Alert-driven Approach. In R. Tailor (Ed.), *Application and Adoption of Robotic Process Automation for Smart Cities*. IGI Global.

Chiu, D. K. W., & Ho, K. K. W. (2022a). Editorial: 40th anniversary: contemporary library research. *Library Hi Tech, 40*(6), 1525–1531. doi:10.1108/LHT-12-2022-517

Chiu, D. K. W., & Ho, K. K. W. (2022b). Special selection on contemporary digital culture and reading. *Library Hi Tech, 40*(5), 1204–1209. doi:10.1108/LHT-10-2022-516

Chung, C., Chiu, D. K. W., Ho, K. K. W., & Au, C. H. (2020). Applying social media to environmental education: Is it more impactful than traditional media? *Information Discovery and Delivery, 48*(4), 255–266. doi:10.1108/IDD-04-2020-0047

Clark, L., & Perry, K. A. (2015). *After access: libraries & digital empowerment.* https://alair.ala.org/bitstream/handle/11213/7559/ALA%20DI%20After%20Access_final_12%2017%2015.pdf?sequence=1

Coleman, J. S. (1988). Social capital in the creation of human capital. *American Journal of Sociology, 94*, S95–S120. doi:10.1086/228943

Connaway, L. S. (2017). Research methods in library and information science (6th ed.). Libraries Unlimited.

Creswell, J. W., & Guetterman, T. C. (2019). *Educational Research: Planning, Conducting, and Evaluating Quantitative and Qualitative Research* (6th ed.) Pearson.

Davis, A., & Bertot, J. (n.d.). *New Study: Today's Public Libraries Are Thriving Technology Hubs That Millions Rely on for First or Only Choice for Internet Access.* Bill & Melinda Gates Foundation. https://www.gatesfoundation.org/ideas/media-center/press-releases/2006/09/todays-public-libraries-are-thriving-technology-hubs-that-millions-rely-on-for-first-or-only-choice-for-internet-access

Deng, S., & Chiu, D. K. W. (2023). Analyzing Hong Kong Philharmonic Orchestra's Facebook Community Engagement with the Honeycomb Model. In M. Dennis & J. Halbert (Eds.), *Community Engagement in the Online Space* (pp. 31–47). IGI Global. doi:10.4018/978-1-6684-5190-8.ch003

Deng, W., Chin, G. Y.-l., Chiu, D. K. W., & Ho, K. K. W. (2022). Contribution of Literature Thematic Exhibition to Cultural Education: A Case Study of Jin Yong's Gallery. *Micronesian Educators*, *32*, 14–26.

Ding, S. J., Lam, E. T. H., Chiu, D. K. W., Lung, M. M., & Ho, K. K. W. (2021). Changes in reading behavior of periodicals on mobile devices: A comparative study. *Journal of Librarianship and Information Science*, *53*(2), 233–244. doi:10.1177/0961000620938119

Dukic, Z., Chiu, D. K. W., & Lo, P. (2015). How useful are smartphones for learning? Perceptions and practices of Library and Information Science students from Hong Kong and Japan. *Library Hi Tech*, *33*(4), 545–561. doi:10.1108/LHT-02-2015-0015

Ezeamuzie, N. M., Rhim, A. H. R., Chiu, D. K. W., & Lung, M. M. (2022). Mobile Technology Usage by Foreign Domestic Helpers: Exploring Gender Differences. *Library Hi Tech*. Advance online publication. doi:10.1108/LHT-07-2022-0350

Fan, K. Y. K., Lo, P., Ho, K. K. W., So, S., Chiu, D. K. W., & Ko, K. H. T. (2020). Exploring the mobile learning needs amongst performing arts students. *Information Discovery and Delivery*, *48*(2), 103–112. doi:10.1108/IDD-12-2019-0085

Ferguson, S. (2012). Are Public Libraries Developers of Social Capital? A Review of Their Contribution and Attempts to Demonstrate It. *The Australian Library Journal*, *61*(1), 22–33. doi:10.1080/00049670.2012.10722299

Fong, K. C. H., Au, C. H., Lam, E. T. H., & Chiu, D. K. W. (2020). Social network services for academic libraries: A study based on social capital and social proof. *Journal of Academic Librarianship*, *46*(1), 102091. doi:10.1016/j.acalib.2019.102091

Fung, R. H. Y., Chiu, D. K. W., Ko, E. H. T., Ho, K. K., & Lo, P. (2016). Heuristic usability evaluation of university of Hong Kong Libraries' mobile website. *Journal of Academic Librarianship*, *42*(5), 581–594. doi:10.1016/j.acalib.2016.06.004

Gerner Nielsen, B. (2014). Public libraries and lifelong learning. *Perspectives of Innovations, Economics and Business*, *14*(2), 94–102. doi:10.15208/pieb.2014.11

Gong, J. Y., Schumann, F., Chiu, D. K. W., & Ho, K. K. W. (2017). Tourists' mobile information seeking behavior: An investigation on China's youth. *International Journal of Systems and Service-Oriented Engineering, 7*(1), 58–76. doi:10.4018/IJSSOE.2017010104

Halpern, D. (2005). *Social capital.* Polity Press.

Hanifan, L. J. (1920). *The community center.* Silver, Burdett.

Hart, G. (2013). Social capital: A fresh vision for public libraries in South Africa? *South African Journal of Library and Information Science, 73*(1), 14–24. doi:10.7553/73-1-1331

He, Z., Chiu, D. K. W., & Ho, K. K. W. (2022). Weibo Analysis on Chinese Cultural Knowledge for Gaming. In Z. Sun (Ed.), *Handbook of Research on Foundations and Applications of Intelligent Business Analytics* (pp. 320–349). IGI Global. doi:10.4018/978-1-7998-9016-4.ch015

Ho, K. K. W., Takagi, T., Ye, S., Au, C. K., & Chiu, D. K. W. (2018). *The Use of Social Media for Engaging People with Environmentally Friendly Lifestyle – A Conceptual Model.* SIG Green Pre ICIS Workshop. https://aisel.aisnet.org/sprouts_proceedings_siggreen_2018/2/

Huang, P. S., Paulino, Y. C., So, S., Chiu, D. K. W., & Ho, K. K. W. (2021). Editorial. *Library Hi Tech, 39*(3), 693–695. doi:10.1108/LHT-09-2021-324

Huang, P.-S., Paulino, Y. C., So, S., Chiu, D. K. W., & Ho, K. K. W. (2022). Guest editorial: COVID-19 Pandemic and Health Informatics Part 2. *Library Hi Tech, 40*(2), 281–285. doi:10.1108/LHT-04-2022-447

Huang, P.-S., Paulino, Y. C., So, S., Chiu, D. K. W., & Ho, K. K. W. (2023). Guest editorial: COVID-19 Pandemic and Health Informatics Part 3. *Library Hi Tech, 41*(3), 1–6. doi:10.1108/LHT-02-2023-585

Hui, S. C., Kwok, M. Y., Kong, E. W. S., & Chiu, D. K. W. (2023). Information Security and Technical Issues of Cloud Storage Services: A Qualitative Study on University Students in Hong Kong. *Library Hi Tech.* Advance online publication. doi:10.1108/LHT-11-2022-0533

James, R. (2018, October 8). *Libraries as Community Hubs.* The Social History Society. https://socialhistory.org.uk/shs_exchange/libraries-as-community-hubs/

Jiang, M., Lam, A. H. C., Chiu, D. K. W., & Ho, K. K. W. (2023). Social media aids for business learning: A quantitative evaluation with the 5E instructional model. *Education and Information Technologies*. Advance online publication. doi:10.100710639-023-11690-z PMID:37361768

Jiang, T., Lo, P., Cheuk, M. K., Chiu, D. K. W., Chu, M. Y., Zhang, X., Zhou, Q., Liu, Q., Tang, J., Zhang, X., Sun, X., Ye, Z., Yang, M., & Lam, S. K. (2019). 文化新語:兩岸四地傑出圖書館、檔案館及博物館傑出工作者訪談 [New Cultural Dialog: Interviews with Outstanding Librarians, Archivists, and Curators in Greater China]. Hong Kong: Systech Publications.

Jiang, X., Chiu, D. K. W., & Chan, C. T. (2022). Application of the AIDA model in social media promotion and community engagement for small cultural organizations: A case study of the Choi Chang Sau Qin Society. In M. Dennis & J. Halbert (Eds.), *Community Engagement in the Online Space* (pp. 48–70). IGI Global.

Johnson, C. A. (2010). Do public libraries contribute to social capital? A preliminary investigation into the relationship. *Library & Information Science Research*, *32*(2), 147–155. doi:10.1016/j.lisr.2009.12.006

Johnson, C. A. (2012). How do public libraries create social capital? An analysis of interactions between library staff and patrons. *Library & Information Science Research*, *34*(1), 52–62. doi:10.1016/j.lisr.2011.07.009

Johnson, C. A. (2015). Social Capital and Library and Information Science Research: Definitional Chaos or Coherent Research Enterprise? *Information Research*, *20*(4), n4.

Johnson, C. A., & Griffis, M. R. (2009). A place where everybody knows your name? Investigating the relationship between public libraries and social capital. *Canadian Journal of Information and Library Science*, *33*(3-4), 159–191.

Johnson, C. A., & Griffis, M. R. (2014). The effect of public library use on the social capital of rural communities. *Journal of Librarianship and Information Science*, *46*(3), 179–190. doi:10.1177/0961000612470278

Khoir, S., Du, J. T., Davison, R. M., & Koronios, A. (2017). Contributing to social capital: An investigation of Asian immigrants' use of public library services. *Library & Information Science Research*, *39*(1), 34–45. doi:10.1016/j.lisr.2017.01.005

Khosrow-Pour, M. (2014). Academic Libraries in the Digital Age. In C. O. Jefferson (Ed.), *Encyclopedia of Information Science and Technology* (3rd ed., pp. 4815–4822). IGI Global. doi:10.4018/978-1-4666-5888-2

Kranich, N. (2001). Libraries create social capital: A unique, if fleeting, opportunity to carve out a new library mission. (United States)(Brief Article). *Library Journal*, *126*(19), 40.

Lam, A. H. C., Ho, K. K. W., & Chiu, D. K. W. (2022). Instagram for student learning and library promotions? A quantitative study using the 5E Instructional Model. *Aslib Journal of Information Management*, *75*(1), 112–130. doi:10.1108/ AJIM-12-2021-0389

Lam, E. T. H., Au, C. H., & Chiu, D. K. W. (2019). Analyzing the use of Facebook among university libraries in Hong Kong. *Journal of Academic Librarianship*, *45*(3), 175–183. doi:10.1016/j.acalib.2019.02.007

Lau, K. P., Chiu, D. K., Ho, K. K., Lo, P., & See-To, E. W. (2017). Educational Usage of Mobile Devices: Differences between Postgraduate and Undergraduate Students. *Journal of Academic Librarianship*, *43*(3), 201–208. doi:10.1016/j. acalib.2017.03.004

Lau, K. S. N., Lo, P., Chiu, D. K. W., Ho, K. K. W., Jiang, T., Zhou, Q., Percy, P., & Allard, B. (2020). Library, Learning, and Recreational Experiences Turned Mobile: A Comparative Study between LIS and non-LIS students. *Journal of Academic Librarianship*, *46*(2), 102103. doi:10.1016/j.acalib.2019.102103

Lei, S. Y., Chiu, D. K. W., Lung, M. M., & Chan, C. T. (2021). Exploring the aids of social media for musical instrument education. *International Journal of Music Education*, *39*(2), 187–201. doi:10.1177/0255761420986217

Leung, T.N., Hui, Y.M., Luk, C.K.L., Chiu, D.K.W., & Kevin, K.K.W. (2023) Evaluating Facebook as aids for learning Japanese: learners' perspectives. *Library Hi Tech*. . doi:10.1108/LHT-11-2021-0400

Leung, T. N., Luk, C. K. L., Chiu, D. K. W., & Kevin, K. K. W. (2022). User perceptions, academic library usage, and social capital: A correlation analysis under COVID-19 after library renovation. *Library Hi Tech*, *40*(2), 304–322. doi:10.1108/ LHT-04-2021-0122

Li, Q., Wong, J., & Chiu, D. K. W. (2023). School library reading support for students with dyslexia: A qualitative study in the digital age. *Library Hi Tech*. Advance online publication. doi:10.1108/LHT-03-2023-0086

Li, S., Xie, Z., Chiu, D. K. W., & Ho, K. K. W. (2023). Sentiment Analysis and Topic Modeling Regarding Online Classes on the Reddit Platform: Educators versus Learners. *Applied Sciences (Basel, Switzerland)*, *13*(4), 2250. doi:10.3390/ app13042250

Li, S. M., Lam, A. H. C., & Chiu, D. K. W. (2023). Digital transformation of ticketing services: A value chain analysis of POPTICKET in Hong Kong. In J. D. Santos & I. V. Pereira (Eds.), *Management and Marketing for Improved Retail Competitiveness and Performance*. IGI. Global.

Lightfoot, M. (2016). The Emergence of Digital Social Capital in Education. In I. R. Haslam & M. S. Khine (Eds.), *Leveraging Social Capital in Systemic Education Reform* (pp. 43–66). Brill Sense. doi:10.1007/978-94-6300-651-4_3

Lin, C.-H., Chiu, D.K.W., & Lam, K. T. (2022). Hong Kong Academic Librarians' Attitudes Towards Robotic Process Automation. *Library Hi Tech*. . doi:10.1108/LHT-03-2022-0141

Liu, Y., Chiu, D. K. W., & Ho, K. K. W. (2023). Short-Form Videos for Public Library Marketing: Performance Analytics of Douyin in China. *Applied Sciences (Basel, Switzerland)*, *13*(6), 3386. doi:10.3390/app13063386

Lo, P., Allard, B., Anghelescu, H. G. B., Xin, Y., Chiu, D. K. W., & Stark, A. J. (2020b). Transformational Leadership and Library Management in World's Leading Academic Libraries. *Journal of Librarianship and Information Science*, *52*(4), 972–999. doi:10.1177/0961000619897991

Lo, P., Allard, B., Wang, N., & Chiu, D. K. W. (2020a). Servant leadership theory in practice: North America's leading public libraries. *Journal of Librarianship and Information Science*, *52*(1), 249–270. doi:10.1177/0961000618792387

Lo, P., & Chiu, D. K. W. (2015). Enhanced and Changing Roles of School Librarians under the Digital Age. *New Library World*, *116*(11/12), 696–710. doi:10.1108/NLW-05-2015-0037

Lo, P., Chiu, D. K. W., & Chu, W. (2014). Modeling Your College Library after a Commercial Bookstore? The Hong Kong Design Institute Library Experience. *Community & Junior College Libraries*, *19*(3-4), 59–76. doi:10.1080/02763915.2014.915186

Lo, P., Cho, A., & Chiu, D. K. W. (2017). *World's Leading National, Public, Monastery and Royal Library Directors: Leadership, Management, Future of Libraries*. Walter de Gruyter GmbH. doi:10.1515/9783110533347

Lo, P., Cho, A., Law, B. K. K., Chiu, D. K. W., & Allard, B. (2017). Progressive trends in electronic resources management among academic libraries in Hong Kong. *Library Collections, Acquisitions & Technical Services*, *40*(1-2), 28–37. doi:10.1080/14649055.2017.1291243

Lo, P., Cho, A., Law, B. K. K., Chiu, D. K. W., & Allard, B. (2017). Progressive Trends in Electronic Resources Management among Academic Libraries in Hong Kong. *Library Collections, Acquisitions & Technical Services*, *40*(1-2), 28–37. doi:10.1080/14649055.2017.1291243

Lo, P., Hsu, W.-E., Wu, S. H. S., Travis, J., & Chiu, D. K. W. (2021). *Creating a Global Cultural City via Public Participation in the Arts: Conversations with Hong Kong's Leading Arts and Cultural Administrators*. Nova Science Publishers.

Lo, P., Yu, K., & Chiu, D. K. W. (2015). A Research Agenda for the Enhancing Roles and Practice of School Librarians in Hong Kong's 21st Century Learning Environments. *School Libraries Worldwide*, *21*(1), 19–37. doi:10.29173lw6881

Lu, S. S., Tian, R., & Chiu, D. K. W. (2023). Why Do People Not Attend Public Library Programs in the Current Digital Age? *Library Hi Tech*. Advance online publication. doi:10.1108/LHT-04-2022-0217

Lybeck, R., Koiranen, I., & Koivula, A. (2023). From digital divide to digital capital: The role of education and digital skills in social media participation. *Universal Access in the Information Society*. Advance online publication. doi:10.100710209-022-00961-0

Ma, J., Chiu, D. K. W., & Tang, J. K. (2016). Exploring the Use of Social Media to Advance K12 Science Education. *International Journal of Systems and Service-Oriented Engineering*, *6*(4), 47–59. doi:10.4018/IJSSOE.2016100104

Mak, M. Y. C., Poon, A. Y. M., & Chiu, D. K. W. (2022). Using Social Media as Learning Aids and Preservation: Chinese Martial Arts in Hong Kong. In S. Papadakis & A. Kapaniaris (Eds.), *The Digital Folklore of Cyberculture and Digital Humanities* (pp. 171–185). IGI Global. doi:10.4018/978-1-6684-4461-0.ch010

Meng, Y., Chu, M. Y., & Chiu, D. K. W. (2022). The impact of COVID-19 on museums in the digital era: Practices and Challenges in Hong Kong. *Library Hi Tech*, *41*(1), 130–151. doi:10.1108/LHT-05-2022-0273

Ni, J., Chiu, D. K. W., & Ho, K. K. W. (2022). Information search behavior among Chinese self-drive tourists in the smartphone era. *Information Discovery and Delivery*, *50*(3), 285–296. doi:10.1108/IDD-05-2020-0054

Ni, Y., Lam, A. H. C., & Chiu, D. K. W. (2023). Leveraging Online Communities for Building Social Capital in University Libraries: A Case Study of Fudan University Medical Library. In *Balance and Boundaries in Creating Meaningful Relationships in Online Higher Education*. IGI Global. Retrieved from https://peterboroughtownlibrary.org/history/

Putnam, R. D. (1995). Turning in, Turning out: The Strange Disappearance of social capital in America. *PS, Political Science & Politics*, *28*(4), 1–20. doi:10.2307/420517

Putnam, R. D. (Ed.). (2002). *Democracies in Flux: The Evolution of Social Capital in Contemporary Society*. Oxford University Press. doi:10.1093/0195150899.001.0001

Sharma, J. (2019). Role of Public Library to growing Digital Literacy in our Society. *International Research Journal of Multidisciplinary Science & Technology*, *1*(6), 295–297.

Stevens, C. R., & Campbell, P. J. (2006). Collaborating to connect global citizenship, information literacy, and lifelong learning in the global studies classroom. *RSR. Reference Services Review*, *34*(4), 536–556. doi:10.1108/00907320610716431

Suen, R. L. T., Tang, J., & Chiu, D. K. W. (2020). Virtual reality services in academic libraries: Deployment experience in Hong Kong. *The Electronic Library*, *38*(4), 843–858. doi:10.1108/EL-05-2020-0116

Sun, X., Chiu, D. K. W., & Chan, C. T. (2022). Recent Digitalization Development of Buddhist Libraries: A Comparative Case Study. In S. Papadakis & A. Kapaniaris (Eds.), *The Digital Folklore of Cyberculture and Digital Humanities* (pp. 251–266). IGI Global. doi:10.4018/978-1-6684-4461-0.ch014

Sung, Y. Y. C., & Chiu, D. K. W. (2022). E-book or print book: Parents' Current View in Hong Kong. *Library Hi Tech*, *40*(5), 1289–1304. doi:10.1108/LHT-09-2020-0230

Thompson, K. M., Jaeger, P. T., Taylor, N. G., Subramaniam, M. M., & Bertot, J. C. (2014). *Digital literacy and digital inclusion: information policy and the public library*. Rowman & Littlefield.

Tsang, A. L. Y., & Chiu, D. K. W. (2022). Effectiveness of Virtual Reference Services in Academic Libraries: A Qualitative Study based on the 5E Learning Model. *Journal of Academic Librarianship*, *48*(4), 012533. doi:10.1016/j.acalib.2022.102533

Tse, H. L., Chiu, D. K., & Lam, A. H. (2022). From Reading Promotion to Digital Literacy: An Analysis of Digitalizing Mobile Library Services With the 5E Instructional Model. In A. Almeida & S. Esteves (Eds.), *Modern Reading Practices and Collaboration Between Schools, Family, and Community* (pp. 239–256). IGI Global. doi:10.4018/978-1-7998-9750-7.ch011

Vårheim, A. (2007). Social capital and public libraries: The need for research. *Library & Information Science Research*, *29*(3), 416–428. doi:10.1016/j.lisr.2007.04.009

Vårheim, A. (2011). Gracious space: Library programming strategies towards immigrants as tools in the creation of social capital. *Library & Information Science Research*, *33*(1), 12–18. doi:10.1016/j.lisr.2010.04.005

Vårheim, A., Steinmo, S., & Ide, E. (2008). Do libraries matter? Public libraries and the creation of social capital. *The Journal of Documentation*, *64*(6), 877–892. doi:10.1108/00220410810912433

Wai, I. S. H., Ng, S. S. Y., Chiu, D. K. W., Ho, K. K., & Lo, P. (2018). Exploring undergraduate students' usage pattern of mobile apps for education. *Journal of Librarianship and Information Science*, *50*(1), 34–47. doi:10.1177/0961000616662699

Wang, J., Deng, S., Chiu, D. K. W., & Chan, C. T. (2022). Social Network Customer Relationship Management for Orchestras: A Case Study on Hong Kong Philharmonic Orchestra. In N. B. Ammari (Ed.), *Social Customer Relationship Management (Social-CRM) in the Era of Web 4.0*. IGI Global. doi:10.4018/978-1-7998-9553-4.ch012

Wang, P., Chiu, D. K. W., Ho, K. K., & Lo, P. (2016). Why read it on your mobile device? Change in reading habit of electronic magazines for university students. *Journal of Academic Librarianship*, *42*(6), 664–669. doi:10.1016/j.acalib.2016.08.007

Wang, W., Lam, E. T. H., Chiu, D. K. W., Lung, M. M., & Ho, K. K. W. (2021). Supporting Higher Education with Social Networks: Trust and Privacy vs. Perceived Effectiveness. *Online Information Review*, *45*(1), 207–219. doi:10.1108/OIR-02-2020-0042

White, B. (2012). Guaranteeing Access to Knowledge: The Role of Libraries. *WIPO Magazine, 2012*(4). https://www.wipo.int/wipo_magazine/en/2012/04/article_0004.html

Wong, A. K.-k., & Chiu, D. K. W. (2023). Digital Transformation of Museum Conservation Practices: A Value Chain Analysis of Public Museums in Hong Kong. In R. Pettinger, B. B. Gupta, A. Roja, & D. Cozmiuc (Eds.), *Handbook of Research on the Digital Transformation Digitalization Solutions for Social and Economic Needs*. IGI Global. doi:10.4018/978-1-6684-4102-2.ch010

Wong, J., Chiu, D. K. W., Leung, T. N., & Kevin, K. K. W. (2023). Exploring the Associations of Addiction as a Motive for Using Facebook with Social Capital Perceptions. *Online Information Review*, *47*(2), 283–298. doi:10.1108/OIR-06-2021-0300

Wong, K. C., & Chiu, D. K. W. (2023). Promoting The Use of Electronic Resources of International Schools in Hong Kong: A Case Study of ESF King George V School. In E. Meletiadou (Ed.), *Handbook of Research on Redesigning Teaching, Learning, and Assessment in the Digital Era* (pp. 123–134). IGI Global. doi:10.4018/978-1-6684-8292-6.ch007

World Bank. (2001). *Understanding and measuring social capital: A synthesis of findings and recommendations from the Social Capital Initiative*. Retrieved from http://siteresources.worldbank.org/INTSOCIALCAPITAL/Resources/Social-Capital-Initiative-Working-Paper-Series/SCI-WPS-24.pdf

Wu, M., Lam, A. H. C., & Chiu, D. K. W. (2023). Transforming and Promoting Reference Services with Digital Technologies: A Case Study on Hong Kong Baptist University Library. In B. Holland (Ed.), Handbook of Research on Advancements of Contactless Technology and Service Innovation in Library and Information Science (pp. 128-145). IGI Global. doi:10.4018/978-1-6684-7693-2.ch007

Xie, Z., Chiu, D. K. W., & Ho, K. K. W. (2023). The Role of Social Media as Aids for Accounting Education and Knowledge Sharing: Learning Effectiveness and Knowledge Management Perspectives in Mainland China. *Journal of the Knowledge Economy*. Advance online publication. doi:10.100713132-023-01262-4

Xie, Z., Wong, G. K. W., Chiu, D. K. W., & Lei, J. (2023). Bridging K-12 Mathematics and Computational Thinking in the Scratch Community: A Big Data Analysis. *IT Professional*, *25*(2), 64–70. doi:10.1109/MITP.2023.3243393

Xue, B., Lam, A. H. C., & Chiu, D. K. W. (2023). Redesigning Library Information Literacy Education with the BOPPPS Model: A Case Study of the HKUST. In R. Taiwo, B. Idowu-Faith, & S. Ajiboye (Eds.), *Transformation of Higher Education Through Institutional Online Spaces*. IGI. Global. doi:10.4018/978-1-6684-8122-6.ch017

Yang, Z., Zhou, Q., Chiu, D. K. W., & Wang, Y. (2022). Exploring the factors influencing continuance usage intention of academic social network sites. *Online Information Review*, *46*(7), 1225–1241. doi:10.1108/OIR-01-2021-0015

Yao, L., Lei, J., Chiu, D. K. W., & Xie, Z. (2023). Adult Learners' Perception of Online Language English Learning Platforms in China. In A. Garcés-Manzanera & M. E. C. García (Eds.), *New Approaches to the Investigation of Language Teaching and Literature* (pp. 123–140). IGI Global. doi:10.4018/978-1-6684-6020-7.ch007

Yip, K. H. T., Chiu, D. K. W., Ho, K. K. W., & Lo, P. (2021). Adoption of Mobile Library Apps as Learning Tools in Higher Education: A Tale between Hong Kong and Japan. *Online Information Review*, *45*(2), 389–405. doi:10.1108/OIR-07-2020-0287

Yip, T., Chiu, D. K. W., Cho, A., & Lo, P. (2019). Behavior and informal learning at night in a 24-hour space: A case study of the Hong Kong Design Institute Library. *Journal of Librarianship and Information Science*, *51*(1), 171–179. doi:10.1177/0961000617726120

Yu, H. H. K., Chiu, D. K. W., & Chan, C. T. (2023). Resilience of symphony orchestras to challenges in the COVID-19 era: Analyzing the Hong Kong Philharmonic Orchestra with Porter's five force model. In W. Aloulou (Ed.), *Handbook of Research on Entrepreneurship and Organizational Resilience During Unprecedented Times* (pp. 586–601). IGI Global.

Yu, H. Y., Tsoi, Y. Y., Rhim, A. H. R., Chiu, D. K., & Lung, M. M. W. (2022). Changes in habits of electronic news usage on mobile devices in university students: A comparative survey. *Library Hi Tech*, *40*(5), 1322–1336. doi:10.1108/LHT-03-2021-0085

Yu, P. Y., Lam, E. T. H., & Chiu, D. K. W. (2023). Operation management of academic libraries in Hong Kong under COVID-19. *Library Hi Tech*, *41*(1), 108–129. doi:10.1108/LHT-10-2021-0342

Zhang, X., Lo, P., So, S., Chiu, D. K. W., Leung, T. N., Ho, K. K. W., & Stark, A. (2021). Medical students' attitudes and perceptions towards the effectiveness of mobile learning: A comparative information-need perspective. *Journal of Librarianship and Information Science*, *53*(1), 116–129. doi:10.1177/0961000620925547

Zhang, Y., Lo, P., So, S., & Chiu, D. K. W. (2020). Relating library user education to business students' information needs and learning practices: A comparative study. *RSR. Reference Services Review*, *48*(4), 537–558. doi:10.1108/RSR-12-2019-0084

Zheng, J., Lam, A. H. C., & Chiu, D. K. W. (2023). Evaluating the Effectiveness of Learning Commons as Third Space with the 5E Usability Model: The Case of Hong Kong University of Science and Technology Library. In C. Kaye, J. H. Writer, & J. Batsaikhan (Eds.), *Third-Space Exploration in Education*. IGI Global.

Zhou, J., Lam, E. T. H., Au, C. H., Lo, P., & Chiu, D. K. W. (2022). Library café or elsewhere: Usage of study space by different majors under contemporary technological environment. *Library Hi Tech*, *40*(6), 1567–1581. doi:10.1108/LHT-03-2021-0103

Zuo, Y., Lam, A. H. C., & Chiu, D. K. W. (2023). Digital protection of traditional villages for sustainable heritage tourism: A case study on Qiqiao Ancient Village, China. In A. Masouras, C. Papademetriou, D. Belias, & S. Anastasiadou (Eds.), *Sustainable Growth Strategies for Entrepreneurial Venture Tourism and Regional Development* (pp. 121–195). IGI Global. doi:10.4018/978-1-6684-6055-9.ch009

APPENDIX

Table 1. Library programs that help build social capital (listed in the order in the book)

Names of Libraries	Regions / Countries	Library Services for Underprivileged Children	Library Services for Underprivileged Elderly	Library Services for Homeless	Library Services for Unemployed People	Library Services for New Immigrants	STEM	Other Non-Traditional Library Services	Services for Building Connections with community
Anythink Libraries (AL)	Adams County (Colorado)						● bring goats to the library every year to give people a chance to feed goats	● Setting up unique "experience zones" to make information and learning more attractive within the library setting	● Block parties ● Watching a local football game in a local park ● Dog party ● Community conversation with the Aspen Institute Citizenship and American Identity Program
New York Public Library (NYPL)	New York	● After-school programming			● Comprehensive job search resources	● English language and literacy classes			
Los Angeles Public Library (LAPL)	Los Angeles	● Free lunches to students				● "Your Path to Citizenship Begins at the Los Angeles Public Library"			● LAPL Book Bike ● LGBTQ library staff committee ● Huge photo and the Feathers Map Collection
Boston Public Library (BPL)	Boston					● ESL conversation groups	● Research & technology classes		
Cleveland Public Library (CPL)	Cleveland	● Provide meals to school-age children	● Free legal aid services ● Financial aid services	● Free legal aid services ● Financial aid services			● TechCentral downtown & technology labs		
Names of Libraries	Regions / Countries	Library services for underprivileged children	Library services for underprivileged elderly	Library services for homeless	Library services for unemployed people	Library Services for new immigrants	STEM	Other non-traditional library services	Services for building connections with community or community building

Names of Libraries	Regions / Countries	Library Services for Underprivileged Children	Library Services for Underprivileged Elderly	Library Services for Homeless	Library Services for Unemployed People	Library Services for New Immigrants	STEM	Other Non-Traditional Library Services	Services for Building Connections with community
Orange County Library System (OCLS)	Orlando, Florida	● Right Service Right Time database	● Right Service Right Time database	● Right Service Right Time database	● Job finding, resume and interview preparation classes	● Computer classes taught in both English and Spanish and Haitian Creole ● Classes for immigrants who wish to prepare for the US Naturalization test/interview, and people who need English as a Second Language instruction and support in pronunciation, conversion, and writing	● Hands-on class in technology and software ● Hands-on STEM programs	● Writers programs which help local authors to develop their writing skills	● Orlando Memory ● Operate EPOCH for the community to save local personal histories, photo, documents of loved ones ● Set up its own e-book platform, ePulp, to host self-published e-books reviewed by the local community
San Diego Public Library (SDPL)	San Diego	● READ/San Diego literacy program					● IDEA Lab ● Bio-Safety lab ● I^3 Lab		
The Seattle Public Library System (SPLS)	Seattle City		● Books on Bikes program			● English-as-a-second-language classes for immigrants		● In-person homework help for students ● Cooperation with museums: use library cards to reserve free tickets for Seattle museums	● Mini-libraries pop up at community events like farmers markets, concerts and summer festivals
Names of Libraries	Regions / Countries	Library services for underprivileged children	Library services for underprivileged elderly	Library services for homeless	Library services for unemployed people	Library Services for new immigrants	STEM	Other non-traditional library services	Services for building connections with community or community building

Names of Libraries	Regions / Countries	Library Services for Underprivileged Children	Library Services for Underprivileged Elderly	Library Services for Homeless	Library Services for Unemployed People	Library Services for New Immigrants	STEM	Other Non-Traditional Library Services	Services for Building Connections with community
McAllen Public Library (MAPL)	Texas	● Summer Reading Program ● Homework help ● Provide free lunches and snacks to children and teens ● Provide the annual McAllen Book Festival for children and teens in the Rio Grande Valley			● Look for jobs and learn to write resumes	● Classes for GED or ESL exams		● Transact business or study in free study rooms ● Rent space in Meeting Center for performance, meeting, or presentation	● The McAllen Book Festival

Compilation of References

Abbas, N., Whitfield, J., Atwell, E., Bowman, H., Pickard, T., & Walker, A. (2022). Online chat and chatbots to enhance mature student engagement in higher education. *International Journal of Lifelong Education*, *41*(3), 1–19. doi:10.1080/02601370.2022.2066213

Abbate, T., Vecco, M., Vermiglio, C., Zarone, V., & Perano, M. (2022). Blockchain and art market: Resistance or adoption? *Consumption Markets & Culture*, *25*(2), 105–123. doi:10.108 0/10253866.2021.2019026

Abrams, Z. (2022). *Beyond OpenSea: Our guide to to the biggest NFT marketplaces.* https://www.businessofbusiness.com/articles/plenty-of-fish-besides-opensea-here-is-our-guide-to-the-most-popular-nft-marketplaces/

AccessEngineering. (2015). *Making a makerspace? Guidelines for accessibility and universal design.* https://www.washington.edu/doit/sites/default/files/atoms/files/Making_a_Makerspace_8_03_15.pdf

Adam, M., Wessel, M., & Benlian, A. (2021). AI-based chatbots in customer service and their effects on user compliance. *Electronic Markets*, *31*(2), 427–445. doi:10.100712525-020-00414-7

Adams Becker, S., Cummins, M., Davis, A., Freeman, A., Giesinger Hall, C., Ananthanarayanan, V., Langley, K., & Wolfson, N. (2017). *NMC horizon report: 2017 library edition.* The New Media Consortium.

Afrane, D. A., Donkor, A. B., & Yamson, G. C. (2022). Libraries for Tomorrow: The Use of ICT and Space Transformation in Some Academic Libraries in Ghana. *Mousaion: South African Journal of Information Science*, *40*(2). Advance online publication. doi:10.25159/2663-659X/9896

Afrane, D. A., Van Der Walt, T., & Donkor, A. B. (2022). Student's assessment of Balme Library's use of information technology in providing quick and efficient library services. *South African Journal of Library and Information Science*, *88*(1). Advance online publication. doi:10.7553/88-1-1975

Ahmad, K., Iqbal, W., El-Hassan, A., Qadir, J., Benhaddou, D., Ayyash, M ., & Al-Fuqaha, A. (2021). *Artificial Intelligence in Education: A Panoramic Review.* Academic Press.

Compilation of References

Ahmad, N. A., Che, M. H., Zainal, A., Abd Rauf, M. F., & Adnan, Z. (2018). Review of chatbots design techniques. *International Journal of Computer Applications*, *181*(8), 7–10. doi:10.5120/ijca2018917606

Aittola, M., Ryhänen, T., & Ojala, T. (2003, September). SmartLibrary-Location-aware mobile library service. Academic Press.

Akinwalere, S. N., & Ivanov, V. (2022). Artificial intelligence in higher education and opportunities. *Boarder Crosssing*, *12*(1), 1–15. doi:10.33182/bc.v12i1.2015

Al Aklabi, A. T. (2021). An Artificial Intelligence (AI) and Digital Services at King Saud University Libraries in COVID-19 Pandemic. *Journal of Engineering and Applied Sciences Technology*, 1–8. doi:10.47363/JEAST/2021(3)131

Alajmi, Q. A., Kamaludin, A., Arshah, R. A., & Al-Sharafi, M. A. (2018). The effectiveness of cloud-based e-learning towards quality of academic services: An Omanis' expert view. *International Journal of Advanced Computer Science and Applications*, *9*(4). Advance online publication. doi:10.14569/IJACSA.2018.090425

Al-athwari, B., & Hossain, M. A. (2022). IoT Architecture: Challenges and Open Research Issues. In *Proceedings of 2nd International Conference on Smart Computing and Cyber Security*. Springer. 10.1007/978-981-16-9480-6_39

AlDhaen, F. (2022). The Use of Artificial Intelligence in Higher Education–Systematic Review. *COVID-19 Challenges to University Information Technology Governance*, 269-285.

Alhojailan, M. I. (2012). Thematic analysis: A critical review of its process and evaluation. *West East Journal of Social Sciences*, *1*(1), 39–47.

Ali, G. (n.d.). *Everything about NFT - According to a Hong Kong NFT* artist. Retrieved from https://www.aligstudios.com/everything-about-nft

Ali, M. Y., Naeem, S. B., & Bhatti, R. (2020). Artificial intelligence tools and perspectives of university librarians: An overview. *Business Information Review*, *37*(3), 116–124. doi:10.1177/0266382120952016

Ali, Z. H., Ali, H. A., & Badawy, M. M. (2015). Internet of Things (IoT): Definitions, challenges and recent research directions. *International Journal of Computer Applications*, *128*(1), 37–47. doi:10.5120/ijca2015906430

Alkhariji, L., De, S., Rana, O., & Perera, C. (2023). Semantics-based privacy by design for Internet of Things applications. *Future Generation Computer Systems*, *138*, 280–295. doi:10.1016/j.future.2022.08.013

Almada, A., Yu, Q., & Patel, P. (2023). Proactive chatbot framework based on the PS2CLH model: an AI-Deep Learning chatbot assistant for students. In *Proceedings of SAI Intelligent Systems Conference* (pp. 751-770). Springer. 10.1007/978-3-031-16072-1_54

271

Almansor, E. H., & Hussain, F. K. (2021). Fuzzy Prediction Model to Measure Chatbot Quality of Service. *2021 IEEE International Conference on Fuzzy Systems (FUZZ-IEEE),* 1–4. 10.1109/FUZZ45933.2021.9494346

Almuhaya, M. A., Jabbar, W. A., Sulaiman, N., & Abdulmalek, S. (2022). A survey on Lorawan technology: Recent trends, opportunities, simulation tools and future directions. *Electronics (Basel), 11*(1), 164. doi:10.3390/electronics11010164

Al-Sharafi, M. A., Al-Emran, M., Iranmanesh, M., Al-Qaysi, N., Iahad, N. A., & Arpaci, I. (2022). Understanding the impact of knowledge management factors on the sustainable use of AI-based chatbots for educational purposes using a hybrid SEM-ANN approach. *Interactive Learning Environments,* 1–20. doi:10.1080/10494820.2022.2075014

Altrabsheh, N. (2016). *Sentiment analysis and students' real-time feedback* [Doctor of Philosophy Thesis]. The University of Portsmouth.

American Library Association. (2017). *The state of America's libraries 2017.* Retrieved from https://www.ala.org/news/sites/ala.org.news/files/content/State-of-Americas-Libraries-Report-2017.pdf

Amini, M., Vakilimofrad, H., & Saberi, M. K. (2021). Human factors affecting information security in libraries. *The Bottom Line (New York, N.Y.), 34*(1), 45–67. doi:10.1108/BL-04-2020-0029

Anand, R., & Beel, J. (2020). *Auto-Surprise: An Automated Recommender-System (AutoRecSys) Library with Tree of Parzens Estimator (TPE) Optimization.* doi:10.1145/3383313.3411467

Angal, Y., & Gade, A. (2017, February). Development of library management robotic system. In *2017 International Conference on Data Management, Analytics and Innovation (ICDMAI)* (pp. 254-258). IEEE. 10.1109/ICDMAI.2017.8073520

Anoop, A., & Ubale, N. A. (2020, August). Cloud Based Collaborative Filtering Algorithm for Library Book Recommendation System. In *2020 Third International Conference on Smart Systems and Inventive Technology (ICSSIT)* (pp. 695-703). IEEE. 10.1109/ICSSIT48917.2020.9214243

Ansari, N., Vakilimofrad, H., Mansoorizadeh, M., & Amin, M. (2021). Using data mining techniques to predict user's behavior and create recommender systems in libraries and information centers. *Global Knowledge Memory and Communication, 70*(6–7), 538–557. doi:10.1108/GKMC-04-2020-0058

Antevski, K., Redondi, A. E., & Pitic, R. (2016, July). A hybrid BLE and Wi-Fi localization system for the creation of study groups in smart libraries. In *2016 9th IFIP wireless and mobile networking conference (WMNC)* (pp. 41-48). IEEE. 10.1109/WMNC.2016.7543928

An, Y., & Yan, Y. (2022). Intelligent retrieval method of library document information based on hidden topic mining. *WEB INTELLIGENCE, 20*(2), 93–102. doi:10.3233/WEB-210484

Arshad, M., Khan, A., Ahmed, P., Abbas Shah, N., & Ahmad, P. (2020). Next Generation Data Analytics: Text Mining in Library Practice and Research. *Library Philosophy and Practice (e-Journal).*

Asemi, A., & Asemi, A. (2018). Artificial Intelligence (AI) application in Library Systems in Iran: A taxonomy study. *Library Philosophy and Practice (e-Journal)*. https://digitalcommons.unl.edu/libphilprac/1840

Asemi, A., & Ko, A. (2021). A Novel Combined Business Recommender System Model Using Customer Investment Service Feedback. *34th Bled EConference Digital Support from Crisis to Progressive Change: Conference Proceedings*, 223–237. 10.18690/978-961-286-485-9.17

Asemi, A., Ko, A., & Nowkarizi, M. (2021). Intelligent libraries: A review on expert systems, artificial intelligence, and robots. *Library Hi Tech*, *39*(2), 412–434. doi:10.1108/LHT-02-2020-0038

Ashton, K. (2009). That ''Internet of Things'' thing: In the Real World Things Matter More than Ideas. *RFiD Journal*. Retrieved from http://kevinjashton.com/2009/06/22/the-internet-of-things/

Au, C. H., Law, K. M. Y., Chiu, D. K. W., & Ho, K. K. W. (2022). Investigating mainstreaming strategies of hot cryptocurrencies-wallet. *Proceedings of Australasian Conference on Information Systems (ACIS), 1*. https://aisel.aisnet.org/acis2022/1

Ayanwale, M. A., Sanusi, I. T., Adelana, O. P., Aruleba, K. D., & Oyelere, S. S. (2022). Teachers' Readiness and Intention to teach artificial intelligence in Schools. *Computers and Education: Artificial Intelligence*, *3*, 100099. doi:10.1016/j.caeai.2022.100099

Bagal, D., & Saindane, P. (2019, July). Librany-A Face Recognition and QR Code Technology based Smart Library System. In *2019 International Conference on Communication and Electronics Systems (ICCES)* (pp. 253-258). IEEE. 10.1109/ICCES45898.2019.9002530

Bagley, C. A. (2014). *Makerspaces top trailblazing projects: a LITA guide*. American Library Association.

Bahja, M., Hammad, R., & Butt, G. (2020, July). A user-centric framework for educational chatbots design and development. In *International Conference on Human-Computer Interaction* (pp. 32-43). Springer. 10.1007/978-3-030-60117-1_3

Bailin, K. (2011). Changes in academic library space: A case study at the University of New South Wales. *Australian Academic and Research Libraries*, *42*(4), 342–359. doi:10.1080/0004 8623.2011.10722245

Bai, R., Zhao, J., Li, D., Lv, X., Wang, Q., & Zhu, B. (2020). RNN-based demand awareness in smart library using CRFID. *China Communications*, *17*(5), 284–294. doi:10.23919/JCC.2020.05.021

Bajeh, A. O., Mojeed, H. A., Ameen, A. O., Abikoye, O. C., Salihu, S. A., Abdulraheem, M., & Awotunde, J. B. (2021). Internet of robotic things: its domain, methodologies, and applications. In *Emergence of Cyber Physical System and IoT in Smart Automation and Robotics: Computer Engineering in Automation* (pp. 135–146). Springer International Publishing. doi:10.1007/978-3-030-66222-6_9

Bak, A., Kozik, V., Walczak, M., Fraczyk, J., Kaminski, Z., Kolesinska, B., Smolinski, A., & Jampilek, J. (2018). Towards Intelligent Drug Design System: Application of Artificial Dipeptide Receptor Library in QSAR-Oriented Studies. *Molecules (Basel, Switzerland)*, *23*(8), 1964. Advance online publication. doi:10.3390/molecules23081964 PMID:30082652

Baker, R. S. (2021). *Artificial intelligence and Bringing it all together. OECD Digital Education Outlook: Pushing the frontier with Artificial intelligence, Blockchain and robots*. IECD.

Bandyopadhyay, S., Balamuralidhar, P., & Pal, A. (2013). Interoperations among IoT standards. *Journal of ICT Standardization*, *1*(2), 253–270. doi:10.13052/jicts2245-800X.12a9

Bansal, A., Arora, D., & Suri, A. (2018). Internet of things: Beginning of new era for libraries. *Library Philosophy and Practice*, 1.

Barik, L., Barukab, O., & Ahmed, A. A. (2020). Employing artificial intelligence techniques for student Performance evaluation and teaching strategy enrichment: An innovative approach. *International Journal of Advanced and Applied Sciences*, *7*(11), 10–24. doi:10.21833/ijaas.2020.11.002

Barile, L. (2011). Mobile technologies for libraries: A list of mobile applications and resources for development. *College & Research Libraries News*, *72*(4), 222–228. doi:10.5860/crln.72.4.8545

Barron-Estrada, M., Zatarain-Cabada, & Oramas-Bustillos, R. (2019). Emotion recognition for education using sentiment analysis. *Research in Computing Science, 148*(5).

Beckman, M. (2021). *The Comprehensive Guide: NFTS, Digital Artwork, Blockchain Technology*. Skyhorse Pubishing.

Begum, B. A., & Nandury, S. V. (2023). Data Aggregation Protocols for WSN and IoT Applications–A Comprehensive Survey. *Journal of King Saud University-Computer and Information Sciences*.

Behera, T. M., Mohapatra, S. K., Samal, U. C., Khan, M. S., Daneshmand, M., & Gandomi, A. H. (2019). I-SEP: An improved routing protocol for heterogeneous WSN for IoT-based environmental monitoring. *IEEE Internet of Things Journal*, *7*(1), 710–717. doi:10.1109/JIOT.2019.2940988

Belpaeme, T., Kennedy, J., Ramachandran, A., Scassellati, B., & Tanaka, F. (2018). Social robots for education: *A review. Science Robotics*, *3*(21), eaat5954. Advance online publication. doi:10.1126cirobotics.aat5954 PMID:33141719

Ben-Daya, M., Hassini, E., & Bahroun, Z. (2017). Internet of Things and supply chain management: *A literature review. International Journal of Production Research*, ●●●, 1–24. doi:10.1080/00207543.2017.1402140

Benotti, L., Martínez, M. C., & Schapachnik, F. (2014, June). Engaging high school students using chatbots. In *Proceedings of the 2014 conference on Innovation & technology in computer science education* (pp. 63-68). 10.1145/2591708.2591728

Beqiri, R. (2016). *Artificial intelligence Architecture*. Longread.

Bieraugel, M. (2019). Do your library spaces help entrepreneurs? Space planning for boosting creative thinking. *Supporting Entrepreneurship and Innovation*, *40*, 21–32. doi:10.1108/S0732-067120190000040001

Bieraugel, M., & Neill, S. (2017). Ascending bloom's pyramid:fostering student creativity and innovation in academic library spaces. *College & Research Libraries*, *78*(1), 35–52. doi:10.5860/crl.78.1.35

Bi, S., Wang, C., Zhang, J., Huang, W., Wu, B., Gong, Y., & Ni, W. (2022). A Survey on Artificial Intelligence Aided Internet-of-Things Technologies in Emerging Smart Libraries. *Sensors (Basel)*, *22*(8), 2991. doi:10.339022082991 PMID:35458974

Bischell, J. C. (2018). *Examining parents' perceptions of and preferences toward the use of comics in the classroom* [Doctoral dissertation, The University of Tennessee]. UTC Scholar. https://scholar.utc.edu/cgi/viewcontent.cgi?article=1695&context=theses

Bitstamp. (2022). *What is the Bored Ape Yacht Club? (BAYC)*. https://www.bitstamp.net/learn/web3/what-is-the-bored-ape-yacht-club-bayc/

Bogdanowicz, M. (2015). *Digital Entrepreneurship Barriers and Drivers*. JRC Technical Report European Commission.

Boston Public Library. (2019). *BPL History*. Retrieved from https://www.bpl.org/bpl-history/

Bottorff, T., Glaser, R., Todd, A., & Alderman, B. (2008). Branching out: Communication and collaboration among librarians at multi-campus institutions. *Journal of Library Administration*, *48*(4), 329–363. doi:10.1080/01930820802289391

Botz, A. (n.d.). *Is blockchain the future of art?* https://artbasel.com/news/blockchain-artworld-cryptocurrency-cryptokitties

Botzakis, S., Savitz, R., & Low, D. E. (2017). Adolescents reading graphic novels and comics: What we know from research. In K. A. Hinchman & D. A. Appleman (Eds.), *Adolescent literacies: A handbook of practice-based research* (pp. 310–322). The Guilford Press.

Bourdieu, P. (1992). *An invitation to reflexive sociology*. University of Chicago Press.

Bradley, F. (2022). *Representation of Libraries in Artificial Intelligence Regulations and Implications for Ethics and Practice*. doi:10.1080/24750158.2022.2101911

Brindley, L. (2006). Re-defining the library. *Library Hi Tech*, *24*(4), 484–495. doi:10.1108/07378830610715356

Bucher, K. T., & Manning, M. L. (2004). Bringing graphic novels into a school's curriculum. *The Clearing House: A Journal of Educational Strategies, Issues and Ideas*, *78*(2), 67–72. doi:10.3200/TCHS.78.2.67-72

Bunn, V. (2012). Researching the Tintin effect: How can the active promotion of graphic novels support and enhance boys' enthusiasm for leisure reading? *School Librarian*, *60*(2), 74–76.

Burke, J. (2015). *Making sense: Can makerspaces work in academic libraries?* https://www.ala.org/acrl/sites/ala.org.acrl/files/content/conferences/confsandpreconfs/2015/Burke.pdf

Burrus, D. (2014). *The Internet of Things is Far Bigger than Anyone Realizes*. Retrieved from https://www.wired.com/insights/2014/11/the-internet-of-things-bigger/

Butcher, W., & Street, P.-A. (2009). Lifelong learning with older adults. *Australasian Public Libraries and Information Services*, *22*(2), 64–70.

Buyya, R., Yeo, C. S., Venugopal, S., Broberg, J., & Brandic, I. (2009). Cloud computing and emerging IT platforms: Vision, hype, and reality for delivering computing as the 5th utility. *Future Generation Computer Systems*, *25*(6), 599–616. doi:10.1016/j.future.2008.12.001

Bybee, R. W., Taylor, J. A., Gardner, A., Van Scotter, P., Powell, J. C., Westbrook, A., & Landes, N. (2006). *The BSCS 5E instructional model: Origins and effectiveness*. BSCS. https://media.bscs.org/bscsmw/5es/bscs_5e_full_report.pdf

Bybee, R. W., Taylor, J. A., Gardner, A., Van Scotter, P., Powell, J. C., Westbrook, A., & Landes, N. (2006). *The BSCS 5E instructional model: Origins and effectiveness*. BSCS.

Cao, G., Liang, M., & Li, X. (2018). How to make the library smart? The conceptualization of the smart library. *The Electronic Library*, *36*(5), 811–825. doi:10.1108/EL-11-2017-0248

Castiglione, J. (2008). Facilitating employee creativity in the library environment: An important managerial concern for library administrators. *Library Management*, *29*(3), 159–172. doi:10.1108/01435120810855296

Celik, I., Dindar, M., Muukkonen, H., & Järvelä, S. (2022). The Promises and Challenges of Artificial Intelligence for Teachers: A Systematic Review of Research. *TechTrends*, *66*(4), 616–630. doi:10.100711528-022-00715-y

Cerf, V. G. (2019). Libraries considered hazardous. *Communications of the ACM*, *62*(2), 5–5. doi:10.1145/3302508

Chan, L. (2021). *Virtual Masterpieces: The Dawn of Digital Art*. https://research.hktdc.com/en/article/ODQ1OTg4MjUx

Chan, A. W. Y., Chiu, D. K. W., & Ho, K. K. (2022). Workforce Information Needs for Vocational Guidance System Design. *International Journal of Systems and Service-Oriented Engineering*, *12*(1), 1–16. doi:10.4018/IJSSOE.297134

Chan, A. W. Y., Chiu, D. K. W., Ho, K. K. W., & Wang, M. (2019). Information Needs of Vocational Training from Training Providers' Perspectives. *International Journal of Systems and Service-Oriented Engineering*, *8*(4), 26–42. doi:10.4018/IJSSOE.2018100102

Chan, D. L. H., & Spodick, E. (2014). Space development: A case study of HKUST Library. *New Library World*, *115*(5/6), 250–262. doi:10.1108/NLW-04-2014-0042

Chang, V., Chiu, D. K. W., Ramachandran, M., & Li, C.-S. (2018). Internet of Things, Big Data and Complex Information Systems: Challenges, solutions and outputs from IoTBD 2016, COMPLEXIS 2016 and CLOSER 2016 selected papers and CLOSER 2015 keynote. *Future Generation Computer Systems*, *79*(3), 973–974. doi:10.1016/j.future.2017.09.013

Chan, M. K. Y., Chiu, D. K. W., & Lam, E. T. H. (2020). Effectiveness of overnight learning commons: A comparative study. *Journal of Academic Librarianship*, *46*(6), 102253. Advance online publication. doi:10.1016/j.acalib.2020.102253 PMID:34173399

Chan, M. M. W., & Chiu, D. K. W. (2022). Alert Driven Customer Relationship Management in Online Travel Agencies: Event-Condition-Actions rules and Key Performance Indicators. In A. Naim & S. Kautish (Eds.), *Building a Brand Image Through Electronic Customer Relationship Management*. IGI Global. doi:10.4018/978-1-6684-5386-5.ch012

Chan, T. T. W., Lam, A. H. C., & Chiu, D. K. W. (2020). From Facebook to Instagram: Exploring user engagement in an academic library. *Journal of Academic Librarianship*, *46*(6), 102229. doi:10.1016/j.acalib.2020.102229 PMID:34173399

Chan, V. H. Y., & Chiu, D. K. W. (2023). Integrating the 6C's motivation into reading promotion curriculum for disadvantaged communities with technology tools: A case study of Reading Dreams Foundation in rural China. In A. S. Etim (Ed.), *Adoption and use of technology tools and services by economically disadvantaged communities: Implications for growth and sustainability*. IGI Global.

Chan, V. H. Y., & Chiu, D. K. W. (2023). Integrating the 6C's motivation into reading promotion curriculum for disadvantaged communities with technology tools: A Case Study of reading dreams foundation in rural China. In A. S. Etim (Ed.), *Adoption and use of technology tools and services by economically disadvantaged communities: Implications for growth and sustainability*. IGI Global.

Chan, V. H. Y., & Chiu, D. K. W. (2023). Integrating the 6C's Motivation into Reading Promotion Curriculum for Disadvantaged Communities with Technology Tools: A Case Study of Reading Dreams Foundation in Rural China. In A. Etim & J. Etim (Eds.), *Adoption and Use of Technology Tools and Services by Economically Disadvantaged Communities: Implications for Growth and Sustainability*. IGI Global.

Chan, V. H. Y., Ho, K. K. W., & Chiu, D. K. W. (2022). Mediating effects on the relationship between perceived service quality and public library app loyalty during the COVID-19 era. *Journal of Retailing and Consumer Services*, *67*, 102960. doi:10.1016/j.jretconser.2022.102960

Cheng, J., Yuen, A. H., & Chiu, D. K. (2022). Systematic review of MOOC research in mainland China. *Library Hi Tech*. doi:10.1108/LHT-02-2022-0099

Cheng, W. W. H., Lam, E. T. H., & Chiu, D. K. W. (2020). Social media as a platform in academic library marketing: A comparative study. *The Journal of Academic Librarianship*, *46*(5). doi:10.1016/j.acalib.2020.102188

Cheng, T., & Hou, H. (2022). Intelligent Educational Evaluation of Research Performance between Digital Library and Open Government Data (vol 12, 791, 2022). *Applied Sciences-Basel*, *12*(21). Advance online publication. doi:10.3390/app122110925

Cheng, W., Tian, R., & Chiu, D. K. W. (2023). Travel vlogs influencing tourist decisions: Information preferences and gender differences. *Aslib Journal of Information Management*. Advance online publication. doi:10.1108/AJIM-05-2022-0261

Cheng, Y. (2021). *Improving students' academic performance with artificial intelligence and Semantic Technologies*. The Australian National University.

Chen, M., McNab, A., & Zhang, W. (2021, September 21). How to improve the university library intelligent knowledge service: A system dynamics model. *Journal of Information Science*. Advance online publication. doi:10.1177/01655515211042240

Chen, M., & Shen, C. (2019). The correlation analysis between the service quality of intelligent library and the behavioral intention of users. *The Electronic Library*, *38*(1), 95–112. doi:10.1108/EL-07-2019-0163

Chen, T.-T., & Ke, H.-R. (2017). Public library as a place and breeding ground of social capital: A case of Singang Library. *Malaysian Journal of Library and Information Science*, *22*(1), 45–58. doi:10.22452/mjlis.vol22no1.4

Chen, X. P. (2018). Application of Blockchain Technology in the Library Smart Services. *Modern Information*, *38*(11), 66–71.

Chen, X., Zou, D., Xie, H., Cheng, G., & Liu, C. (2022). Two Decades of Artificial Intelligence in Education: Contributors, Collaborations, Research Topics, Challenges, and Future Directions. *Journal of Educational Technology & Society*, *25*(1), 28–47.

Chen, Y., Chiu, D. K. W., & Ho, K. K. W. (2018). Facilitating the learning of the art of Chinese painting and calligraphy at Chao Shao-an Gallery in Hong Kong. *Micronesian Educators*, *26*, 45–58.

Chen, Y., Chiu, D. K. W., & Ho, K. K. W. (2018). Facilitating the learning of the art of Chinese painting and calligraphy at Chao Shao-an Gallery. *Micronesian Educators*, *26*, 45–58.

Chen, Y., & Xu, L. (2015). The Construction of Smart Library Aimed at Ubiquitous Consumers' Smart Service. *Library Journal*, *34*(08), 4–9.

Cheung, L. S. N., Chiu, D. K. W., & Ho, K. K. W. (2022). A quantitative study on utilizing electronic resources to engage children's reading and learning: Parents' perspectives through the 5E instructional model. *The Electronic Library*, *40*(6), 662–679. doi:10.1108/EL-09-2021-0179

Cheung, T. Y., Ye, Z., & Chiu, D. K. W. (2021). Value chain analysis of information services for the visually impaired: A case study of contemporary technological solutions. *Library Hi Tech*, *39*(2), 625–642. doi:10.1108/LHT-08-2020-0185

Cheung, V. S. Y., Lo, J. C. Y., Chiu, D. K. W., & Ho, K. K. W. (2023). Predicting Facebook's influence on travel products marketing based on the AIDA model. *Information Discovery and Delivery*, *51*(1), 66–73. doi:10.1108/IDD-10-2021-0117

Chin, G. Y. L., & Chiu, D. K. W. (2023). RFID-based Robotic Process Automation for Smart Museums with an Alert-driven Approach. In R. Tailor (Ed.), *Application and Adoption of Robotic Process Automation for Smart Cities*. IGI Global.

Chin, G. Y. L., & Chiu, D. K. W. (2023). RFID-based robotic process automation for smart museums with an alert-driven approach. In R. K. Tailor (Ed.), *Application and adoption of robotic process automation for smart cities*. IGI Global.

Chiu, D. K. W., & Ho, K. K. W. (2022). Special selection on contemporary digital culture and reading. *Library Hi Tech*, *40*(5), 1204–1209. doi:10.1108/LHT-10-2022-516

Chiu, D. K. W., & Ho, K. K. W. (2022b). Editorial: 40th anniversary: Contemporary library research. *Library Hi Tech*, *40*(6), 1525–1531. doi:10.1108/LHT-12-2022-517

Chiu, D. K. W., & Wong, S. W. S. (2023). Reevaluating remote library storage in the digital age: A comparative study. *portal. Portal (Baltimore, Md.)*, *23*(1), 89–109. doi:10.1353/pla.2023.0009

Choi, Y., & Joo, S. (2019, June). Topic detection of online book reviews: preliminary results. In *2019 ACM/IEEE Joint Conference on Digital Libraries (JCDL)* (pp. 418-419). IEEE. 10.1109/JCDL.2019.00098

Chopra, K., Gupta, K., & Lambora, A. (2019, February). Future internet: The internet of things-a literature review. In *2019 International Conference on Machine Learning, Big Data, Cloud and Parallel Computing (COMITCon)* (pp. 135-139). IEEE. 10.1109/COMITCon.2019.8862269

Choy, F. C., & Goh, S. N. (2016). A framework for planning academic library spaces. *Library Management*, *37*(1/2), 13–28. doi:10.1108/LM-01-2016-0001

Christensen, L. L. (2006). Graphic global conflict: Graphic novels in the high school social studies classroom. *Social Studies*, *97*(6), 227–230. doi:10.3200/TSSS.97.6.227-230

Chu, J. L., & Duan, M. Z. (2018). Smart Library and Smart Services. *Library Development*, (04), 85–90.

Chung, C., Chiu, D. K. W., Ho, K. K. W., & Au, C. H. (2020). Applying social media to environmental education: Is it more impactful than traditional media? *Information Discovery and Delivery*, *48*(4), 255–266. doi:10.1108/IDD-04-2020-0047

Clark, L., & Perry, K. A. (2015). *After access: libraries & digital empowerment.* https://alair.ala.org/bitstream/handle/11213/7559/ALA%20DI%20After%20Access_final_12%2017%2015.pdf?sequence=1

Clark, M. (2022). *People are spending millions on NFTs. What? Why?* The Verge.

Clark, R. C., & Mayer, R. E. (2016). *E-learning and the science of instruction: Proven guidelines for consumers and designers of multimedia learning.* John Wiley & Sons. doi:10.1002/9781119239086

Coba, L., Confalonieri, R., & Zanker, M. (2022). RecoXplainer: A Library for Development and Offline Evaluation of Explainable Recommender Systems. *IEEE Computational Intelligence Magazine*, *17*(1), 46–58. doi:10.1109/MCI.2021.3129958

Cointelegraph. (2021). *How to create an NFT: A guide to creating a non-fungible token.* https://cointelegraph.com/nonfungible-tokens-for-beginners/how-to-create-an-nft

Coleman, J. S. (1988). Social capital in the creation of human capital. *American Journal of Sociology*, *94*, S95–S120. doi:10.1086/228943

Collins, A., Tkaczyk, D., Aizawa, A., & Beel, J. (2018). Position Bias in Recommender Systems for Digital Libraries. doi:10.1007/978-3-319-78105-1_37

Connaway, L. S. (2017). Research methods in library and information science (6th ed.). Libraries Unlimited.

Cooper, S., Nesmith, S., & Schwarz, G. (2011). Exploring graphic novels for elementary science and mathematics. *School Library Media Research*, *14*, 1–17.

Coslor, E. (2016). Transparency in an opaque market: Evaluative frictions between "thick" valuation and "thin" price data in the art market. *Accounting, Organizations and Society*, *50*, 13–26. doi:10.1016/j.aos.2016.03.001

Cox, A. (2022). How artificial intelligence might change academic library work: Applying the competencies literature and the theory of the professions. *Journal of the Association for Information Science and Technology*. Advance online publication. doi:10.1002/asi.24635

Cox, A. M., & Mazumdar, S. (2022). Defining artificial intelligence for librarians. *Journal of Librarianship and Information Science*. Advance online publication. doi:10.1177/09610006221142029

Cox, A. M., Pinfield, S., & Rutter, S. (2019). The intelligent library: Thought leaders' views on the likely impact of artificial intelligence on academic libraries. *Library Hi Tech*, *37*(3), 418–435. doi:10.1108/LHT-08-2018-0105

Crawford, P. (2004). Using graphic novels to attract reluctant readers. *Library Media Connection*, *27*, 26–28.

Creswell, J. W., & Guetterman, T. C. (2019). *Educational Research: Planning, Conducting, and Evaluating Quantitative and Qualitative Research* (6th ed.) Pearson.

Crompton, H., & Song, D. (2021). The potential of artificial intelligence in higher education. *Revista Virtual Universidad Católica Del Norte*, *62*(62), 1–4. doi:10.35575/rvucn.n62a1

Cronbach, L. J. (1951). Coefficient alpha and the internal structure of tests. *Psychometrika*, *16*(3), 297–334. doi:10.1007/BF02310555

Compilation of References

Crowley, J. (2015). Graphic novels in the school library: Using graphic novels to encourage reluctant readers and improve literacy. *School Librarian*, *63*(3), 140–142.

Cruz-Cunha, M. M., Miranda, I. M., & Gonçalves, P. (2013). What is Fuzzy Logic? In *Handbook of Research on ICTs and Management Systems for Improving Efficiency in Healthcare and Social Care* (Vols. 1–2). IGI Global. doi:10.4018/978-1-4666-3990-4

Cui, L., Zhang, Z., Gao, N., Meng, Z., & Li, Z. (2019). Radio frequency identification and sensing techniques and their applications—A review of the state-of-the-art. *Sensors (Basel)*, *19*(18), 4012. doi:10.339019184012 PMID:31533321

Cummings, W. R. (2020, March 6). Why are graphic novels so popular amongst students who have ADHA? *Psych Central*. https://blogs.psychcentral.com/childhood-behavioral/2020/03/why-are-graphic-novels-so-popular-amongst-students-who-have-adhd/

Cunningham-Nelson, S., Boles, W., Trouton, L., & Margerison, E. (2019). A review of chatbots in education: practical steps forward. In *30th Annual Conference for the Australasian Association for Engineering Education (AAEE 2019): Educators Becoming Agents of Change: Innovate, Integrate, Motivate* (pp. 299-306). Engineers Australia.

Curry, R. (2017). Makerspaces: A beneficial new service for academic libraries? *Library Review*, *66*(4/5), 201–212. doi:10.1108/LR-09-2016-0081

Cushing, A. L., & Osti, G. (2022). "So how do we balance all of these needs?": How the concept of AI technology impacts digital archival expertise. *The Journal of Documentation*, *79*(7), 12–29. Advance online publication. doi:10.1108/JD-08-2022-0170

Dai, C., & Chiu, D. K. W. (2023). Impact of COVID-19 on reading behaviors and preferences: Investigating high school students and parents with the 5E instructional model. *Library Hi Tech*. Advance online publication. doi:10.1108/LHT-10-2022-0472

Dai, Y. (2018). Research on the Intelligent Service System of Library in the Ubiquitous Information Society. *Journal of Library Science*, *40*(09), 52–55.

Daniel, O. C., Ramsurrun, V., & Seeam, A. K. (2019, September). Smart library seat, occupant and occupancy information system, using pressure and RFID sensors. In *2019 Conference on Next Generation Computing Applications (NextComp)* (pp. 1-5). IEEE. 10.1109/NEXTCOMP.2019.8883610

Davidson, P., & Piché, R. (2016). A survey of selected indoor positioning methods for smartphones. *IEEE Communications Surveys and Tutorials*, *19*(2), 1347–1370. doi:10.1109/COMST.2016.2637663

Davis, A., & Bertot, J. (n.d.). *New Study: Today's Public Libraries Are Thriving Technology Hubs That Millions Rely on for First or Only Choice for Internet Access.* Bill & Melinda Gates Foundation. https://www.gatesfoundation.org/ideas/media-center/press-releases/2006/09/todays-public-libraries-are-thriving-technology-hubs-that-millions-rely-on-for-first-or-only-choice-for-internet-access

De Sarkar, T. (2022). Internet of Things (IOT) and library services. *Library Hi Tech News*, *39*(9), 18–22. doi:10.1108/LHTN-06-2022-0079

Delgado, L., Galvez, D., Hassan, A., Palominos, P., & Morel, L. (2020). Innovation spaces in universities: Support for collaborative learning. *Journal of Innovation Economics and Management*, *31*(1), 123–153. doi:10.3917/jie.pr1.0064

Deng, Q., Allard, B., Lo, P., Chiu, D. K. W., See-To, E. W. K., & Bao, A. Z. R. (2019). The role of the library café as a learning space: A comparative analysis of three universities. *Journal of Librarianship and Information Science*, *51*(3), 823–842. doi:10.1177/0961000617742469

Deng, S., & Chiu, D. K. W. (2022). Analyzing Hong Kong Philharmonic Orchestra's Facebook community engagement with the Honeycomb Model. In M. Dennis & J. Halbert (Eds.), *Community Engagement in the Online Space* (pp. 31–47). IGI Global.

Deng, S., & Chiu, D. K. W. (2023). Analyzing Hong Kong Philharmonic Orchestra's Facebook Community Engagement with the Honeycomb Model. In M. Dennis & J. Halbert (Eds.), *Community Engagement in the Online Space*. IGI Global. doi:10.4018/978-1-6684-5190-8.ch003

Deng, W., Chin, G. Y.-l., Chiu, D. K. W., & Ho, K. K. W. (2022). Contribution of literature thematic exhibition to cultural education: A case study of Jin Yong's Gallery. *Micronesian Educators*, *32*, 14–26.

Deng, W., Chin, G. Y.-l., Chiu, D. K. W., & Ho, K. K. W. (2022). Contribution of Literature Thematic Exhibition to Cultural Education: A Case Study of Jin Yong's Gallery. *Micronesian Educators*, *32*, 14–26.

Determe, J. F., Azzagnuni, S., Singh, U., Horlin, F., & De Doncker, P. (2022). Monitoring large crowds with WiFi: A privacy-preserving approach. *IEEE Systems Journal*, *16*(2), 2148–2159. doi:10.1109/JSYST.2021.3139756

Devi, J. V., Yamini, U. M., Sanjuri, K., & Virinchi, G. (2022). Traditional sentiment and emotion identification system to improve teaching and learning. *Journal of Engineering Sciences*, *13*(5).

Dewi, H. K., Rahim, N. A., Putri, R. F., Wardani, T. I., & Pandin, M. G. R. (2021). *The use of artificial intelligence in English learning among university students: A case study in English Department*. Univerrsitas Airlagga.

Diebold, G., & Han, C. (2022). *How Artificial intelligence can improve K-12 Education in the United States*. Center for Data Innovations.

Ding, S. J., Lam, E. T. H., Chiu, D. K. W., Lung, M. M., & Ho, K. K. W. (2021). Changes in reading behavior of periodicals on mobile devices: A comparative study. *Journal of Librarianship and Information Science*, *53*(2), 233–244. doi:10.1177/0961000620938119

Dokukina, I., & Gumanova, J. (2020). The rise of chatbots–new personal assistants in foreign language learning. *Procedia Computer Science*, *169*, 542–546. doi:10.1016/j.procs.2020.02.212

Compilation of References

Dong, G., Chiu, D. K. W., Huang, P.-S., Lung, M. M., Ho, K. K. W., & Geng, Y. (2021). Relationships between research supervisors and students from coursework-based Master's degrees: Information usage under social media. *Information Discovery and Delivery*, *49*(4), 319–327. doi:10.1108/IDD-08-2020-0100

Downey, E. M. (2011). Graphic novels in curriculum and instruction collections. *Reference and User Services Quarterly*, *49*(2), 181–188. doi:10.5860/rusq.49n2.181

Dsouza, L. (2023). *Are NFTs the Future for the Art World?* Motiva. https://www.motiva.art/blog/are-nfts-the-future-for-the-art-world/

Dukic, Z., Chiu, D. K. W., & Lo, P. (2015). How useful are smartphones for learning? Perceptions and practices of Library and Information Science students from Hong Kong and Japan. *Library Hi Tech*, *33*(4), 545–561. doi:10.1108/LHT-02-2015-0015

Duursma, E., Augustyn, M., & Zuckerman, B. (2008). Reading aloud to children: The evidence. *Archives of Disease in Childhood*, *93*(7), 554–557. doi:10.1136/adc.2006.106336 PMID:18477693

Efe, R. T. (2021). Awareness of the concept of makerspace: the scenario of university libraries in Nigeria. *Library Philosophy and Practice*, 1-17. https://digitalcommons.unl.edu/libphilprac/4952/

El-Keiey, S., ElMenshawy, D., & Hassanein, E. (2022). Student's Performance Prediction based on Personality Traits and Intelligence Quotient using Machine Learning. *International Journal of Advanced Computer Science and Applications*, *13*(9). Advance online publication. doi:10.14569/IJACSA.2022.0130934

Engel, M. M., & Lie, H. D. (2022). Library Self Service System Using Nfc And 2fa Google Authenticator. *Jurnal Teknik Informatika (Jutif)*, *3*(3), 753–761.

Erinfolami, K. (2022, May 7). *Bored ape yacht club: What is it & why are they so expensive?* https://bitkan.com/learn/what-is-bored-ape-and-what-makes-bored-ape-expensive-8695

Ezeamuzie, N. M., Rhim, A. H. R., Chiu, D. K. W., & Lung, M. M. W. (2022). Mobile Technology Usage by Foreign Domestic Helpers: Exploring Gender Differences. *Library Hi Tech*. Advance online publication. doi:10.1108/LHT-07-2022-0350

Fang, J., Chiu, D. K. W., & Ho, K. K. W. (2021). Exploring Cryptocurrency Sentimental with Clustering Text Mining on Social Media. In Z. Sun (Ed.), *Handbook of Research on Intelligent Analytics with Multi-Industry Applications* (pp. 157–171). IGI Global. doi:10.4018/978-1-7998-4963-6.ch007

Fan, K. Y. K., Lo, P., Ho, K. K. W., So, S., Chiu, D. K. W., & Ko, K. H. T. (2020). Exploring the mobile learning needs amongst performing arts students. *Information Discovery and Delivery*, *48*(2), 103–112. doi:10.1108/IDD-12-2019-0085

Farag, H. A., Mahfouz, S. N., & Alhajri, S. (2021). *Artificial Intelligence Investing in Academic Libraries: Reality and Artificial Intelligence Investing in Academic Libraries: Reality and Challenges Challenges*. https://digitalcommons.unl.edu/libphilprac

Farritor, S. (2017). University-based makerspaces: A source of innovation. *Technology and Innovation*, *19*(1), 389–395. doi:10.21300/19.1.2017.389

Farrokhnia, M., Banihashem, S. K., Noroozi, O., & Wals, A. (2023). A SWOT analysis of ChatGPT: Implications for educational practice and research. *Innovations in Education and Teaching International*, 1–15. Advance online publication. doi:10.1080/14703297.2023.2195846

Fei, J. (2021). Research on the Management Mode of Library Security Crisis Early-Warning Based on Competitive Intelligence. *New Century Library*, *42*(06), 45–49.

Feng, Z., Chiu, D., Peng, R., Gong, P., He, K., & Huang, Y. (2016). Facilitating Cloud Process Family Co-evolution by Reusable Process Plug-in: An Open-source Prototype. *IEEE Transactions on Services Computing*, *10*(6), 854–867. doi:10.1109/TSC.2015.2504973

Ferguson, S. (2012). Are Public Libraries Developers of Social Capital? A Review of Their Contribution and Attempts to Demonstrate It. *The Australian Library Journal*, *61*(1), 22–33. doi:10.1080/00049670.2012.10722299

Fletcher-Spear, K., Jenson-Benjamin, M., & Copeland, T. (2005). The truth about graphic novels: A format, not a genre. *The ALAN Review*, *32*(2), 37–44. doi:10.21061/alan.v32i2.a.7

Fong, K. C. H., Au, C. H., Lam, E. T. H., & Chiu, D. K. W. (2020). Social network services for academic libraries: A study based on social capital and social proof. *Journal of Academic Librarianship*, *46*(1), 102091. doi:10.1016/j.acalib.2019.102091

Fortnow, M., Terry, Q., & Nguyen, K. (2021). *The NFT Handbook: How to Create, Sell and Buy Non-Fungible Tokens*. Ascent Audio.

Frangoudes, F., Hadjiaros, M., Schiza, E. C., Matsangidou, M., Tsivitanidou, O., & Neokleous, K. (2021, July). An Overview of the Use of Chatbots in Medical and Healthcare Education. In *International Conference on Human-Computer Interaction* (pp. 170-184). Springer. 10.1007/978-3-030-77943-6_11

Frederick, A., Run, Y., Frederick, A., & Run, Y. (2019). The Role of Academic Libraries in Research Data Management: A Case in Ghanaian University Libraries. *OAlib*, *6*(3), 1–16. doi:10.4236/oalib.1105286

Fu, J., Lv, J., & Li, X. (2011). The Research about Intelligent Search Engine and in Digital Library Personalization Services. *Advanced Materials Research*, *143–144*, 333–337. doi:10.4028/WWW. SCIENTIFIC.NET/AMR.143-144.333

Fung, R. H. Y., Chiu, D. K. W., Ko, E. H. T., Ho, K. K. W., & Lo, P. (2016). Heuristic usability evaluation of University of Hong Kong Libraries' mobile website. *Journal of Academic Librarianship*, *42*(5), 581–594. doi:10.1016/j.acalib.2016.06.004

Fu, P., Zou, X. Z., & Wu, D., etal. (2018). Retrospective Analysis and Prediction: Artificial Intelligence and Its Applications in Libraries. *Tushu Qingbao Zhishi*, *35*(02), 50–60.

Fu, Y. X. (2018). Research on Application of Artificial Intelligence in Library Construction. *Library Work and Study*, *40*(09), 47–51.

Gade, A., & Angal, Y. (2018). Real-Time Intelligent NI myRIO-Based Library Management Robotic System Using LabVIEW. doi:10.1007/978-981-10-5520-1_36

Gao, W., Lam, K. M., Chiu, D. K. W., & Ho, K. K. W. (2021). A Big data Analysis of the Factors Influencing Movie Box Office in China. In Z. Sun (Ed.), *Handbook of Research on Intelligent Analytics with Multi-Industry Applications* (pp. 232–249). IGI Global. doi:10.4018/978-1-7998-4963-6.ch011

Gao, Y. (2018). Implementation of an Intelligent Library System Based on WSN and RFID. *International Journal of Online Engineering*, *14*(5), 211–224. doi:10.3991/ijoe.v14i05.8601

Gao, Y. (2020). Intelligent library knowledge innovation service system based on multimedia technology. *Personal and Ubiquitous Computing*, *24*(3), 333–345. doi:10.100700779-019-01269-2

Gao, Z., & Zou, G. (2021). Intelligent Data Mining of Computer-Aided Extension Residential Building Design Based on Algorithm Library. *Complexity*, *2021*, 1–9. Advance online publication. doi:10.1155/2021/6690746

Gardner, J., O'Leary, M., & Yuan, L. (2021). Artificial intelligence in educational assessment: Breakthrough? Or buncombe and ballyhoo? Journal of Computer Assisted Learning. doi:10.1111/jcal.12577

Garrido-Hidalgo, C., Roda-Sanchez, L., Ramírez, F. J., Fernández-Caballero, A., & Olivares, T. (2023). Efficient online resource allocation in large-scale LoRaWAN networks: A multi-agent approach. *Computer Networks*, *221*, 109525. doi:10.1016/j.comnet.2022.109525

Gasparini, A., & Kautonen, H. (2022). Understanding Artificial Intelligence in Research Libraries: An Extensive Literature Review. *LIBER Quarterly*, *32*(1), 1–36. doi:10.53377/lq.10934

Gerner Nielsen, B. (2014). Public libraries and lifelong learning. *Perspectives of Innovations, Economics and Business*, *14*(2), 94–102. doi:10.15208/pieb.2014.11

Ghasempour, A. (2019). Internet of things in smart grid: Architecture, applications, services, key technologies, and challenges. *Inventions (Basel, Switzerland)*, *4*(1), 22. doi:10.3390/inventions4010022

Gigli, M., & Koo, S. (2011). Internet of Things: Services and applications categorization. *Advances in Internet of Things*, *1*(2), 27–31. doi:10.4236/ait.2011.12004

Gocen, A & Aydemir, F. (2020). AI in education and schools. *Research on Education and Media, 12*(1).

Goddard, J. (2020). Public Libraries Respond to the COVID-19 Pandemic, Creating a New Service Model. *Information Technology and Libraries*, *39*(4). Advance online publication. doi:10.6017/ital.v39i4.12847

Gong, J. Y., Schumann, F., Chiu, D. K. W., & Ho, K. K. W. (2017). Tourists' mobile information seeking behavior: An investigation on China's youth. *International Journal of Systems and Service-Oriented Engineering*, 7(1), 58–76. doi:10.4018/IJSSOE.2017010104

González-Calatayud, V., Prendes-Espinosa, P., & Roig-Vila, R. (2021). Artificial Intelligence for Student Assessment: A Systematic Review. *Applied Sciences (Basel, Switzerland)*, 11(12), 5467. doi:10.3390/app11125467

Gopalakrishnan, K. (2020). Security vulnerabilities and issues of traditional wireless sensors networks in IoT. *Principles of internet of things (IoT) ecosystem: Insight paradigm*, 519-549.

Griffith, P. E. (2010). Graphic novels in the secondary classroom and school libraries. *Journal of Adolescent & Adult Literacy*, 54(3), 181–189. doi:10.1598/JAAL.54.3.3

Grush, M. (2015, December 1). *Rethinking the makerspace*. Campus Technology. https://campustechnology.com/Articles/2015/12/01/Rethinking-the-Makerspace.aspx?p=1

Gstalder, S. H. (2017). *Understanding library space planning* (Order No. 10289537). Available from ProQuest Dissertations & Theses A&I. (1954088754). https://repository.upenn.edu/dissertations/AAI10289537

Gu, B., & Tanoue, K. (2022). Research on Library Space Layout and Intelligent Optimization Oriented to Readers' Needs. *Mathematical Problems in Engineering*, 2022, 1–12. Advance online publication. doi:10.1155/2022/4426091

Guo, Y., Lam, A. H. C., Chiu, D. K. W., & Ho, K. K. W. (2022). Perceived quality of reference service with WhatsApp: A quantitative study from user perspectives. *Information Technology and Libraries*, 41(3). Advance online publication. doi:10.6017/ital.v41i3.14325

Gupta, B. B., & Quamara, M. (2018). An overview of Internet of Things (IoT): Architectural aspects, challenges, and protocols. *Concurrency and Computation*, e4946(21). doi:10.1002/cpe.4946

Gupta, T., Tripathi, R., Shukla, M. K., & Mishra, S. (2020). Design and development of IoT based smart library using line follower robot. *Int. J. Emerg. Technol*, 11(2), 1105–1109. https://d1wqtxts1xzle7.cloudfront.net/82644987/Design_20and_20Development_20of_20IOT_20Based_20Smart_20Library_20using_20Line_20Follower_20Robot_20Rohit_20Tripathi_202798-libre.pdf

Haken, H. (2005). *Secrets of Nature's Success Synergetics: the Teaching of Interaction*. Shanghai Translation Publishing House.

Halaweh, M. (2023). ChatGPT in education: Strategies for responsible implementation. *Contemporary Educational Technology*, 15(2), ep421. doi:10.30935/cedtech/13036

Halpern, D. (2005). *Social capital*. Polity Press.

Hanifan, L. J. (1920). *The community center*. Silver, Burdett.

Compilation of References

Hart, G. (2013). Social capital: A fresh vision for public libraries in South Africa? *South African Journal of Library and Information Science*, *73*(1), 14–24. doi:10.7553/73-1-1331

Hashkey Group. (n.d.). *Combing NFT Summer: What are the future directions of NFT worth looking forward to?* https://www.hashkey.com/en/insights/combing-nft-summer-what-are-the-future-directions-of-nft-worth-looking-forward-to.html

HashKey Group. (n.d.). *HashKey Releases Commemorative NFT for Hong Kong FinTech Week.* https://www.hashkey.com/en/insights/hashkey-releases-commemorative-nft-for-hong-kong-fintech-week.html

Hatch, M. (2013). *The maker movement manifesto: rules for innovation in the new world of crafters, hackers, and tinkerers*. McGraw Hill Professional.

Haugeland, J. (Ed.). (1985). *Artificial intelligence.The very idea*. MIT Press.

Herron, J., & Kaneshiro, K. (2017). A university-wide collaborative effort to designing a makerspace at an academic health sciences library. *Medical Reference Services Quarterly*, *36*(1), 1–8. doi:10.1080/02763869.2017.1259878 PMID:28112641

He, Z., Chiu, D. K. W., & Ho, K. K. W. (2022). Weibo Analysis on Chinese Cultural Knowledge for Gaming. In Z. Sun (Ed.), *Handbook of Research on Foundations and Applications of Intelligent Business Analytics* (pp. 320–349). doi:10.4018/978-1-7998-9016-4.ch015

Ho, K. K. W., Takagi, T., Ye, S., Au, C. K., & Chiu, D. K. W. (2018). *The Use of Social Media for Engaging People with Environmentally Friendly Lifestyle – A Conceptual Model*. SIG Green Pre ICIS Workshop. https://aisel.aisnet.org/sprouts_proceedings_siggreen_2018/2/

Hobert, S. (2019). How are you, chatbot? evaluating chatbots in educational settings–results of a literature review. *DELFI 2019.* https://www.chatcompose.com/chatbot-learning.html https://www.investopedia.com/terms/c/chatbot.asp https://yellow.ai/chatbots/use-cases-of-chatbots-in-education industry/#:~:text=Chatbots%20act%20as%20a%20data,all%20their%20students%20at%20onc

Ho, C. Y., Chiu, D. K. W., & Ho, K. K. W. (2023). Green space development in academic libraries: A case study in Hong Kong. In V. Okojie & M. O. Igbinovia (Eds.), *Global perspectives on sustainable library practices* (pp. 142–156). IGI Global. doi:10.4018/978-1-6684-5964-5.ch010

Ho, C. Y., Chiu, D. K., & Ho, K. K. (2023). Green space development in academic libraries: a case study in Hong Kong. In *Global Perspectives on Sustainable Library Practices* (pp. 142–156). IGI Global.

Huang, P. S., Paulino, Y. C., So, S., Chiu, D. K. W., & Ho, K. K. W. (2021). Editorial. *Library Hi Tech*, *39*(3), 693–695. doi:10.1108/LHT-09-2021-324

Huang, P. S., Paulino, Y. C., So, S., Chiu, D. K. W., & Ho, K. K. W. (2022). Guest editorial: COVID-19 Pandemic and Health Informatics Part 2. *Library Hi Tech*, *40*(2), 281–285. doi:10.1108/LHT-04-2022-447

Huang, P.-S., Paulino, Y. C., So, S., Chiu, D. K. W., & Ho, K. K. W. (2023). Guest editorial: COVID-19 pandemic and health informatics part 3. *Library Hi Tech, 41*(1), 1–6. doi:10.1108/LHT-02-2023-585

Huang, X., Craig, P., Lin, H., & Yan, Z. (2016). SecIoT: A security framework for theInternet of Things. *Security and Communication Networks, 9*(16), 3083–3094. doi:10.1002ec.1259

Huang, Z., Chung, W., Ong, T.-H., & Chen, H. (2002). A graph-based recommender system for digital library. *Proceedings of the 2nd ACM/IEEE-CS Joint Conference on Digital Libraries*, 65–73. 10.1145/544220.544231

Hu, B., Tang, W., & Xie, Q. (2022). A two-factor security authentication scheme for wireless sensor networks in IoT environments. *Neurocomputing, 500*, 741–749. doi:10.1016/j.neucom.2022.05.099

Hui, S. C., Kwok, M. Y., Kong, E. W. S., & Chiu, D. K. W. (2023). Information security and technical issues of cloud storage services: A qualitative study on university students in Hong Kong. *Library Hi Tech*. Advance online publication. doi:10.1108/LHT-11-2022-0533

Hunt, J. M., Goodner, R. E., & Jay, A. (2020). *Comparing male and female student responses on MIT maker survey: understanding the implications and strategies for more inclusive spaces.* https://ijamm.pubpub.org/pub/m71isalo/release/1?readingCollection=ee0f734d

Hussain, A. (2023). Use of artificial intelligence in the library services: Prospects and challenges. *Library Hi Tech News, 40*(2), 15–17. doi:10.1108/LHTN-11-2022-0125

Huvila, I., Lloyd, A., Budd, J. M., Palmer, C., & Toms, E. (2016). Information work in information science research and practice. *Proceedings of the Association for Information Science and Technology, 53*(1), 1–5. doi:10.1002/pra2.2016.14505301004

Iantovics, L., Rotar, C., & Nechita, E. (2018). Intelligent University Library Information Systems to Support Students Efficient Learning. doi:10.1007/978-3-030-04224-0_17

IFLA FAIFE. (2020). *IFLA Statement on Libraries and Artificial Intelligence.* https://repository.ifla.org/handle/123456789/1646

Igbinovia, M. O. (2021). Internet of things in libraries and focus on its adoption in developing countries. *Library Hi Tech News, 38*(4), 13–17. doi:10.1108/LHTN-05-2021-0020

Igbinovia, M. O., & Okuonghae, O. (2021). Internet of Things in contemporary academic libraries: Application and challenges. *Library Hi Tech News, 38*(5), 1–4. doi:10.1108/LHTN-05-2021-0019

Ilkka, T. (2018). *The impact of AI on learning, teaching and education: policies for the future. JRS Science for Policy Report.* European Commission.

Irankunda, D. (2021). *Development of radio frequency identification based library management and anti-theft system: a case of east African community region* [Doctoral dissertation]. NM-AIST.

Ismail, M., & Ade-Ibijola, A. (2019, November). Lecturer's apprentice: A chatbot for assisting novice programmers. In *2019 International Multidisciplinary Information Technology and Engineering Conference (IMITEC)* (pp. 1-8). IEEE. 10.1109/IMITEC45504.2019.9015857

James, B. (2014, March 11). Redefining literacy with graphic novels. *Minnesota English Journal*. Retrieved from https://minnesotaenglishjournalonline.org/2014/03/11/praxis-strategies-for-teaching-literature-3-graphic-novels/

James, R. (2018, October 8). *Libraries as Community Hubs*. The Social History Society. https://socialhistory.org.uk/shs_exchange/libraries-as-community-hubs/

Jamil, H. (2021). Architecture of an Intelligent Personal Health Library for Improved Health Outcomes. doi:10.1109/ICDH52753.2021.00012

Jennings, K. A., Rule, A. C., & Vander Zanden, S. M. (2014). Fifth graders' enjoyment, interest, and comprehension of graphic novels compared to heavily-illustrated and traditional novels. *International Electronic Journal of Elementary Education*, 6(2), 257–274.

Jeon, K. E., She, J., Soonsawad, P., & Ng, P. C. (2018). BLE beacons for internet of things applications: Survey, challenges, and opportunities. *IEEE Internet of Things Journal*, 5(2), 811–828. doi:10.1109/JIOT.2017.2788449

Jiang, M., Lam, A.H.C., Chiu, D.K.W., & Ho, K.K.W. (2023). Social media aids for business learning: A quantitative evaluation with the 5E instructional model. *Education and Information Technology*. doi:10.1007/s10639-023-11690-z

Jiang, J. (2019). The main purpose and principles of Artificial Intelligence ethics under the perspective of risk. *Information and Communications Technology and Policy*, 45(06), 13–16.

Jiang, M., Hu, Y., Worthey, G., Dubnicek, R. C., Underwood, T., & Downie, J. S. (2021, September). Evaluating BERT's Encoding of Intrinsic Semantic Features of OCR'd Digital Library Collections. In *2021 ACM/IEEE Joint Conference on Digital Libraries (JCDL)* (pp. 308-309). IEEE. 10.1109/JCDL52503.2021.00045

Jiang, S., & Yang, X. (2022). Analysis of the Influence of Library Information on the Utilization of Regional Environmental and Ecological Resources: From the Perspective of Intelligent Adaptive Learning. *Journal of Environmental and Public Health*, 2022, 1–12. Advance online publication. doi:10.1155/2022/1110105 PMID:36213048

Jiang, X., Chiu, D. K. W., & Chan, C. T. (2023). Application of the AIDA model in social media promotion and community engagement for small cultural organizations: A case study of the Choi Chang Sau Qin Society. In M. Dennis & J. Halbert (Eds.), *Community Engagement in the Online Space* (pp. 48–70). IGI Global. doi:10.4018/978-1-6684-5190-8.ch004

Jing, Z. (2019). *Logical Construction and Realization of Intelligent Service in Ai Library*. doi:10.1109/ICSGEA.2019.00069

Johnson, E. D. M. (2017). The right place at the right time: creative spaces in libraries. *The Future of Library Space, 36*, 1-35. doi:10.1108/S0732-067120160000036001

Johnson, C. A. (2010). Do public libraries contribute to social capital? A preliminary investigation into the relationship. *Library & Information Science Research*, *32*(2), 147–155. doi:10.1016/j. lisr.2009.12.006

Johnson, C. A. (2012). How do public libraries create social capital? An analysis of interactions between library staff and patrons. *Library & Information Science Research*, *34*(1), 52–62. doi:10.1016/j.lisr.2011.07.009

Johnson, C. A. (2015). Social Capital and Library and Information Science Research: Definitional Chaos or Coherent Research Enterprise? *Information Research*, *20*(4), n4.

Johnson, C. A., & Griffis, M. R. (2009). A place where everybody knows your name? Investigating the relationship between public libraries and social capital. *Canadian Journal of Information and Library Science*, *33*(3-4), 159–191.

Johnson, C. A., & Griffis, M. R. (2014). The effect of public library use on the social capital of rural communities. *Journal of Librarianship and Information Science*, *46*(3), 179–190. doi:10.1177/0961000612470278

Jomsri, P. (2018). FUCL mining technique for book recommender system in library service. doi:10.1016/j.promfg.2018.03.081

Jones, M. (1985). Application of artificial intelligence within education. *Computers & Mathematics with Applications (Oxford, England)*, *11*(5), 517–526. doi:10.1016/0898-1221(85)90054-9

Julian, K. D., & Parrott, D. J. (2017). Makerspaces in the library: Science in a student's hands. *Journal of Learning Spaces*, *6*(2), 13–21. https://eric.ed.gov/?id=EJ1152687

Jung, S., & Kim, S. (2020). A Study of Promoting Method a traditional market by implementing RFID technology and 6W1H context awareness. *Journal of Convergence for Information Technology*, *10*(10), 9–14.

Kaba, A., & Ramaiah, C. K. (2019). The internet of things: opportunities and challenges for libraries. *Library Philosophy and Practice (e-Journal),* 3704. Retrieved from https://digitalcommons.unl. edu/libphilprac/3704

Kaliraj, P., & Devi, T. (2022). *Artificial Intelligence Theory, Models, And Applications* (P. Kaliraj & T. Devi, Eds.; 1st ed.). CRC Press.

Karthikeyan, D., Arumbu, V. P., Surendhirababu, K., Selvakumar, K., Divya, P., Suhasini, P., & Palanisamy, R. (2021). Sophisticated and modernized library running system with OCR algorithm using IoT. *Indonesian Journal of Electrical Engineering and Computer Science*, *24*(3), 1680–1691. doi:10.11591/ijeecs.v24.i3.pp1680-1691

Kaushal, V., & Yadav, R. (2022). The Role of Chatbots in Academic Libraries: An Experience-based Perspective. *Journal of the Australian Library and Information Association*, *71*(3), 215–232. doi:10.1080/24750158.2022.2106403

Kee, H. C. L., & Chiu, D. K. W. (2023). Building social capital in major U.S. public libraries. In D. K. W. Chiu & K. K. W. Ho (Eds.), *Emerging technology-based services and systems in libraries, educational institutions, and non-profit organizations*. IGI Global.

Kengam, J. (2020). *Artificial intelligence in education*. Researchgate.

Khan, I. M., Ahmad, A. R., Jabeur, N & Mahdi, N. (2021) An Artificial intelligence approach to monitoring student performance and devise preventive measures. *Smart Learning Environments, 17*.

Khan, R., Khan, S. U., Zaheer, R., & Khan, S. (2012, December). Future internet: the internet of things architecture, possible applications and key challenges. In *2012 10th international conference on frontiers of information technology* (pp. 257-260). IEEE. 10.1109/FIT.2012.53

Khan, A. U., Zhang, Z., Chohan, S. R., & Rafique, W. (2022). Factors fostering the success of IoT services in academic libraries: A study built to enhance the library performance. *Library Hi Tech, 40*(6), 1976–1995. doi:10.1108/LHT-06-2021-0179

Khanna, A., & Kaur, S. (2020). Internet of things (IoT), applications and challenges: A comprehensive review. *Wireless Personal Communications, 114*(2), 1687–1762. doi:10.100711277-020-07446-4

Khemani, D. (2013). *A First Course in Artificial Intelligence*. McGraw Hill Education.

Khoir, S., Du, J. T., Davison, R. M., & Koronios, A. (2017). Contributing to social capital: An investigation of Asian immigrants' use of public library services. *Library & Information Science Research, 39*(1), 34–45. doi:10.1016/j.lisr.2017.01.005

Khosrow-Pour, M. (2014). Academic Libraries in the Digital Age. In C. O. Jefferson (Ed.), *Encyclopedia of Information Science and Technology* (3rd ed., pp. 4815–4822). IGI Global. doi:10.4018/978-1-4666-5888-2

Kim, J., & Myers, R. (2012). Discovering greatness: YALSA's great graphic novels for teens list. *Young Adult Library Services, 10*(3), 39–41.

Koedinger, K. R., Brunskill, E., Baker, R. S. J. D., McLaughlin, E. A., & Stamper, J. (2013). *New Potentials for Data-Driven Intelligent Tutoring System Development and Optimization*. Association for the Advancement of Artificial Intelligence. doi:10.1609/aimag.v34i3.2484

Koh, K., & Abbas, J. (2015). Competencies for information professionals in learning labs and makerspaces. *Journal of Education for Library and Information Science, 56*(2), 114–129. Advance online publication. doi:10.12783/issn.2328-2967/56/2/3

Koliarakis, A., Krouska, A., Troussas, C., & Sgouropoulou, C. (2022). *Modified collaborative filtering for hybrid recommender systems and personalized search: The case of digital library*. doi:10.1109/SMAP56125.2022.9942020

Kranich, N. (2001). Libraries create social capital: A unique, if fleeting, opportunity to carve out a new library mission. (United States)(Brief Article). *Library Journal, 126*(19), 40.

Kroski, E. (Ed.). (2017). *The makerspace librarian's sourcebook*. Facet Publishing.

Kufakunesu, R., Hancke, G. P., & Abu-Mahfouz, A. M. (2020). A survey on adaptive data rate optimization in lorawan: Recent solutions and major challenges. *Sensors (Basel)*, *20*(18), 5044. doi:10.339020185044 PMID:32899454

Kuhail, M. A., Alturki, N., Alramlawi, S., & Alhejori, K. (2022). Interacting with educational chatbots: A systematic review. *Education and Information Technologies*, 1–46.

Kumar, V. (2018). *Role of artificial intelligence in education systems, its challenges and opportunities in India*. Academic Press.

Kumar, J. A. (2021). Educational chatbots for project-based learning: Investigating learning outcomes for a team-based design course. *International Journal of Educational Technology in Higher Education*, *18*(1), 1–28. doi:10.118641239-021-00302-w PMID:34926790

Kumar, V. (2018). Application of internet of things for smart libraries: An overview. *International Journal of Multidisciplinary Education Research*, *7*(2). https://www.researchgate.net/publication/328475575_APPLICATIONS_OF_INTERNET_OF_THINGS_FOR_SMART_LIBRARIES_AN_OVERVIEW

Kurzweil, R. (1990). *The age of intelligent machines*. MIT Press.

Lam, S. Y., & Li, Y. (2023, May 30). *By embracing NFTs and the metaverse, Hong Kong can remodel its identity and global role*. South China Morning Post. https://www.scmp.com/comment/letters/article/3222159/embracing-nfts-and-metaverse-hong-kong-can-remodel-its-identity-and-global-role

Lam, A. H. C., Ho, K. K. W., & Chiu, D. K. W. (2023). Instagram for student learning and library promotions? A quantitative study using the 5E Instructional Model. *Aslib Journal of Information Management*, *75*(1), 112–130. doi:10.1108/AJIM-12-2021-0389

Lam, E. T. H., Au, C. H., & Chiu, D. K. W. (2019). Analyzing the use of Facebook among university libraries in Hong Kong. *Journal of Academic Librarianship*, *45*(3), 175–183. doi:10.1016/j.acalib.2019.02.007

Landbot. (2021). *Product Recommendation Chatbot: No-Code Tutorial*. Landbot.Io. https://landbot.io/blog/product-recommendation-chatbot-for-ecommerce

Lau, K. P., Chiu, D. K. W., Ho, K. K. W., Lo, P., & See-To, E. W. K. (2017). Educational usage of mobile devices: Differences between postgraduate and undergraduate students. *Journal of Academic Librarianship*, *43*(3), 201–208. doi:10.1016/j.acalib.2017.03.004

Lau, K. S. N., Lo, P., Chiu, D. K. W., Ho, K. K. W., Jiang, T., Zhou, Q., Percy, P., & Allard, B. (2020). Library and learning experiences turned mobile: A comparative study between LIS and non-LIS students. *Journal of Academic Librarianship*, *46*(2), 102103. Advance online publication. doi:10.1016/j.acalib.2019.102103

Law, T. Y., Leung, F. C. W., Chiu, D. K. W., Lo, P., Lung, M. M.-W., Zhou, Q., Xu, Y., Lu, Y., & Ho, K. K. W. (2019). Mobile learning usage of LIS students in Mainland China. *International Journal of Systems and Service-Oriented Engineering*, *9*(2), 12–34. doi:10.4018/IJSSOE.2019040102

Lee, C. T., Pan, L. Y., & Hsieh, S. H. (2021). Artificial intelligent chatbots as brand promoters: A two-stage structural equation modelling-artificial neural network approach. *Internet Research.*

Leedy, P., & Ormrod, J. (2015). *Practical research planning and design* (11th ed.). Pearson Harlow.

Lee, I., & Lee, K. (2015). The Internet of Things (IoT): Applications, investments, and challenges for enterprises. *Business Horizons*, *58*(4), 431–440. doi:10.1016/j.bushor.2015.03.008

Lee, R. J. (2017). Campus-library collaboration with makerspaces. *Public Services Quarterly*, *13*(2), 108–116. doi:10.1080/15228959.2017.1303421

Lei, S. Y., Chiu, D. K. W., Lung, M. M., & Chan, C. T. (2021). Exploring the aids of social media for musical instrument education. *International Journal of Music Education*, *39*(2), 187–201. doi:10.1177/0255761420986217

Leonardi, S., & Torchiano, M. (2023). Educational Chatbot to Support Question Answering on Slack. In *International Conference in Methodologies and intelligent Systems for Technology Enhanced Learning* (pp. 20-25). Springer. 10.1007/978-3-031-20617-7_4

Leorke, D., Wyatt, D., & McQuire, S. (2018). More than just a library: Public libraries in the smart city. *City. Cultura e Scuola*, *15*, 37–44. doi:10.1016/j.ccs.2018.05.002

Leung, L. M., Chiu, D. K. W., & Lo, P. (2020) School librarians' view of cooperation with public libraries: A win-win situation in Hong Kong. *School Library Research, 23*. www.ala.org/aasl/slr/volume23/leung-chiu-lo

Leung, T.N., Hui, Y.M., Luk, C.K.L., Chiu, D.K.W., & Kevin, K.K.W. (2022). Evaluating Facebook as aids for learning Japanese: learners' perspectives, *Library Hi Tech.* . doi:10.1108/LHT-11-2021-0400

Leung, G. S. M. (2007). A study of the effects of parental support and children's resourcefulness on the academic stress of senior primary school students in Hong Kong (China). *Dissertation Abstracts International. A, The Humanities and Social Sciences*, *67*(8-A), 3180.

Leung, T. N., Chiu, D. K. W., Ho, K. K. W., & Luk, C. K. L. (2022). User perceptions, academic library usage, and social capital: A correlation analysis under COVID-19 after library renovation. *Library Hi Tech*, *40*(2), 304–322. doi:10.1108/LHT-04-2021-0122

Liang, X. (2018). Internet of Things and its applications in libraries: A literature review. *Library Hi Tech*, *38*(1), 67–77. doi:10.1108/LHT-01-2018-0014

Liao, P. K., & Shieh, J. C. (2015, July). The development of library mobile book-finding system based on NFC. In *2015 IIAI 4th International Congress on Advanced Applied Informatics* (pp. 148-153). IEEE. 10.1109/IIAI-AAI.2015.198

Library of Congress. (2014, June 19). *Cataloging Graphic Novels*. https://www.loc.gov/aba/cyac/graphicnovels.html

Li, D.-Y., Xie, S.-D., Chen, R.-J., & Tan, H.-Z. (2016). Design of Internet of Things System for Library Materials Management using UHF RFID. *2016 IEEE International Conference on RFID Technology and Applications (RFID-TA)*, 44-48. 10.1109/RFID-TA.2016.7750755

Lightfoot, M. (2016). The Emergence of Digital Social Capital in Education. In I. R. Haslam & M. S. Khine (Eds.), *Leveraging Social Capital in Systemic Education Reform* (pp. 43–66). Brill Sense. doi:10.1007/978-94-6300-651-4_3

Li, K. K., & Chiu, D. K. W. (2022). A worldwide quantitative review of the iSchools' archival education. *Library Hi Tech*, *40*(5), 1497–1518. doi:10.1108/LHT-09-2021-0311

Lin, C.-H., Chiu, D.K.W., Lam, K. T. (2022). Hong Kong Academic Librarians' Attitudes Towards Robotic Process Automation. *Library Hi Tech.* . doi:10.1108/LHT-03-2022-0141

Lin, C. J., & Mubarok, H. (2021). Learning analytics for investigating the mind map-guided AI Chatbot approach in an EFL flipped speaking classroom. *Journal of Educational Technology & Society*, *24*(4), 16–35.

Lin, W. H., Chang, S. S., Li, P., Chiu, T. T., & Lou, S. J. (2019, May). Exploration of usage behavioral model construction for university library electronic resources from Deep Learning Multilayer perceptron. In *2019 IEEE International Conference on Consumer Electronics-Taiwan (ICCE-TW)* (pp. 1-2). IEEE. 10.1109/ICCE-TW46550.2019.8991756

Li, Q., Wong, J., & Chiu, D. K. W. (2023). (in press). School library reading support for students with dyslexia: A qualitative study in the digital age. *Library Hi Tech*. Advance online publication. doi:10.1108/LHT-03-2023-0086

Li, S. M., Lam, A. H. C., & Chiu, D. K. W. (2023). Digital transformation of ticketing services: A value chain analysis of POPTICKET in Hong Kong. In J. D. Santos & I. V. Pereira (Eds.), *Management and Marketing for Improved Retail Competitiveness and Performance*. IGI. Global.

Li, S. M., Lam, A. H. C., & Chiu, D. K. W. (2023). Digital transformation of ticketing services: A value chain analysis of POPTICKET in Hong Kong. In J. D. Santos, I. V. Pereira, & P. B. Pires (Eds.), *Management and marketing for improved retail competitiveness and performance*. IGI. Global.

Li, S., Chiu, D. K. W., & Ho, K. K. W. (2023). Social Media Analytics for Non-governmental Organizations: A Case Study of Hong Kong Next Generation Arts. In Z. Sun (Ed.), *Driving Socioeconomic Development with Big Data: Theories, Technologies, and Applications*. IGI Global. doi:10.4018/978-1-6684-5959-1.ch013

Li, S., Xie, Z., Chiu, D. K. W., & Ho, K. K. W. (2023). Sentiment analysis and topic modeling regarding online classes on the Reddit Platform: Educators versus learners. *Applied Sciences (Basel, Switzerland)*, *13*(4), 2250. Advance online publication. doi:10.3390/app13042250

Liu, G. R., Zhang, L., & Huang, X. Q. (2017). Research on the Dynamic Mechanism of Cooperation between Library and Society—Based on the Synergy Theory. *Library Work and Study*, *39*(07), 26–30.

Liu, H., & Gao, Z. Q. (2017). Construction of Agricultural Science and Technology Service System with Multi-agent Coordination. *Agricultural Engineering*, 7(02), 146–149.

Liu, N., & Ye, Z. (2021). Empirical research on the blockchain adoption - based on TAM. *Applied Economics*, 53(37), 4263–4275. doi:10.1080/00036846.2021.1898535

Liu, Y. (2018). Opinions on the Construction of Smart Libraries Based on Cloud Computing and the Internet of Things. *Journal of Library Science*, 40(11), 124–127.

Liu, Y. (2021). Research on Data Ethical Risk Identification and Interest Coordination Strategy in Smart Library. *New Century Library*, 42(12), 11–15.

Liu, Y. (2022). Investigating users' willingness of acceptance for background music service in intelligent library. *Library Hi Tech*, 40(1), 33–44. doi:10.1108/LHT-02-2019-0052

Liu, Y. (2022). The Risk of Personal Information Breach of Library Readers and Its Scenario-Based Goverance. *Tushuguanxue Yanjiu*, 44(06), 18–26.

Liu, Y., Chiu, D. K. W., & Ho, K. K. W. (2023). Short-form videos for public library marketing: Performance analytics of Douyin in China. *Applied Sciences (Basel, Switzerland)*, 13(6), 3386. Advance online publication. doi:10.3390/app13063386

Liu, Y., Ye, H., & Sun, H. (2021). Mobile phone library service: Seat management system based on WeChat. *Library Management*, 42(6/7), 421–435. doi:10.1108/LM-09-2020-0132

Li, X. (2006). Library as incubating space for innovations: Practices, trends and skill sets. *Library Management*, 27(6/7), 370–378. doi:10.1108/01435120610702369

Li, X. (2022). Application Analysis of RFID in Library Automation Management. *International Journal of e-Collaboration*, 18(2), 1–10. doi:10.4018/IJeC.304036

Li, Y. K. (2019). *Research on Quality Improvement of Information System Based on Defect Analysis*. Dalian Maritime University.

Lo, Z. (2022). *Hong Kong NFT experts decode why the blockchain isn't as safe as you think.* https://www.tatlerasia.com/culture/arts/hong-kong-art-tech-screens-guru-new-blockchain-solution-interview

Lo, C. K. (2023). *What is the impact of ChatGPT on education? A rapid review of the literature.* Education Sciences & MDPI. doi:10.3390/educsci13040410

Long, J., & Hicks, J. (2020). *Makerspaces for adults: Best practices and great projects.* Rowman & Littlefield.

Lo, P., Allard, B., Anghelescu, H. G. B., Xin, Y., Chiu, D. K. W., & Stark, A. J. (2020). Transformational leadership practice in the world's leading academic libraries. *Journal of Librarianship and Information Science*, 52(4), 972–999. doi:10.1177/0961000619897991

Lo, P., Allard, B., Wang, N., & Chiu, D. K. W. (2020a). Servant leadership theory in practice: North America's leading public libraries. *Journal of Librarianship and Information Science*, *52*(1), 249–270. doi:10.1177/0961000618792387

Lo, P., Chan, H. H. Y., Tang, A. W. M., Chiu, D. K. W., Cho, A., Ho, K. K. W., See-To, E., & He, J. (2019). Visualising and Revitalising Traditional Chinese Martial Arts – Visitors' Engagement and Learning Experience at the 300 Years of Hakka KungFu. *Library Hi Tech*, *37*(2), 273–292. doi:10.1108/LHT-05-2018-0071

Lo, P., & Chiu, D. K. W. (2015). Enhanced and Changing Roles of School Librarians under the Digital Age. *New Library World*, *116*(11/12), 696–710. doi:10.1108/NLW-05-2015-0037

Lo, P., Chiu, D. K. W., Cho, A., & Allard, B. (2018). *Conversations with leading academic and research library directors: International perspectives on library management*. Chandos Publishing.

Lo, P., Chiu, D. K. W., & Chu, W. (2013). Modeling your college library after a commercial bookstore? The Hong Kong Design Institute Library experience. *Community & Junior College Libraries*, *19*(3-4), 59–76. doi:10.1080/02763915.2014.915186

Lo, P., Cho, A., & Chiu, D. K. W. (2017). *World's Leading National, Public, Monastery and Royal Library Directors: Leadership, Management, Future of Libraries*. Walter de Gruyter GmbH. doi:10.1515/9783110533347

Lo, P., Cho, A., Law, B. K. K., Chiu, D. K. W., & Allard, B. (2017). Progressive Trends in Electronic Resources Management among Academic Libraries in Hong Kong. *Library Collections, Acquisitions & Technical Services*, *40*(1-2), 28–37. doi:10.1080/14649055.2017.1291243

Lo, P., Hsu, W. E., Wu, S. H. S., Travis, J., & Chiu, D. K. W. (2021). *Creating a Global Cultural City via Public Participation in the Arts: Conversations with Hong Kong's Leading Arts and Cultural Administrators*. Nova Science Publishers.

Lo, P., Yu, K., & Chiu, D. K. W. (2015). A Research Agenda for the Enhancing Roles and Practice of School Librarians in Hong Kong's 21st Century Learning Environments. *School Libraries Worldwide*, *21*(1), 19–37. doi:10.29173lw6881

Lu, S.S., Tian, R., & Chiu, D.K.W. (2023). Why do people not attend public library programs in the current digital age? *Library Hi Tech*. doi:10.1108/LHT-04-2022-0217

Lu, K., Pan, X. Y., Liu, H., & Ren, B. B. (2020). Research on Ethics Norms of Library Smart Service from the Perspective of AI Governance. *The Library*, *48*(06), 21–28.

Lund, B. D., Omame, I., Tijani, S., & Agbaji, D. (2020). Perceptions toward Artificial Intelligence among Academic Library Employees and Alignment with the Diffusion of Innovations' Adopter Categories. In *College & Research Libraries* (Vol. 81, Issue 5). https://crl.acrl.org/index.php/crl/rt/printerFriendly/24516/32350

Compilation of References

Lund, D., MacGillivray, C., Turner, V., & Morales, M. (2014). Worldwide and Regional Internet of Things (IoT) 2014-2020 Forecast: A Virtuous Circle of Proven Value and Demand. *International Data Corporation (IDC)*. Retrieved from http://branden.biz/wp-content/uploads/2017/06/IoT-worldwide_regional_2014-2020-forecast.pdf

Lund, B. D., Wang, T., Mannuru, N. R., Nie, B., Shimray, S., & Wang, Z. (2023). ChatGPT and a New Academic Reality: AI-Written Research Papers and the Ethics of the Large Language Models in Scholarly Publishing. *Journal of the Association for Information Science and Technology*. Advance online publication. doi:10.1002/asi.24750

Lu, X. H. (2016). Research on Smart Library Management and Service Based on Wearable Technology. *Journal of Library Science*, *38*(10), 114–117.

Lybeck, R., Koiranen, I., & Koivula, A. (2023). From digital divide to digital capital: The role of education and digital skills in social media participation. *Universal Access in the Information Society*. Advance online publication. doi:10.100710209-022-00961-0

Mabunda, K. (2020). *An Intelligent Chatbot for Guiding Visitors and Locating Venues* [Doctoral dissertation]. University of Johannesburg.

Madakam, S., Ramaswamy, R., & Tripathi, S. (2015). Internet of Things (IoT): A literature review. *Journal of Computer and Communications*, *3*(5), 164–173. doi:10.4236/jcc.2015.35021

Maepa, M. R., & Moeti, M. N. (2021, December). IoT-Based Smart Library Seat Occupancy and Reservation System using RFID and FSR Technologies for South African Universities of Technology. In *Proceedings of the International Conference on Artificial Intelligence and Its Applications* (pp. 1-8). 10.1145/3487923.3487933

Mahey, M., Al-Abdulla, A., Ames, S., Bray, P., Candela, G., Chambers, S., Derven, C., Dobreva-McPherson, M., Gasser, K., Karner, S., Kokegei, K., Laursen, D., Potter, A., Straube, A., Wagner, S., & Wilms, L. (2019). *Open a GLAM Lab*. Qatar University Press.

Ma, J., Chiu, D. K. W., & Tang, J. K. (2016). Exploring the Use of Social Media to Advance K12 Science Education. *International Journal of Systems and Service-Oriented Engineering*, *6*(4), 47–59. doi:10.4018/IJSSOE.2016100104

Mak, M. Y. C., Poon, A. Y. M., & Chiu, D. K. W. (2022). Using Social Media as Learning Aids and Preservation: Chinese Martial Arts in Hong Kong. In S. Papadakis & A. Kapaniaris (Eds.), *The Digital Folklore of Cyberculture and Digital Humanities* (pp. 171–185). IGI Global. doi:10.4018/978-1-6684-4461-0.ch010

Malathy, S., & Kantha, P. (2013). Application of mobile technologies to libraries. *DESIDOC Journal of Library and Information Technology*, *33*(5).

Malipatil, N., Roopashree, V., Gowda, R. S., Shobha, M. R., & Kumar, H. S. (2020). RFID based library management system. *Int. J. Res. Eng. Sci. Manag*, *3*(7), 112–115.

Mann, H. B., & Whitney, D. R. (1947). On a test of whether one of two random variables is stochastically larger than the other. *Annals of Mathematical Statistics*, *18*(1), 50–60. https://www.jstor.org/stable/2236101. doi:10.1214/aoms/1177730491

Maric, J. (2018). The gender-based digital divide in maker culture: Features, challenges and possible solutions. *Journal of Innovation Economics and Management*, *3*(27), 147–168. doi:10.3917/jie.027.0147

Martinez-Martin, E., Ferrer, E., Vasilev, I., & Del Pobil, A. P. (2021). The uji aerial librarian robot: A quadcopter for visual library inventory and book localisation. *Sensors (Basel)*, *21*(4), 1079. doi:10.339021041079 PMID:33557363

Mathew, A. N., & Paulose, J. (2021). NLP-based personal learning assistant for school education. *International Journal of Electrical & Computer Engineering, 11*(5).

Ma, X. T. (2016). Design of Library Risk Alarm System Based on Big Data Analysis. *Library Theory and Practice*, *38*(08), 81–84.

Ma, Y. J., Gam, H. J., & Banning, J. (2017). Perceived ease of use and usefulness of sustainability labels on apparel products: Application of the technology acceptance model. *Fashion and Textiles*, *4*(1), 1–20. doi:10.118640691-017-0093-1

Ma, Y., & Zheng, J. M. (2015). Analysis of Main and Branch Library System Mode Based on Synergistic and Cluster Theory. *The Library*, *43*(07), 80–84.

McClanahan, B. J., & Nottingham, M. (2019). A suite of strategies for navigating graphic novels: A dual coding approach. *The Reading Teacher*, *73*(1), 39–50. doi:10.1002/trtr.1797

Melo, M., & Nichols, J. T. (Eds.). (2020). *Re-making the library makerspace: Critical theories, reflections, and practices*. Library Juice Press.

Mendoza, S., Hernández-León, M., Sánchez-Adame, L. M., Rodríguez, J., Decouchant, D., & Meneses-Viveros, A. (2020, July). Supporting student-teacher interaction through a chatbot. In *International conference on human-computer interaction* (pp. 93-107). Springer.

Mendoza-Silva, G. M., Matey-Sanz, M., Torres-Sospedra, J., & Huerta, J. (2019). BLE RSS measurements dataset for research on accurate indoor positioning. *Data*, *4*(1), 12. doi:10.3390/data4010012

Meng, Y., Chu, M. Y., & Chiu, D. K. W. (2023). The impact of COVID-19 on museums in the digital era: Practices and Challenges in Hong Kong. *Library Hi Tech*, *41*(1), 130–151. doi:10.1108/LHT-05-2022-0273

Merelo, J. J., Castillo, P. A., Mora, A. M., Barranco, F., Abbas, N., Guillén, A., & Tsivitanidou, O. (2022). Exploring the role of chatbots and messaging applications in higher education: a teacher's perspective. In *International Conference on Human-Computer Interaction* (pp. 205-223). Springer. 10.1007/978-3-031-05675-8_16

Michalak, R., & Rysavy, M. D. T. (2019). Academic libraries in 2018: A comparison of makerspaces within academic research libraries. *Supporting Entrepreneurship and Innovation*, *40*, 67–88. doi:10.1108/S0732-067120190000040008

Miller, R. A., & Shortliffe, E. H. (2022). Corrigendum to: The roles of the US National Library of Medicine and Donald A.B. Lindberg in revolutionizing biomedical and health informatics. *Journal of the American Medical Informatics Association : JAMIA*, *29*(5), 1025. doi:10.1093/jamia/ocac026 PMID:35226056

Minhaj, S. U., Mahmood, A., Abedin, S. F., Hassan, S. A., Bhatti, M. T., Ali, S. H., & Gidlund, M. (2023). Intelligent Resource Allocation in LoRaWAN using Machine Learning Techniques. *IEEE Access : Practical Innovations, Open Solutions*, *11*, 10092–10106. doi:10.1109/ACCESS.2023.3240308

Mo, Q. (2021). *How NFTs are transforming culture, society and financial investments*. https://www.asiaglobalonline.hku.hk/how-nfts-are-transforming-culture-society-and-financial-investments

Moeller, R. A. (2011). "Aren't these boy books?": High school students' readings of gender in graphic novels. *Journal of Adolescent & Adult Literacy*, *54*(7), 476–484. doi:10.1598/JAAL.54.7.1

Moeller, R. A., & Becnel, K. E. (2021). Recommended reading: Comparing elementary/middle school graphic novel collections to recommended reading lists. *Children & Libraries*, *19*(2), 6–13. doi:10.5860/cal.19.2.6

Moersch, C. M. (2014). *Improving achievement with digital age best practices*. Corwin. doi:10.4135/9781506374482

Mohd Rahim, N. I., & Iahad, N., Yusof, A. F., & Al-Sharafi, M. (2022). AI-Based chatbots adoption model for higher-education institutions: A hybrid PLS-SEM-Neural network modelling approach. *Sustainability*, *14*(19), 12726. doi:10.3390u141912726

Molnár, G., & Szuts, Z. (2018, September). The role of chatbots in formal education. In *2018 IEEE 16th International Symposium on Intelligent Systems and Informatics (SISY)* (pp. 197-202). IEEE. 10.1109/SISY.2018.8524609

Mol, S. E., & Bus, A. (2011). To read or not to read: A meta-analysis of print exposure from infancy to early adulthood. *Psychological Bulletin*, *137*(2), 267–296. doi:10.1037/a0021890 PMID:21219054

Monti, L., Mirri, S., Prandi, C., & Salomoni, P. (2019, December). Preservation in smart libraries: An experiment involving iot and indoor environmental sensing. In *2019 IEEE Global Communications Conference (GLOBECOM)* (pp. 1-6). IEEE. 10.1109/GLOBECOM38437.2019.9014149

Moorefield-Lang, H. (2015). Change in the making: makerspaces and the ever-changing landscape of libraries. *TechTrends, 59*(3), 107-112. doi:10.1007/s11528-015-0860-z

Moorefield-lang, H. (2019). Lesson learned: Intentional implementation of second makerspaces. *RSR. Reference Services Review*, *47*(1), 37–47. Advance online publication. doi:10.1108/RSR-07-2018-0058

Morgan, J. (2014). A Simple Explanation of 'The Internet of Things'. *Forbes*. Retrieved from https://www.forbes.com/sites/jacobmorgan/2014/05/13/simple-explanation-internet-things-that-anyone-can-understand/

Morley, J., Floridi, L., Kinsey, L., & Elhalal, A. (2020). From what to how: An initial review of publicly available AI ethics tools, methods and research to translate principles into practices. *Science and Engineering Ethics*, 26(4), 2141–2168. doi:10.100711948-019-00165-5 PMID:31828533

Morrison, L. (2017, April 14). *The research behind graphic novels and young learners* [Blog post]. https://www.ctd.northwestern.edu/blog/research-behind-graphic-novels-and-young-learners

Mouha, R. (2021). Internet of Things (IoT). *Journal of Data Analysis and Information Processing*, 9(2), 77–101. doi:10.4236/jdaip.2021.92006

Mplus. (n.d.). *Museums and digital culture: A Field of productive tension*. https://www.mplus.org.hk/en/events/museums-and-digital-culture-a-field-of-productive-tension/

Mueller, J. P., & Massaron, L. (2018). *Artificial intelligence for dummies*. John Wiley & Sons.

Murphy, R. F. (2019). *Artificial intelligence application to suppirtK-12 teachers and teaching, A review of promising applications, opportunities and challenges. Perspectives, Expert insights on tertiary timely*. Rand Corporation.

Nakamoto, S. (2008). Bitcoin: A peer-to-peer electronic cash system. *Decentralized Business Review*, 21260.

Negi, D. S. (2014). Using mobile technologies in libraries and information centers. *Library Hi Tech News*, 31(5), 14–16. doi:10.1108/LHTN-05-2014-0034

Negnevitsky, M. (2005). *Artificial intelligence: a guide to intelligent systems* (2nd ed.). Addison-Wesley.

Nesmith, S., Cooper, S., Schwarz, G., & Walker, A. (2016). Are graphic novels always "cool"? Parent and student perspectives on elementary mathematics and science graphic novels: The need for action research by school leaders. *Planning and Changing*, 47(3-4), 228–245.

Ngoc, H. V., & Thi, M. N. (2021). Sentiment analysis of students' review on online course. A transfer learning method. *Proceedings of the International Conference on Industrial Engineering and Operations Management*.

Ng, T. C. W., Chiu, D. K. W., & Li, K. K. (2022). Motivations of choosing archival studies as major in the i-Schools: Viewpoint between two universities across the Pacific Ocean. *Library Hi Tech*, 40(5), 1483–1496. doi:10.1108/LHT-07-2021-0230

Ngu, A. H. H., Gutierrez, M., Metsis, V., Nepal, S., & Sheng, M. Z. (2017). IoT Middleware: A Survey on Issues and Enabling technologies. *IEEE Internet of Things Journal*, 4(1), 1–20. doi:10.1109/JIOT.2016.2615180

Nguyen, P., Di Ruscio, D., Di Rocco, J., Di Sipio, C., & Di Penta, M. (2021). *Adversarial Machine Learning: On the Resilience of Third-party Library Recommender Systems.* doi:10.1145/3463274.3463809

Ni, Y., Lam, A. H. C., & Chiu, D. K. W. (2023). Leveraging Online Communities for Building Social Capital in University Libraries: A Case Study of Fudan University Medical Library. In *Balance and Boundaries in Creating Meaningful Relationships in Online Higher Education.* IGI Global. Retrieved from https://peterboroughtownlibrary.org/history/

Ni, Y., Lam, A. H. C., & Chiu, D. K. W. (2023). Leveraging online communities for building social capital in university libraries: A case study of Fudan University Medical Library. In S. H. Jarvie, & C. Metz (Eds.), Balance and boundaries in creating meaningful relationships in online higher education. IGI Global.

Nie, Y. (2018). The Way to Construct the Intelligent Library-Taking Nanyang College Library as an example. Academic Press.

Ni, J., Chiu, D. K. W., & Ho, K. K. W. (2022). Information search behavior among Chinese self-drive tourists in the smartphone era. *Information Discovery and Delivery*, *50*(3), 285–296. doi:10.1108/IDD-05-2020-0054

Nilsson, N. J. (1998). *Artificial intelligence: A new synthesis.* Morgan Kaufman.

Nord, J. H., Koohang, A., & Paliszkiewicz, J. (2019). The Internet of Things: Review and theoretical framework. *Expert Systems with Applications*, *133*, 97–108. doi:10.1016/j.eswa.2019.05.014

Noura, H., Hatoum, T., Salman, O., Yaacoub, J. P., & Chehab, A. (2020). LoRaWAN security survey: Issues, threats and possible mitigation techniques. *Internet of Things*, *12*, 100303. doi:10.1016/j.iot.2020.100303

Nugroho, P. A., Anna, N. E. V., & Ismail, N. (2023). The shift in research trends related to artificial intelligence in library repositories during the coronavirus pandemic. *Library Hi Tech.* Advance online publication. doi:10.1108/LHT-07-2022-0326

Nunez, J. M., & Lantada, A. D. (2020). Artificial Intelligence Aided Engineering Education: State of the Art, Potentials and Challenges. *International Journal of Engineering Education, 36*(6), 1740–1751.

OECED. (2020). *Trustworthy Ai in Education: Promises and Challenges.* Background paper for the G20 AI Dialogue, Digital Economy Task Force Meeting. Riyadh, Saudi Arabia.

Okonkwo, C. W., & Ade-Ibijola, A. (2021). Chatbots applications in education: A systematic review. *Computers and Education: Artificial Intelligence*, *2*, 100033. doi:10.1016/j.caeai.2021.100033

Okpala, H. N. (2016). Making a makerspace case for academic libraries in Nigeria. *New Library World*, *117*(9/10), 568–586. doi:10.1108/NLW-05-2016-0038

Okunlaya, R. O., Syed Abdullah, N., & Alias, R. A. (2022). Artificial intelligence (AI) library services innovative conceptual framework for the digital transformation of university education. *Library Hi Tech*, *40*(6), 1869–1892. doi:10.1108/LHT-07-2021-0242

Ola, A. C. (2020). Funding and ICT use as determinants of sustainable library development. *Covenant Journal of Library and Information Science*, *3*(1), 32–49.

Onyancha, O. B. (2015). An informetrics view of the relationship between internet ethics, computer ethics and cyberethics. *Library Hi Tech*, *33*(3), 387–408. doi:10.1108/LHT-04-2015-0033

Ornes, S. (2016). The Internet of Things and the explosion of interconnectivity. *Proceedings of the National Academy of Sciences of the United States of America*, *113*(40), 11059–11060. doi:10.1073/pnas.1613921113 PMID:27702874

Osawaru, K. E., Dime, A. I., & Okonjo, E. H. (2020). The right time for makerspaces in Nigerian academic libraries: Perceived benefits and challenges. *International Journal on Integrated Education*, *3*(10), 103–115. doi:10.31149/ijie.v3i10.694

Ouatu, B. I., & Gifu, D. (2021). Chatbot, the Future of Learning? In *Ludic, Co-design and Tools Supporting Smart Learning Ecosystems and Smart Education* (pp. 263–268). Springer. doi:10.1007/978-981-15-7383-5_23

Ouyang, F., & Jiao, P. (2021). *Artificial intelligence in education: The three paradigms. Computers and Education: Artificial intelligence*. Elsevier.

Ozeer, A., Sungkur, Y., & Nagowah, S. D. (2019, December). Turning a traditional library into a smart library. In *2019 International Conference on Computational Intelligence and Knowledge Economy (ICCIKE)* (pp. 352-358). IEEE. 10.1109/ICCIKE47802.2019.9004242

Oz, H., & Efecioglu, E. (2015). Graphic novels: An alternative approach to teach English as a foreign language. *Journal of Language and Linguistic Studies*, *11*(1), 75–90.

Palo, T., & Tähtinen, J. (2013). Networked business model development for emerging technology-based services. *Industrial Marketing Management*, *42*(5), 773–782. doi:10.1016/j.indmarman.2013.05.015

Panahi, U., & Bayılmış, C. (2023). Enabling secure data transmission for wireless sensor networks based IoT applications. *Ain Shams Engineering Journal*, *14*(2), 101866. doi:10.1016/j.asej.2022.101866

Pang, L. (2022). Intelligent Big Information Retrieval of Smart Library Based on Graph Neural Network (GNN) Algorithm. *Computational Intelligence and Neuroscience*, *2022*, 1–12. Advance online publication. doi:10.1155/2022/1475069 PMID:35875784

Pannucci, C. J., Wilkins, E. G., & Pannucci, C. (2010). Identifying and Avoiding Bias in Research. *Plastic and Reconstructive Surgery*, *126*(2), 619–625. doi:10.1097/PRS.0b013e3181de24bc PMID:20679844

Park, S. (2019). Analyzing Library Space Use Patterns in a Public Library through Smartphone WiFi. *Jeongbo Gwan'ri Hag'hoeji, 36*(1), 295–313.

Patrickson, B. (2021). What do blockchain technologies imply for digital creative industries? *Creativity and Innovation Management, 30*(3), 585–595. doi:10.1111/caim.12456

Paullier, A., & Sotelo, R. (2020). *A Recommender Systems' algorithm evaluation using the Lenskit library and MovieLens databases.* doi:10.1109/BMSB49480.2020.9379914

Pence, H. E. (2019). Artificial intelligence in education: New wine in old wineskins. *Journal of Educational Technology Systems, 48*(1), 5–13. doi:10.1177/0047239519865577

Pera, M. (2015). *Libraries and the "Internet of Things". OCLC Symposium shows benefits, raises questions.* Available at: https://americanlibrariesmagazine.org/blogs/thescoop/libraries-and-the-internet-of-things/

Pera, M. (2015). *Libraries and the "Internet of Things": OCLC Symposium shows benefits, raises questions.* Retrieved from https://americanlibrariesmagazine.org/blogs/thescoop/libraries-and-the-internet-of-things/

Pérez, J. Q., Daradoumis, T., & Puig, J. M. M. (2020). Rediscovering the use of chatbots in education: A systematic literature review. *Computer Applications in Engineering Education, 28*(6), 1549–1565. doi:10.1002/cae.22326

Perles, A., Pérez-Marín, E., Mercado, R., Segrelles, J. D., Blanquer, I., Zarzo, M., & Garcia-Diego, F. J. (2018). An energy-efficient internet of things (IoT) architecture for preventive conservation of cultural heritage. *Future Generation Computer Systems, 81*, 566–581. doi:10.1016/j.future.2017.06.030

Perrier, L., Blondal, E., Ayala, A. P., Dearborn, D., Kenny, T., Lightfoot, D., Reka, R., Thuna, M., Trimble, L., & MacDonald, H. (2017). Research data management in academic institutions: A scoping review. *PLoS One, 12*(5), e0178261. Advance online publication. doi:10.1371/journal.pone.0178261 PMID:28542450

Pham, H. T., & Tanner, K. (2015). Collaboration Between Academics and Library Staff: A Structurationist Perspective. *Australian Academic and Research Libraries, 46*(1), 2–18. doi:10.1080/00048623.2014.989661

Pluzhnikova, N. N. (2020, October). Technologies of artificial intelligence in educational management. In *2020 International Conference on Engineering Management of Communication and Technology (EMCTECH)* (pp. 1-6). IEEE. 10.1109/EMCTECH49634.2020.9261561

Poole, D., Mackworth, A. K., & Goebel, R. (1998). *Computational intelligence. A logical approach.* Oxford University Press.

Prabhu, V. S., Abinaya, R. M., Archana, G., & Aishwarya, R. (2021). IoT-based automatic library management robot. In *Advances in Smart System Technologies: Select Proceedings of ICFSST 2019* (pp. 567-575). Springer Singapore. 10.1007/978-981-15-5029-4_46

Preuveneers, D., & Berbers, Y. (2008). Internet of things: A context-awareness perspective. In *The Internet of things* (pp. 287–308). Auerbach Publications.

Private Museum. (n.d.). *Art NFTs: a Revolution in the Art Industry.* https://www.privatemuseum.art/blog/guide/art-nfts-revolution-art-industry

Pujara, S. M., & Satyanarayanab, K. V. (2015). Internet of Things and libraries. *Annals of Library and Information Studies*, *62*, 186–190.

Putnam, R. D. (1995). Turning in, Turning out: The Strange Disappearance of social capital in America. *PS, Political Science & Politics*, *28*(4), 1–20. doi:10.2307/420517

Putnam, R. D. (Ed.). (2002). *Democracies in Flux: The Evolution of Social Capital in Contemporary Society.* Oxford University Press. doi:10.1093/0195150899.001.0001

Qian, Y. (2022). *The Semantic Framework of Library Intelligent Question Answering System Based on Exploratory Search Behavior.* doi:10.1109/CCAI55564.2022.9807737

Qi, Y. F. (2013). A New Way of Urban Public Service Supply in the Acceleration of Urbanization: Multi-coordinated Network Supply. *Contemporary World and Socialism*, *34*(01), 165–168.

Radniecki, T. (2017). Supporting 3D modeling in the academic library. *Library Hi Tech*, *35*(2), 240–250. doi:10.1108/LHT-11-2016-0121

Radniecki, T., & Winterman, M. (2020). Leveraging student expertise for niche services. *RSR. Reference Services Review*, *48*(2), 287–306. doi:10.1108/RSR-11-2019-0083

Rajan, S. S., Esmail, M., & Musthafa, K. M. (2022). Repositioning Academic Libraries as a Hub of technology enhanced learning space: Innovations and Challenges. *Library Philosophy and Practice*, 1-14. Retrieved from https://media.proquest.com/media/hms/PFT/1/DOO6M?_s=KnuGNUzf73G%2BpYJcmo%2BmkOrneZ8%3D

Rashid, T. A., & Aziz, N. K. (2016). *Student academic performance using artificial Intelligence.* ZANCO Jou.

Renaissance Learning. (2021). *What kids are reading 2021.* https://www.renaissance.com/wkar-report/?int=https://www.renaissance.com/wkar/

Rhanoui, M., Mikram, M., Yousfi, S., Kasmi, A., & Zoubeidi, N. (2022). A hybrid recommender system for patron driven library acquisition and weeding. *Journal of King Saud University-Computer and Information Sciences*, *34*(6), 2809–2819. doi:10.1016/j.jksuci.2020.10.017

Richardson, E. M. (2017). "Graphic novels are real books": Comparing graphic novels to traditional text novels. *Delta Kappa Gamma Bulletin*, *83*(5), 24–31.

Rich, E., & Knight, K. (1991). *Artificial intelligence* (2nd ed.). McGraw Hill.

Robins, D. (2002). Information Architecture in Library and Information Science Curricula. *Bulletin of the American Society for Information Science*, *28*(2), 20–22. doi:10.1002/bult.231

Rodriguez, S., & Mune, C. (2022). Uncoding library chatbots: Deploying a new virtual reference tool at the San Jose State University library. *RSR. Reference Services Review*, *50*(3/4), 392–405. doi:10.1108/RSR-05-2022-0020

Roos, S. (2018). Chatbots in education: A passing trend or a valuable pedagogical tool? IEEE.

Ross, L. (2022). *How to make and sell NFTs.* https://www.benzinga.com/money/how-to-make-your-own-nft

Rubei, R., Di Sipio, C., Di Rocco, J., Di Ruscio, D., & Nguyen, P. (2022). *Endowing third-party libraries recommender systems with explicit user feedback mechanisms.* doi:10.1109/SANER53432.2022.00099

Rubei, R., Di Ruscio, D., Di Sipio, C., Di Rocco, J., & Nguyen, P. (2022). Providing upgrade plans for third-party libraries: A recommender system using migration graphs. *Applied Intelligence*, *52*(10), 12000–12015. doi:10.100710489-021-02911-4

Sabzalieva, E., & Valentini, A. (2023). *ChatGPT and artificial intelligence in higher education. A quick start guide.* UNESCO.

Said, O., & Masud, M. (2013). Towards internet of things: Survey and future vision. *International Journal of Computer Networks*, *5*(1), 1–17.

Salas-Pilco, S. Z., Xiao, K., & Oshima, J. (2022). Artificial intelligence and new technologies in inclusive education for minority students: A Systematic Review. *Sustainability (Basel)*, *14*(13572), 13572. Advance online publication. doi:10.3390u142013572

Salisu, M. S. (n.d.). *Non-Fungible Tokens (Nfts): Future of Digital Art & Collectibles.* LinkedIn. https://www.linkedin.com/pulse/non-fungible-tokens-nfts-future-digital-art-m-seun-salisu-

Samin, H & Azim, T. (2019). Knowledge-based recommender systems for academia using machine learning: a case study on Higher Education landscape of Pakistan. *IEEE. Research Gate.*

Sanji, M., Behzadi, H., & Gomroki, G. (2022). Chatbot: An intelligent tool for libraries. *Library Hi Tech News*, *39*(3), 17–20. doi:10.1108/LHTN-01-2021-0002

Schachter, K. (2022). *10 best NFT marketplaces.* https://www.fool.com/the-ascent/cryptocurrency/nft-marketplaces/

Scholastic. (2018). *A guide to using graphic novels with children and teens.* https://www.scholastic.com/content/dam/teachers/lesson-plans/18-19/Graphic-Novel-Discussion-Guide-2018.pdf

Schopfel, J. (2018). Smart libraries. *Infrastructures*, *3*(4), 43. doi:10.3390/infrastructures3040043

Schuemie, M. J., van der Straaten, P., Krijn, M., & van der Mast, C. A. P. G. (2004, July 5). *Research on Presence in Virtual Reality: A Survey (world).* Http://Www.Liebertpub.Com/Cpb

Seebacher, S., & Schüritz, R. (2017). Blockchain Technology as an Enabler of Service Systems: A Structured Literature Review. *Exploring Service Science*, *279*, 12–23.

Seo, K., Tang, J., Roll, I., Fels, S., & Yoon, D. (2021). The impact of artificial intelligence on learner-instructor interaction in online learning. *International Journal of Educational Technology in Higher Education*, *18*(1), 54. doi:10.118641239-021-00292-9 PMID:34778540

Shang, S., Yu, Z., Geng, A., Xu, X., Ma, H., & Wang, G. (2022). Intelligent Optimization Method of Resource Recommendation Service of Mobile Library Based on Digital Twin Technology. *Computational Intelligence and Neuroscience*, *2022*, 1–10. Advance online publication. doi:10.1155/2022/3582719 PMID:36065374

Shan, Z., Chen, Y., & Shao, B. (2021). A Case Study of Risk Regulation of Smart Library Transition Abroad. *Tushuguanxue Yanjiu*, *43*(01), 2–8.

Sharma, A. (2015). *The tech behind Internet of things*. Retrieved from https://www.pcquest.com/the-tech-behind-internet-things/

Sharma, R. (2023). *Non-Fungible Token (NFT): What It Means and How It Works*. Investopedia. https://www.investopedia.com/non-fungible-tokens-nft-5115211

Sharma, J. (2019). Role of Public Library to growing Digital Literacy in our Society. *International Research Journal of Multidisciplinary Science & Technology*, *1*(6), 295–297.

Sha, Y. Z. (2004). A Study of Library Professional Ethics. *Journal of Library Science in China*, (4), 22–25.

Shen, X. (2022). *Hong Kong NFT gamers among the world's most enthusiastic*. https://www.scmp.com/tech/tech-trends/article/3179694/hong-kong-nft-gamers-among-worlds-most-enthusiastic-only-behind

Shen, Y. (2018). *Digital Library Systems in Intelligent Infrastructure for Human-Centered Communities: Qualitative Research*. doi:10.1145/3197026.3203894

Sheng, X., & Sun, L. (2007). Developing knowledge innovation culture of libraries. *Library Management*, *28*(1/2), 36–52. doi:10.1108/01435120710723536

Shen, Y. (2019). Emerging scenarios of data infrastructure and novel concepts of digital libraries in intelligent infrastructure for human-centred communities: A qualitative research. *Journal of Information Science*, *45*(5), 691–704. doi:10.1177/0165551518811459

Shimray, S. R., Keerti, C., & Ramaiah, C. K. (2015). An overview of mobile reading habits. *DESIDOC Journal of Library and Information Technology*, *35*(5), 343–354. doi:10.14429/djlit.35.5.8901

Shin, D. (2021). The effects of explainability and causability on perception, trust, and acceptance: Implications for explainable AI. *International Journal of Human-Computer Studies*, *146*, 102551. doi:10.1016/j.ijhcs.2020.102551

Shi, X., Tang, K., & Lu, H. (2021). Smart library book sorting application with intelligence computer vision technology. *Library Hi Tech*, *39*(1), 220–232. doi:10.1108/LHT-10-2019-0211

Short, J. C., Randolph-Seng, B., & McKenny, A. F. (2013). Graphic presentation: An empirical examination of the graphic novel approach to communicate business concepts. *Business Communication Quarterly*, *76*(3), 273–303. doi:10.1177/1080569913482574

Sidorko, P., & Lee, L. (2014). JURA: A collaborative solution to hong kong academic libraries storage challenge. *Library Management*, *35*(1), 46–68. doi:10.1108/LM-03-2013-0025

Simovic, A. (2018). A Big Data smart library recommender system for an educational institution. *Library Hi Tech*, *36*(3), 498–523. doi:10.1108/LHT-06-2017-0131

Singh, N. K. (2020). Near-field Communication (NFC). *Information Technology and Libraries*, *39*(2). Advance online publication. doi:10.6017/ital.v39i2.11811

Sinha, S., Basak, S., Dey, Y., & Mondal, A. (2020). An educational Chatbot for answering queries. In *Emerging technology in modelling and graphics* (pp. 55–60). Springer. doi:10.1007/978-981-13-7403-6_7

Snowball, C. (2005). Teenage reluctant readers and graphic novels. *Young Adult Library Services*, *3*(4), 43–45.

Soleimanzade, N., Asemi, A., Cheshmehsohrabi, M., & Shabani, A. (2019). The Scientific Information Exchange General Model at Digital Library Context: Internet of Things. *Library Philosophy and Practice*, *2019*, 21–38.

Sophia, J. J., & Jacob, T. P. (2021, August). Edubot-a chatbot for education in covid-19 pandemic and vqabot comparison. In *2021 Second International Conference on Electronics and Sustainable Communication Systems (ICESC)* (pp. 1707-1714). IEEE. 10.1109/ICESC51422.2021.9532611

Statista. (n.d.). *NFT - Hong Kong | Statista Market Forecast*. https://www.statista.com/outlook/dmo/fintech/digital-assets/nft/hong-kong

Steele, K. M., Cakmak, M., & Blaser, B. (2018). Accessible making: Designing a makerspace for accessibility. *International Journal of Designs for Learning*, *9*(1), 114–121. doi:10.14434/ijdl.v9i1.22648

Stevens, C. R., & Campbell, P. J. (2006). Collaborating to connect global citizenship, information literacy, and lifelong learning in the global studies classroom. *RSR. Reference Services Review*, *34*(4), 536–556. doi:10.1108/00907320610716431

Stover, M., Jefferson, C., & Santos, I. (2019). Innovation and creativity: A new facet of the traditional mission for university libraries. *Supporting Entrepreneurship and Innovation*, *40*, 135–151. doi:10.1108/S0732-067120190000040006

Suen, R. L. T., Tang, J., & Chiu, D. K. W. (2020). Virtual reality services in academic libraries: Deployment experience in Hong Kong. *The Electronic Library*, *38*(4), 843–858. doi:10.1108/EL-05-2020-0116

Su, H. (2021). Design of the online platform of intelligent library based on machine learning and image recognition. *Microprocessors and Microsystems*, *82*, 103851. Advance online publication. doi:10.1016/j.micpro.2021.103851

Sun, Z. (2022). *Hong Kong's securities and futures commission warns of non-fungible token risks*. https://cointelegraph.com/news/hong-kong-s-securities-and-futures-commission-warn-of-nonfungible-token-risks

Sung, Y. Y. C., & Chiu, D. K. W. (2022). E-book or print book: Parents' Current View in Hong Kong. *Library Hi Tech*, *40*(5), 1289–1304. doi:10.1108/LHT-09-2020-0230

Sun, H. (2020). Research on interest reading recommendation method of intelligent library based on big data technology. *Web Intelligence*, *18*(2), 121–131. doi:10.3233/WEB-200434

Sun, L., Li, Y., & Lu, Y. (2022). Construction of Cloud Library Intelligent Service Platform Relying on Artificial Neural Network. *Mobile Information Systems*, *2022*, 1–11. Advance online publication. doi:10.1155/2022/6259127

Sun, X., Chiu, D. K. W., & Chan, C. T. (2022). Recent Digitalization Development of Buddhist Libraries: A Comparative Case Study. In S. Papadakis & A. Kapaniaris (Eds.), *The Digital Folklore of Cyberculture and Digital Humanities* (pp. 251–266). IGI Global. doi:10.4018/978-1-6684-4461-0.ch014

Su, X., & Chen, N. (2022). Intelligent Information Service System of Smart Library Based on Virtual Reality and Eye Movement Technology. *Scientific Programming*, *2022*, 1–12. Advance online publication. doi:10.1155/2022/9174756

Su, Y. (2018). Research on Library Intelligent Service with Drive of Big Data and Artificial Intelligence. *Library and Information*, *38*(05), 103–106.

Swati, B., Namrata, P., Shikha, S., & Shivani, S. (2019). Digital Entrepreneurship And The Bottom of The Pyramid- A Conceptual Framework. *International Journal on Recent Trends in Business and Tourism*, *3*(1), 33–42.

Swiecki, Z., Khosravi, H., Chen, G., Martinez-Maldonado, R., Lodge, J. M., Milligan, S., Selwyn, N., & Gašević, D. (2022). Assessment in the age of artificial intelligence. *Computers and Education: Artificial Intelligence*, *3*, 100075. doi:10.1016/j.caeai.2022.100075

Taha Osman, G. T. O., Hassan Mohammed, M., & Babiker Al Shafei, I. (2023). The Internet of Things in information institutions: Concept, use and challenges. *BSU-Journal of Pedagogy and Curriculum*, *2*(3), 153–172. doi:10.21608/bsujpc.2023.278442

Tamayo, P. A., Herrero, A., Martín, J., Navarro, C., & Tránchez, J. M. (2020). Design of a chatbot as a distance learning assistant. *Open Praxis*, *12*(1), 145–153. doi:10.5944/openpraxis.12.1.1063

Tang, Y., & Xie, S. M. (2013). Study on Embedded Services of Academic Libraries Based on the Synergetic Theories. *Library and Information Service*, *57*(08), 78–81.

Tan, S. C., Lee, A. V. Y., & Lee, M. (2022). A systematic review of artificial intelligence techniques for collaborative learning over the past two decades. *Computers and Education. Artificial Intelligence*, *3*, 1000097.

Tariq, M. U., Babar, M., Poulin, M., Khattak, A. S., Alshehri, M. D., & Kaleem, S. (2021). Human Behavior Analysis Using Intelligent Big Data Analytics. *Frontiers in Psychology*, *12*, 686610. Advance online publication. doi:10.3389/fpsyg.2021.686610 PMID:34295289

Tejeda-Lorente, A., Bernabe-Moreno, J., Porcel, C., & Herrera-Viedma, E. (2018). Using Bibliometrics and Fuzzy Linguistic Modeling to Deal with Cold Start in Recommender Systems for Digital Libraries. doi:10.1007/978-3-319-66827-7_36

Tella, A. (2020). Robots are coming to the libraries: Are librarians ready to accommodate them? *Library Hi Tech News*, *37*(8), 13–17. doi:10.1108/LHTN-05-2020-0047

Temiz, S., & Salelkar, L. P. (2020). Innovation during crisis: Exploring reaction of Swedish university libraries to COVID-19. *Digital Library Perspectives*, *36*(4), 365–375. doi:10.1108/DLP-05-2020-0029

Tempone, D. (2021). *How Can Artists Tokenize Their Work and Enter the New Crypto Art Market?* https://www.domestika.org/en/blog/7273-how-can-artists-tokenize-their-work-and-enter-the-new-crypto-art-market

Thamizhmaran, K. (2021). RFID for Library Management System. *Journal of Advancement in Communication System*, *4*(1).

Thompson, K. M., Jaeger, P. T., Taylor, N. G., Subramaniam, M. M., & Bertot, J. C. (2014). *Digital literacy and digital inclusion: information policy and the public library*. Rowman & Littlefield.

Tian, L. (2020). Research on the construction of smart library based on the Internet of Things. *Journal of Library Science*, *42*(10), 101–104.

Tilak, S., Abu-Ghazaleh, N., & Heinzelman, W. (2002). A taxonomy of wireless microsensor network models. *ACM Mobile Computing and Communications Review*, *6*(2), 28–36. doi:10.1145/565702.565708

Timoshenko, I. (2020). RFID in Libraries: automatic identification and data collection technology for library documents. In Maintenance Management. IntechOpen. doi:10.5772/intechopen.82032

Trichur Narayanan, R. (2021). Recommender System: Personalizing User Experience or Scientifically Deceiving Users? *2021 the 5th International Conference on Information System and Data Mining,* 138–144. 10.1145/3471287.3471303

Tsabedze, V. W., Mathabela, N. N., & Ademola, S. S. (2022). A framework for integrating artificial intelligence into library and information science curricula. In *Innovative Technologies for Enhancing Knowledge Access in Academic Libraries* (pp. 233–246). IGI Global. doi:10.4018/978-1-6684-3364-5.ch014

Tsang, A. L. Y., & Chiu, D. K. W. (2022). Effectiveness of virtual reference services in academic libraries: A qualitative study based on the 5E Learning Model. *Journal of Academic Librarianship*, *48*(4), 102533. Advance online publication. doi:10.1016/j.acalib.2022.102533

Tse, H. L., Chiu, D. K., & Lam, A. H. (2022). From Reading Promotion to Digital Literacy: An Analysis of Digitalizing Mobile Library Services With the 5E Instructional Model. In A. Almeida & S. Esteves (Eds.), *Modern Reading Practices and Collaboration Between Schools, Family, and Community* (pp. 239–256). IGI Global. doi:10.4018/978-1-7998-9750-7.ch011

UKRI. (2021). *Transforming our world with AI*. Available at: https://www.ukri.org/wp-content/uploads/2021/02/UKRI-120221-TransformingOurWorldWithAI.pdf

Ulrich Beck. (2004). Risk Society: On the Road to a New Modernity. Yilin Press.

UNESCO. (2019). Artificial intelligence in Education: challenges and opportunities for sustainable development. In *Development Working Papers On Education Policy*. United Nations.

Upala, M., & Wong, W. K. (2019, April). IoT solution for smart library using facial recognition. *IOP Conference Series. Materials Science and Engineering*, *495*(1), 012030. doi:10.1088/1757-899X/495/1/012030

Uttarwar, M. L., & Chong, P. H. (2017). Bealib: A beacon enabled smart library system. *Wireless Sensor Network*, *9*(8), 302–310. doi:10.4236/wsn.2017.98017

Vårheim, A. (2007). Social capital and public libraries: The need for research. *Library & Information Science Research*, *29*(3), 416–428. doi:10.1016/j.lisr.2007.04.009

Vårheim, A. (2011). Gracious space: Library programming strategies towards immigrants as tools in the creation of social capital. *Library & Information Science Research*, *33*(1), 12–18. doi:10.1016/j.lisr.2010.04.005

Vårheim, A., Steinmo, S., & Ide, E. (2008). Do libraries matter? Public libraries and the creation of social capital. *The Journal of Documentation*, *64*(6), 877–892. doi:10.1108/00220410810912433

Vekaria, K., Calyam, P., Sivarathri, S. S., Wang, S., Zhang, Y., Pandey, A., Chen, C., Xu, D., Joshi, T., & Nair, S. (2021). Recommender-as-a-service with chatbot guided domain-science knowledge discovery in a science gateway. *Concurrency and Computation*, *33*(19), e6080. doi:10.1002/cpe.6080 PMID:35495546

Viet, D. H. (2019). *Near field communication enabled library* [Doctoral dissertation]. Vietnamese-German University. Retrieved from http://epub.vgu.edu.vn/handle/dlibvgu/545

Vuorikari, R., Ferrari, A., & Punie, Y. (2019). *Makerspaces for education and training: Exploring future implications for Europe*. https://publications.jrc.ec.europa.eu/repository/handle/JRC117481

Wai, I. S. H., Ng, S. S. Y., Chiu, D. K. W., Ho, K. K., & Lo, P. (2018). Exploring undergraduate students' usage pattern of mobile apps for education. *Journal of Librarianship and Information Science*, *50*(1), 34–47. doi:10.1177/0961000616662699

Wang, Y., & Wei, Y. (2019). *Research on Library functional layout based on Intelligent occupying system*. doi:10.1088/1742-6596/1176/3/032010

Wang, Y., He, H., & Liu, T. (2019). *Intelligent Library Information Service Terminal Based on BDS and Wireless Communication*. Academic Press.

Wang, B., Chiu, D. K. W., & Ho, K. K. W. (2022a). A Comparison of Deep Learning Models in Time Series Forecasting of Web Traffic Data from Kaggle. In Z. Sun (Ed.), *Handbook of Refromch on Foundations and Applications of Intelligent Business Analytics* (pp. 301–319). IGI Global. doi:10.4018/978-1-7998-9016-4.ch014

Wang, C. S. (2019). An AR mobile navigation system integrating indoor positioning and content recommendation services. *World Wide Web (Bussum)*, *22*(3), 1241–1262. doi:10.100711280-018-0580-3

Wang, F., Wang, W., Wilson, S., & Ahmed, N. (2016). The state of library makerspaces. *International Journal of Librarianship*, *1*(1), 2–16. doi:10.23974/ijol.2016.vol1.1.12

Wang, H., & Ding, J. (2022). Development Strategy of Intelligent Digital Library without Human Service in the Era of "Internet plus". *Computational Intelligence and Neuroscience*, *2022*, 1–11. Advance online publication. doi:10.1155/2022/7892738 PMID:36262604

Wang, J., Deng, S., Chiu, D. K. W., & Chan, C. T. (2022b). Social Network Customer Relationship Management for Orchestras: A Case Study on Hong Kong Philharmonic Orchestra. In N. B. Ammari (Ed.), *Social Customer Relationship Management (Social-CRM) in the Era of Web 4.0* (pp. 250–268). IGI Global. doi:10.4018/978-1-7998-9553-4.ch012

Wang, J., Song, Y. F., & Du, P. P., etal. (2018). Research on Intelligent Library and Service Process Control Based on Collaborative Theory. *Library Work and Study*, *40*(10), 42–46.

Wang, L., & Xu, X. W. (2017). The Construction of Library Intelligence Service System Based on Collaborative Ecology. *New Century Library*, *39*(4), 28–32.

Wang, P. (2014). Study on Library Information Resources Construction and Sharing Based on Synergetic Theory. *Modern Information*, *34*(4), 33–37.

Wang, P., Chiu, D. K. W., Ho, K. K., & Lo, P. (2016). Why read it on your mobile device? Change in reading habit of electronic magazines for university students. *Journal of Academic Librarianship*, *42*(6), 664–669. doi:10.1016/j.acalib.2016.08.007

Wang, S. W. (2011). New Pattern of Future Libraries: The Smart Library. *Library Development*, *34*(12), 1–5.

Wang, S. W. (2012). On Three Main Features of the Smart Library. *Journal of Library Science in China*, *38*(06), 22–28.

Wang, S. W. (2017a). On the Five Relations of the Smart library. *Library Journal*, *36*(04), 4–10.

Wang, S. W. (2017b). Artificial Intelligence and Library Service Reshaping. *Library and Information*, *37*(06), 6–18.

Wang, S. W. (2019). Research on Artificial Intelligence and Library Renewal. *Tushu Qingbao Zhishi*, *36*(04), 35–42.

Wang, S. W., & Wang, T. N. (2018). Library Space Reengineering and Service Based on Artificial Intelligence. *Library and Information*, *38*(03), 50–55.

Wang, W., Lam, E. T. H., Chiu, D. K. W., Lung, M. M., & Ho, K. K. W. (2021). Supporting higher education with social networks: Trust and privacy vs perceived effectiveness. *Online Information Review*, *45*(1), 207–219. doi:10.1108/OIR-02-2020-0042

Wang, X., Feng, X., & Guo, K. (2022). Research hotspots and prospects of ethics education of science and technology in China based on bibliometrics. *Library Hi Tech*. Advance online publication. doi:10.1108/LHT-06-2022-0298

Wang, Y., & Wang, L. (2013). Construction and Application of the Smart Library Based on RFID Technology. *Journal of Library Science*, *35*(12), 98–100.

Warkentin, M., & Orgeron, C. (2020). Using the Security Triad to Assess Blockchain Technology in Public Sector Applications. *International Journal of Information Management*, *52*, 102090–102098. doi:10.1016/j.ijinfomgt.2020.102090

Webb, K. K. (2018). *Development of creative spaces in academic libraries: a decision maker's guide*. Chandos Publishing.

Wei, Q., & Lu, P. (2018). Innovative Supervision Methods to Prevent AI Risks. *Digital Economy*, *5*(03), 22–27.

White, B. (2012). Guaranteeing Access to Knowledge: The Role of Libraries. *WIPO Magazine*, *2012*(4). https://www.wipo.int/wipo_magazine/en/2012/04/article_0004.html

Wilhelm, J. D. (1995). Reading is seeing: Using visual response to improve the literary reading of reluctant readers. *Journal of Reading Behavior*, *27*(4), 467–503. doi:10.1080/10862969509547896

Williams, V. K., & Peterson, D. V. (2009). Graphic novels in libraries supporting teacher education and librarianship programs. *Library Resources & Technical Services*, *53*(3), 166–173. doi:10.5860/lrts.53n3.166

Willingham, T., & De Boer, J. (2015). *Makerspaces in libraries*. Rowman & Littlefield.

Wilson, S. (2021). User Experience Desires Personalization from Academic Library Websites. *School of Information Student Research Journal*, *11*(1). Advance online publication. doi:10.31979/2575-2499.110109

Winkler, R., & Söllner, M. (2018). Unleashing the potential of chatbots in education: A state-of-the-art analysis. *Academy of Management Annual Meeting (AOM)*. 10.5465/AMBPP.2018.15903abstract

Winston, P. H. (1992). *Artificial Intelligence* (3rd ed.). Addison-Wesley Publishing Company.

Wong, A. K. K., & Chiu, D. K. W. (2023). Digital Transformation of Museum Conservation Practices: A Value Chain Analysis of Public Museums in Hong Kong. In R. Pettinger, B. B. Gupta, A. Roja, & D. Cozmiuc (Eds.), *Handbook of Research on the Digital Transformation Digitalization Solutions for Social and Economic Needs*. IGI Global. doi:10.4018/978-1-6684-4102-2.ch010

Wong, A., & Partridge, H. (2016). Making as learning: Makerspaces in universities. *Australian Academic and Research Libraries*, *47*(3), 143–159. doi:10.1080/00048623.2016.1228163

Wong, I. H. S., Fan, E. C. H., Chiu, D. K. W., & Ho, K. K. W. (2023). Internet Celebrities' Influence on Youths' Diet Behaviors: A Gender Study on Social Media. *Aslib Journal of Information Management*. Advance online publication. doi:10.1108/AJIM-11-2022-0495

Wong, J., Chiu, D. K. W., Leung, T. N., & Kevin, K. K. W. (2023). Exploring the Associations of Addiction as a Motive for Using Facebook with Social Capital Perceptions. *Online Information Review*, *47*(2), 283–298. doi:10.1108/OIR-06-2021-0300

Wong, K. C., & Chiu, D. K. W. (2023). Promoting The Use of Electronic Resources of International Schools in Hong Kong: The Case Study of ESF King George V School. In E. Meletiadou (Ed.), *Handbook of Research on Redesigning Teaching, Learning, and Assessment in the Digital Era*. IGI Global. doi:10.4018/978-1-6684-8292-6.ch007

Wong, S. W. S., & Chiu, D. K. W. (2023). Re-examining the value of remote academic library storage in the mobile digital age: A comparative study. *Portal (Baltimore, Md.)*, *23*(1), 89–109.

Woolf, B.P., Lane, C., Chaudhri, V. K., & Kolodner, J. L. (2013). Artificial Intelligent Grand Challenges for Education. *AI Magazine, 10*(13).

World Bank. (2001). *Understanding and measuring social capital: A synthesis of findings and recommendations from the Social Capital Initiative*. Retrieved from http://siteresources.worldbank.org/INTSOCIALCAPITAL/Resources/Social-Capital-Initiative-Working-Paper-Series/SCI-WPS-24.pdf

Wu, M., Lam, A. H. C., & Chiu, D. K. W. (2023) Transforming and Promoting Reference Services with Digital Technologies: A Case Study on Hong Kong Baptist University Library. In B. Holland (Ed.), Handbook of Research on Advancements of Contactless Technology and Service Innovation in Library and Information Science (pp. 128-145). IGI Global. doi:10.4018/978-1-6684-7693-2.ch007

Wu, M., Lu, T. J., Ling, F. Y., Sun, J., & Du, H. Y. (2010, August). Research on the architecture of Internet of Things. In *2010 3rd international conference on advanced computer theory and engineering (ICACTE)* (Vol. 5, pp. V5-484). IEEE. 10.1109/ICACTE.2010.5579493

Xia, Y. P. (2018). Research on the Cooperative Operation Mechanism of Multiple Subjects in Accurate Poverty Alleviation. *Review of Economic Research*, *40*(37), 72–77.

Xie, X. (2018). Study on the Intelligent Terminal Innovation of Personalized Active Service of Mobile Library-A Case Study of Zhejiang University of Media and Communications. doi:10.23977smme.2018.62224

Xie, K., Liu, Z., Fu, L., & Liang, B. (2019). Internet of Things-based intelligent evacuation protocol in libraries. *Library Hi Tech*, *38*(1), 145–163. doi:10.1108/LHT-11-2017-0250

Xie, Y., Liu, J., Zhu, S., Chong, D., Shi, H., & Chen, Y. (2019). An IoT-based risk warning system for smart libraries. *Library Hi Tech*, *37*(4), 918–912. doi:10.1108/LHT-11-2017-0254

Xie, Z., Chiu, D. K. W., & Ho, K. K. W. (2023). The Role of Social Media as Aids for Accounting Education and Knowledge Sharing: Learning Effectiveness and Knowledge Management Perspectives in Mainland China. *Journal of the Knowledge Economy*. Advance online publication. doi:10.100713132-023-01262-4

Xie, Z., Wong, G. K. W., Chiu, D. K. W., & Lei, J. (2023). Bridging K-12 Mathematics and Computational Thinking in the Scratch Community: A Big Data Analysis. *IT Professional*, *25*(2), 64–70. doi:10.1109/MITP.2023.3243393

Xu, C., Lam, A. H. C., Gao, X., & Chiu, D. K. W. (2023). Antique bookstores marketing strategies as urban cultural landmark: A case analysis for Suzhou Antique Bookstore. In M. A. Rodrigues & M. A. M. Carvalho (Eds.), *Exploring niche tourism business models, marketing, and consumer experience*. IGI Global.

Xue, J., Wang, Y., & Wang, M. (2021). Smart Design of Portable Indoor Shading Device for Visual Comfort—A Case Study of a College Library. *Applied Sciences (Basel, Switzerland)*, *11*(22), 10644. doi:10.3390/app112210644

Xue, Y., Lam, A. H. C., & Chiu, D. K. W. (2023). Redesigning library information literacy education with the BOPPPS model: A case study of the HKUST Library. In R. Taiwo, B. Idowu-Faith, & S. Ajiboye (Eds.), *Transformation of higher education through institutional online spaces* (pp. 279–296). IGI Global. doi:10.4018/978-1-6684-8122-6.ch017

Xu, F. (2019). Regulation on Black Box of AI Algorithms. *Oriental Law*, *12*(06), 78–86.

Xu, K. (2022). Intelligent Library Service and Management Based on IoT Assistance and Text Recommendation. *Journal of Sensors*, *2022*, 1–10. Advance online publication. doi:10.1155/2022/3660135

Yang, C. J., Kang, H. B., Zhang, L., & Zhang, R. Y. (2019). A design of smart library energy consumption monitoring and management system based on IoT. In *Proceedings of the Fifth Euro-China Conference on Intelligent Data Analysis and Applications 5* (pp. 217-224). Springer International Publishing. 10.1007/978-3-030-03766-6_24

Yang, G. (2008). Graphic novels in the classroom. *Language Arts*, *85*(3), 185.

Yang, L. H., & Zhang, L. (2016). Comparative Study on Multi-coordinated Treatment of Air Pollution: Cross-Case Analysis of Typical Countries. *Administrative Tribune*, *23*(05), 24–30.

Yang, S., & Evans, C. (2019, November). Opportunities and challenges in using AI chatbots in higher education. In *Proceedings of the 2019 3rd International Conference on Education and E-Learning* (pp. 79-83). 10.1145/3371647.3371659

Yang, Z., Zhou, Q., Chiu, D. K. W., & Wang, Y. (2022). Exploring the factors influencing continuance usage intention of academic social network sites. *Online Information Review*, *46*(7), 1225–1241. doi:10.1108/OIR-01-2021-0015

Yan, Y. C. (2022). Research on the Infringement Risk and Countermeasures in the Text Data Mining of Smart Libraries in China. *Journal of the National Library of China*, *31*(01), 106–113.

Yan, Z., Zhang, P., & Vasilakos, A. V. (2014). A survey on trust management forInternet of Things. *Journal of Network and Computer Applications*, *42*, 120–134. doi:10.1016/j.jnca.2014.01.014

Yao, E. X., & Tang, R. Q. (2018). The Path Optimization of Urban Risk Prevention and Goverance from the Perspective of Multiple Cooperation. *Journal of Hulunbeier Universities*, *26*(01), 49–52.

Yao, L., Lei, J., Chiu, D. K. W., & Xie, Z. (2023). Adult Learners' Perception of Online Language English Learning Platforms in China. In A. Garcés-Manzanera & M. E. C. García (Eds.), *New Approaches to the Investigation of Language Teaching and Literature* (pp. 123–140). IGI Global. doi:10.4018/978-1-6684-6020-7.ch007

Yao, L., Lei, J., Chiu, D. K. W., & Xie, Z. (2023). Adult learners' perception of online language English learning platforms in China. In A. Garcés-Manzanera & M. E. C. García (Eds.), *New approaches to the investigation of language teaching and literature* (pp. 123–149). IGI Global.

Yao, N., Xi, C. L., & Huang, X. T. (2017). Research on the Construction of the Situational Awareness Micro-Service Model in Smart Libraries. *Journal of Library Science*, *39*(08), 57–60.

Yew, A., Chiu, D. K. W., Nakamura, Y., & Li, K. K. (2022). A quantitative review of LIS programs accredited by ALA and CILIP under contemporary technology advancement. *Library Hi Tech*, *40*(6), 1721–1745. doi:10.1108/LHT-12-2021-0442

Yi, F., & Baumann, M. (2020). Guiding principles for designing an accessible, inclusive, and diverse library makerspace. *International Journal of Academic Makerspaces and Making*. https://ijamm.pubpub.org/pub/op9ltdpf

Yi, Y., & Chiu, D.K.W. (2023). Public information needs during the COVID-19 outbreak: A qualitative study in mainland China. *Library Hi Tech*. doi:10.1108/LHT-08-2022-0398

Yip, K. H. T., Chiu, D. K. W., Ho, K. K. W., & Lo, P. (2021). Adoption of Mobile Library Apps as Learning Tools in Higher Education: A Tale between Hong Kong and Japan. *Online Information Review*, *45*(2), 389–405. doi:10.1108/OIR-07-2020-0287

Yip, T., Chiu, D. K. W., Cho, A., & Lo, P. (2019). Behavior and informal learning at night in a 24-hour space: A case study of the Hong Kong Design Institute Library. *Journal of Librarianship and Information Science*, *51*(1), 171–179. doi:10.1177/0961000617726120

Yoon, A., & Schultz, T. (2017). Research data management services in academic libraries in the US: A content analysis of libraries' websites. *College & Research Libraries*, *78*(7), 920–933. doi:10.5860/crl.78.7.920

Yu, H. H. K., Chiu, D. K. W., & Chan, C. T. (2023). Resilience of symphony orchestras to challenges in the COVID-19 era: Analyzing the Hong Kong Philharmonic Orchestra with Porter's five force model. In W. Aloulou (Ed.), *Handbook of Research on Entrepreneurship and Organizational Resilience During Unprecedented Times* (pp. 586–601). IGI Global.

Yu, H. H. K., Chiu, D. K. W., & Chan, C. T. (2023). Resilience of symphony orchestras to challenges in the COVID-19 era: Analyzing the Hong Kong Philharmonic Orchestra with Porter's five force model. In W. J. Aloulou (Ed.), *Handbook of research on entrepreneurship and organizational resilience during unprecedented times* (pp. 586–601). IGI Global. doi:10.4018/978-1-6684-4605-8.ch026

Yu, H. Y., Tsoi, Y. Y., Rhim, A. H. R., Chiu, D. K. W., & Lung, M. M. W. (2022). Changes in habits of electronic news usage on mobile devices in university students: A comparative survey. *Library Hi Tech*, *40*(5), 1322–1336. doi:10.1108/LHT-03-2021-0085

Yu, P. Y., Lam, E. T. H., & Chiu, D. K. W. (2023). Operation management of academic libraries in Hong Kong under COVID-19. *Library Hi Tech*, *41*(1), 108–129. doi:10.1108/LHT-10-2021-0342

Yu, X. H. (2014). Capability Construction of Risk Management System of Library Knowledge Management. *Library Work and Study*, *36*(09), 60–63.

Zawacki-Richter, O., Marín, V. I., Bond, M., & Gouverneur, F. (2019). A systematic review of research on artificial intelligence applications in higher education – where are the educators? *International Journal of Educational Technology in Higher Education*, *2019*(16), 39. doi:10.118641239-019-0171-0

Zeng, Z. M., & Qin, S. Q. (2018). Decentralized Mobile Visual Search Management System for Smart Library. *Information Science*, *36*(01), 11–15.

Zeng, Z. M., & Song, Y. Y. (2017). Research on Mobile Visual Search Service of Smart Library Based on SoLoMo. *The Library*, *45*(07), 92–98.

Zhai, X. (2023). *ChatGPT user experience: Implications for higher education. National Science Foundation*. NSF.

Zhang, H., & Ye, Y. (2023). Intelligence, Cognition and Insight: Characterizing Smart Library. *Journal of Library Science in China*, 1-11. http://kns.cnki.net/kcms/detail/11.2746.G2.20230131.1544.001.html

Zhang, J., Zhang, Y., & Wu, X. (2018). *Research of Intelligent Library Based on RFID Technology*. doi:10.1109/ITME.2018.00129

Zhang, W. (2019). Research on the Construction of Community Library Intelligent Service System from the Perspective of Smart City. doi:10.2991schd-19.2019.1

Zhang, J., Lam, A. H. C., & Chiu, D. K. W. (2023). Evaluating the effectiveness of learning commons as third space with the 5E usability model: The case of Hong Kong University of Science and Technology Library. In C. Kaye & J. H. Writer (Eds.), *Third-Space Exploration in Education*. IGI Global.

Compilation of References

Zhang, Q., Huang, B., & Chiu, D.K.W., & Ho, K.K.W. (2015). Learning Japanese Through Social Network Sites: A Case Study of Chinese Learners' Perceptions. *Micronesian Educators*, *21*, 55–71.

Zhang, T., & Ma, H. Q. (2021). Research on Algorithm Risk and Regulation in Intelligent Intelligence Analysis. *Library and Information Service*, *65*(12), 47–56.

Zhang, X., Lo, P., So, S., Chiu, D. K. W., Leung, T. N., Ho, K. K. W., & Stark, A. (2021). Medical students' attitudes and perceptions towards the effectiveness of mobile learning: A comparative information-need perspective. *Journal of Librarianship and Information Science*, *53*(1), 116–129. doi:10.1177/0961000620925547

Zhang, Y. J., Wang, J., & Huang, H. Q. (2018). Suggestions on Big Data Risk Management in the Era of AI. *China Economic & Trade Herald*, (06), 63–65.

Zhang, Y., Lo, P., So, S., & Chiu, D. K. W. (2020). Relating library user education to business students' information needs and learning practices: A comparative study. *RSR. Reference Services Review*, *48*(4), 537–558. doi:10.1108/RSR-12-2019-0084

Zhang, Y., & Yan, B. (2019). MIL-61 and Eu3+@MIL-61 as Signal Transducers to Construct an Intelligent Boolean Logical Library Based on Visualized Luminescent Metal-Organic Frameworks. *ACS Applied Materials & Interfaces*, *11*(22), 20125–20133. doi:10.1021/acsami.9b00179 PMID:31088052

Zhang, Z. M. (2022). Risk identification and avoidance strategy of intelligent library in the era of artificial intelligence. *Jiangsu Science & Technology Information*, *39*(06), 29–31.

Zheng, J., Lam, A. H. C., & Chiu, D. K. W. (2023). Evaluating the Effectiveness of Learning Commons as Third Space with the 5E Usability Model: The Case of Hong Kong University of Science and Technology Library. In C. Kaye, J. H. Writer, & J. Batsaikhan (Eds.), *Third-Space Exploration in Education*. IGI Global.

Zhou, D. (2019, October). Case Study on Seat Management of University Library Based on WeChat Public Number Client—Taking Jianghan University Library as an Example. In *2019 4th International Conference on Mechanical, Control and Computer Engineering (ICMCCE)* (pp. 630-6303). IEEE.

Zhou, J., Lam, E., Au, C. H., Lo, P., & Chiu, D. K. W. (2022). Library café or elsewhere: Usage of study space by different majors under contemporary technological environment. *Library Hi Tech*, *40*(6), 1567–1581. doi:10.1108/LHT-03-2021-0103

Zhou, L., Yan, S., & Zhu, X. (2017). Context-aware service model and evaluation research of smart library. *Research on Library Science*, *21*, 23–30.

Zhou, Q. X., & Yang, C. J. (2018). Design of smart library monitoring and management system based on internet of things technology. *Zidonghua Yu Yibiao*, *38*(11), 85–88.

Zhuang, Y., Li, Q., Chiu, D. K., Wu, Z., & Hu, H. (2014). Efficient Personalized Probabilistic Retrieval of Chinese Calligraphic Manuscript Images in Mobile Cloud Environment. *ACM Transactions on Asian Language Information Processing*, *13*(4), 18. doi:10.1145/2629575

Zhu, F. Y. (2019). Security Risk and Solution for Library Applications of Artificial Intelligence. *Tushuguanxue Yanjiu*, *41*(01), 6–11.

Zuo, Y., Lam, A. H. C., & Chiu, D. K. W. (2023). Digital protection of traditional villages for sustainable heritage tourism: A case study on Qiqiao Ancient Village, China. In A. Masouras, C. Papademetriou, D. Belias, & S. Anastasiadou (Eds.), *Sustainable Growth Strategies for Entrepreneurial Venture Tourism and Regional Development* (pp. 121–195). IGI Global. doi:10.4018/978-1-6684-6055-9.ch009

Related References

To continue our tradition of advancing media and communications research, we have compiled a list of recommended IGI Global readings. These references will provide additional information and guidance to further enrich your knowledge and assist you with your own research and future publications.

Abashian, N., & Fisher, S. (2018). Intercultural Effectiveness in Libraries: Supporting Success Through Collaboration With Co-Curricular Programs. In B. Blummer, J. Kenton, & M. Wiatrowski (Eds.), *Promoting Ethnic Diversity and Multiculturalism in Higher Education* (pp. 219–236). Hershey, PA: IGI Global. doi:10.4018/978-1-5225-4097-7.ch012

Adebayo, O., Fagbohun, M. O., Esse, U. C., & Nwokeoma, N. M. (2018). Change Management in the Academic Library: Transition From Print to Digital Collections. In R. Bhardwaj (Ed.), *Digitizing the Modern Library and the Transition From Print to Electronic* (pp. 1–28). Hershey, PA: IGI Global. doi:10.4018/978-1-5225-2119-8.ch001

Adegbore, A. M., Quadri, M. O., & Oyewo, O. R. (2018). A Theoretical Approach to the Adoption of Electronic Resource Management Systems (ERMS) in Nigerian University Libraries. In A. Tella & T. Kwanya (Eds.), *Handbook of Research on Managing Intellectual Property in Digital Libraries* (pp. 292–311). Hershey, PA: IGI Global. doi:10.4018/978-1-5225-3093-0.ch015

Adesola, A. P., & Olla, G. O. (2018). Unlocking the Unlimited Potentials of Koha OSS/ILS for Library House-Keeping Functions: A Global View. In M. Khosrow-Pour (Ed.), *Optimizing Contemporary Application and Processes in Open Source Software* (pp. 124–163). Hershey, PA: IGI Global. doi:10.4018/978-1-5225-5314-4.ch006

Adesola, A. P., & Olla, G. O. (2019). Bridging the Digital Divide in Nigerian Information Landscape: The Role of the Library. *International Journal of Digital Literacy and Digital Competence*, *10*(3), 10–31. doi:10.4018/IJDLDC.2019070102

Adetayo, A. J. (2021). Fake News and Social Media Censorship: Examining the Librarian Role. In R. Blankenship (Ed.), *Deep Fakes, Fake News, and Misinformation in Online Teaching and Learning Technologies* (pp. 69–92). IGI Global. https://doi.org/10.4018/978-1-7998-6474-5.ch004

Adetayo, A. J. (2022). Building Civic Engagement in Smart Cities: Role of Smart Libraries. In M. Taher (Ed.), *Handbook of Research on the Role of Libraries, Archives, and Museums in Achieving Civic Engagement and Social Justice in Smart Cities* (pp. 314–333). IGI Global. https://doi.org/10.4018/978-1-7998-8363-0.ch017

Adigun, G. O., Sobalaje, A. J., & Salau, S. A. (2018). Social Media and Copyright in Digital Libraries. In A. Tella & T. Kwanya (Eds.), *Handbook of Research on Managing Intellectual Property in Digital Libraries* (pp. 19–36). Hershey, PA: IGI Global. doi:10.4018/978-1-5225-3093-0.ch002

Adriyana, L., & Fitrina Cahyaningtyas, D. (2022). The Importance of Rural Library Services Based on Social Inclusion in Indonesia. In M. Taher (Ed.), *Handbook of Research on the Role of Libraries, Archives, and Museums in Achieving Civic Engagement and Social Justice in Smart Cities* (pp. 201–218). IGI Global. https://doi.org/10.4018/978-1-7998-8363-0.ch010

Afolabi, O. A. (2018). Myths and Challenges of Building an Effective Digital Library in Developing Nations: An African Perspective. In A. Tella & T. Kwanya (Eds.), *Handbook of Research on Managing Intellectual Property in Digital Libraries* (pp. 51–79). Hershey, PA: IGI Global. doi:10.4018/978-1-5225-3093-0.ch004

Ahuja, Y., & Kumar, P. (2017). Web 2.0 Tools and Application: Knowledge Management and Sharing in Libraries. In B. Gunjal (Ed.), *Managing Knowledge and Scholarly Assets in Academic Libraries* (pp. 218–234). Hershey, PA: IGI Global. doi:10.4018/978-1-5225-1741-2.ch010

Ajmi, A. (2018). Developing In-House Digital Tools: Case Studies From the UMKC School of Law Library. In L. Costello & M. Powers (Eds.), *Developing In-House Digital Tools in Library Spaces* (pp. 117–139). Hershey, PA: IGI Global. doi:10.4018/978-1-5225-2676-6.ch006

Al-Kharousi, R., Al-Harrasi, N. H., Jabur, N. H., & Bouazza, A. (2018). Soft Systems Methodology (SSM) as an Interdisciplinary Approach: Reflection on the Use of SSM in Adoption of Web 2.0 Applications in Omani Academic Libraries. In M. Al-Suqri, A. Al-Kindi, S. AlKindi, & N. Saleem (Eds.), *Promoting Interdisciplinarity in Knowledge Generation and Problem Solving* (pp. 243–257). Hershey, PA: IGI Global. doi:10.4018/978-1-5225-3878-3.ch016

Alenzuela, R. (2017). Research, Leadership, and Resource-Sharing Initiatives: The Role of Local Library Consortia in Access to Medical Information. In S. Ram (Ed.), *Library and Information Services for Bioinformatics Education and Research* (pp. 199–211). Hershey, PA: IGI Global. doi:10.4018/978-1-5225-1871-6.ch012

Alenzuela, R., & Terry, M. A. (2020). Diversity, Indigenous Knowledge, and LIS Pedagogy: Conceptualizing Formal Education in Library and Information Studies in Vanuatu. In R. Alenzuela, H. Kim, & D. Baylen (Eds.), *Internationalization of Library and Information Science Education in the Asia-Pacific Region* (pp. 50–77). IGI Global. doi:10.4018/978-1-7998-2273-8.ch003

Allison, D. (2017). When Sales Talk Meets Reality: Implementing a Self-Checkout Kiosk. In E. Iglesias (Ed.), *Library Technology Funding, Planning, and Deployment* (pp. 36–54). Hershey, PA: IGI Global. doi:10.4018/978-1-5225-1735-1.ch003

Anglim, C. T., & Rusk, F. (2018). Empowering DC's Future Through Information Access. In A. Burtin, J. Fleming, & P. Hampton-Garland (Eds.), *Changing Urban Landscapes Through Public Higher Education* (pp. 57–77). Hershey, PA: IGI Global. doi:10.4018/978-1-5225-3454-9.ch003

Asmi, N. A. (2017). Social Media and Library Services. *International Journal of Library and Information Services*, 6(2), 23–36. doi:10.4018/IJLIS.2017070103

Attademo, G., & Maccaro, A. (2022). Research Ethics in the Social Sciences. In G. Punziano & A. Delli Paoli (Eds.), *Handbook of Research on Advanced Research Methodologies for a Digital Society* (pp. 54–64). IGI Global. https://doi.org/10.4018/978-1-7998-8473-6.ch005

Awoyemi, R. A. (2018). Adoption and Use of Innovative Mobile Technologies in Nigerian Academic Libraries. In J. Keengwe (Ed.), *Handbook of Research on Digital Content, Mobile Learning, and Technology Integration Models in Teacher Education* (pp. 354–389). Hershey, PA: IGI Global. doi:10.4018/978-1-5225-2953-8.ch019

Awoyemi, R. A. (2018). Adoption and Use of Innovative Mobile Technologies in Nigerian Academic Libraries. In J. Keengwe (Ed.), *Handbook of Research on Digital Content, Mobile Learning, and Technology Integration Models in Teacher Education* (pp. 354–389). Hershey, PA: IGI Global. doi:10.4018/978-1-5225-2953-8.ch019

Awoyemi, R. A., & Awoyemi, R. O. (2021). Beyond the Physical Library Space: Creating a 21st Century Digitally-Oriented Library Environment. In C. Chisita, R. Enakrire, O. Durodolu, V. Tsabedze, & J. Ngoaketsi (Eds.), *Handbook of Research on Records and Information Management Strategies for Enhanced Knowledge Coordination* (pp. 189–203). IGI Global. https://doi.org/10.4018/978-1-7998-6618-3.ch012

Babatope, I. S. (2018). Social Media Applications as Effective Service Delivery Tools for Librarians. In M. Khosrow-Pour, D.B.A. (Ed.), Encyclopedia of Information Science and Technology, Fourth Edition (pp. 5252-5261). Hershey, PA: IGI Global. doi:10.4018/978-1-5225-2255-3.ch456

Bakare, A. A. (2018). Digital Libraries and Copyright of Intellectual Property: An Ethical Practice Management. In A. Tella & T. Kwanya (Eds.), *Handbook of Research on Managing Intellectual Property in Digital Libraries* (pp. 377–395). Hershey, PA: IGI Global. doi:10.4018/978-1-5225-3093-0.ch019

Baker, A. A. (2020). To Whose Benefit? At What Cost?: Consideration for Ethical Issues in Social Science Research. In M. Baran & J. Jones (Eds.), *Applied Social Science Approaches to Mixed Methods Research* (pp. 251–260). IGI Global. https://doi.org/10.4018/978-1-7998-1025-4.ch011

Baker-Gardner, R., & Smart, C. (2017). Ignorance or Intent?: A Case Study of Plagiarism in Higher Education among LIS Students in the Caribbean. In D. Velliaris (Ed.), *Handbook of Research on Academic Misconduct in Higher Education* (pp. 182–205). Hershey, PA: IGI Global. doi:10.4018/978-1-5225-1610-1.ch008

Baker-Gardner, R., & Stewart, P. (2018). Educating Caribbean Librarians to Provide Library Education in a Dynamic Information Environment. In S. Bhattacharyya & K. Patnaik (Eds.), *Changing the Scope of Library Instruction in the Digital Age* (pp. 187–226). Hershey, PA: IGI Global. doi:10.4018/978-1-5225-2802-9.ch008

Baran, M. L., & Jones, J. E. (2020). Developing the Research Study: A Step-by-Step Approach. In M. Baran & J. Jones (Eds.), *Applied Social Science Approaches to Mixed Methods Research* (pp. 262–274). IGI Global. https://doi.org/10.4018/978-1-7998-1025-4.ch012

Baskaran, C. (2020). Altmetircs Research: An Impact and Tools. In C. Baskaran (Ed.), *Measuring and Implementing Altmetrics in Library and Information Science Research* (pp. 1–10). IGI Global. https://doi.org/10.4018/978-1-7998-1309-5.ch001

Bengtson, J. (2017). Funding a Gamification Machine. In E. Iglesias (Ed.), *Library Technology Funding, Planning, and Deployment* (pp. 99–112). Hershey, PA: IGI Global. doi:10.4018/978-1-5225-1735-1.ch006

Bhuda, M., & Koitsiwe, M. (2022). The Importance of Underpinning Indigenous Research Using African Indigenous Philosophies: Perspectives From Indigenous Scholars. In R. Tshifhumulo & T. Makhanikhe (Eds.), *Handbook of Research on Protecting and Managing Global Indigenous Knowledge Systems* (pp. 223–248). IGI Global. https://doi.org/10.4018/978-1-7998-7492-8.ch013

Blummer, B., & Kenton, J. M. (2017). Access and Accessibility of Academic Libraries' Electronic Resources and Services: Identifying Themes in the Literature From 2000 to the Present. In H. Alphin Jr, J. Lavine, & R. Chan (Eds.), *Disability and Equity in Higher Education Accessibility* (pp. 242–267). Hershey, PA: IGI Global. doi:10.4018/978-1-5225-2665-0.ch011

Blummer, B., & Kenton, J. M. (2018). Academic and Research Libraries' Portals: A Literature Review From 2003 to the Present. In R. Bhardwaj (Ed.), *Digitizing the Modern Library and the Transition From Print to Electronic* (pp. 29–63). Hershey, PA: IGI Global. doi:10.4018/978-1-5225-2119-8.ch002

Blummer, B., & Kenton, J. M. (2018). International Students and Academic Libraries: Identifying Themes in the Literature From 2001 to the Present. In B. Blummer, J. Kenton, & M. Wiatrowski (Eds.), *Promoting Ethnic Diversity and Multiculturalism in Higher Education* (pp. 237–263). Hershey, PA: IGI Global. doi:10.4018/978-1-5225-4097-7.ch013

Bohuski, L. (2020). What If Your Library Can't Go Green?: Promoting Wellness in Libraries. In A. Kaushik, A. Kumar, & P. Biswas (Eds.), *Handbook of Research on Emerging Trends and Technologies in Library and Information Science* (pp. 13–26). IGI Global. doi:10.4018/978-1-5225-9825-1.ch002

Boom, D. (2017). The Embedded Librarian: Do More With less. In B. Gunjal (Ed.), *Managing Knowledge and Scholarly Assets in Academic Libraries* (pp. 76–97). Hershey, PA: IGI Global. doi:10.4018/978-1-5225-1741-2.ch004

Bosire-Ogechi, E. (2018). Social Media, Social Networking, Copyright, and Digital Libraries. In A. Tella & T. Kwanya (Eds.), *Handbook of Research on Managing Intellectual Property in Digital Libraries* (pp. 37–50). Hershey, PA: IGI Global. doi:10.4018/978-1-5225-3093-0.ch003

Bradley-Sanders, C., & Rudshteyn, A. (2018). MyLibrary at Brooklyn College: Developing a Suite of Digital Tools. In L. Costello & M. Powers (Eds.), *Developing In-House Digital Tools in Library Spaces* (pp. 140–167). Hershey, PA: IGI Global. doi:10.4018/978-1-5225-2676-6.ch007

Brown, V. (2018). Technology Access Gap for Postsecondary Education: A Statewide Case Study. In M. Yildiz, S. Funk, & B. De Abreu (Eds.), *Promoting Global Competencies Through Media Literacy* (pp. 20–40). Hershey, PA: IGI Global. doi:10.4018/978-1-5225-3082-4.ch002

Browne, N. (2021). The IHS Library and Its Response to the COVID-19 Pandemic. In B. Holland (Eds.), *Handbook of Research on Library Response to the COVID-19 Pandemic* (pp. 298-320). IGI Global. https://doi.org/10.4018/978-1-7998-6449-3. ch016

Chaiyasoonthorn, W., & Suksa-ngiam, W. (2018). Users' Acceptance of Online Literature Databases in a Thai University: A Test of UTAUT2. *International Journal of Information Systems in the Service Sector, 10*(1), 54–70. doi:10.4018/ IJISSS.2018010104

Chaudron, G. (2018). Burst Pipes and Leaky Roofs: Small Emergencies Are a Challenge for Libraries. In K. Strang, M. Korstanje, & N. Vajjhala (Eds.), *Research, Practices, and Innovations in Global Risk and Contingency Management* (pp. 211–231). Hershey, PA: IGI Global. doi:10.4018/978-1-5225-4754-9.ch012

Chemulwo, M. J. (2018). Managing Intellectual Property in Digital Libraries and Copyright Challenges. In A. Tella & T. Kwanya (Eds.), *Handbook of Research on Managing Intellectual Property in Digital Libraries* (pp. 165–183). Hershey, PA: IGI Global. doi:10.4018/978-1-5225-3093-0.ch009

Chen, J., Lan, X., Huang, Q., Dong, J., & Chen, C. (2017). Scholarly Learning Commons. In L. Ruan, Q. Zhu, & Y. Ye (Eds.), *Academic Library Development and Administration in China* (pp. 90–109). Hershey, PA: IGI Global. doi:10.4018/978-1-5225-0550-1.ch006

Chigwada, J. P. (2018). Adoption of Open Source Software in Libraries in Developing Countries. *International Journal of Library and Information Services, 7*(1), 15–29. doi:10.4018/IJLIS.2018010102

Chigwada, J. P. (2020). Librarian Skillsets in the 21st Century: The Changing Role of Librarians in the Digital Era. In N. Osuigwe (Ed.), *Managing and Adapting Library Information Services for Future Users* (pp. 41–58). IGI Global. https://doi. org/10.4018/978-1-7998-1116-9.ch003

Chigwada, J. P. (2020). The Role of the Librarian in the Research Life Cycle: Research Collaboration Among the Library and Faculty. In C. Chisita (Ed.), *Cooperation and Collaboration Initiatives for Libraries and Related Institutions* (pp. 335–346). IGI Global. https://doi.org/10.4018/978-1-7998-0043-9.ch017

Related References

Chigwada, J. P. (2021). Research Data Management Services in Tertiary Institutions in Zimbabwe. In B. Holland (Eds.), *Handbook of Research on Knowledge and Organization Systems in Library and Information Science* (pp. 419-437). IGI Global. https://doi.org/10.4018/978-1-7998-7258-0.ch022

Chigwada, J. P., & Maturure, R. (2019). Advocating for Library and Information Services by National Library Associations of Africa in the Context of Sustainable Development Goals. In P. Ngulube (Ed.), *Handbook of Research on Advocacy, Promotion, and Public Programming for Memory Institutions* (pp. 219–237). IGI Global. doi:10.4018/978-1-5225-7429-3.ch012

Chiparausha, B., & Chigwada, J. P. (2019). Promoting Library Services in a Digital Environment in Zimbabwe. In P. Ngulube (Ed.), *Handbook of Research on Advocacy, Promotion, and Public Programming for Memory Institutions* (pp. 284–296). IGI Global. https://doi.org/10.4018/978-1-5225-7429-3.ch015

Chisita, C. T., & Chinyemba, F. (2017). Utilising ICTs for Resource Sharing Initiatives in Academic Institutions in Zimbabwe: Towards a New Trajectory. In B. Gunjal (Ed.), *Managing Knowledge and Scholarly Assets in Academic Libraries* (pp. 174–187). Hershey, PA: IGI Global. doi:10.4018/978-1-5225-1741-2.ch008

Chu, S., Tu, S., Wang, N., & Zhang, W. (2020). Information Equity and Cultural Sharing: The Service for Migrant Workers in Hangzhou Public Library. *International Journal of Library and Information Services*, 9(1), 10–24. https://doi.org/10.4018/IJLIS.2020010102

Clarance, M. M., & Angeline, X. M. (2019). User Opinion on Library Collections and Services: A Case Study of Branch Library in Karaikudi. In S. Thanuskodi (Ed.), *Literacy Skill Development for Library Science Professionals* (pp. 343–375). IGI Global. https://doi.org/10.4018/978-1-5225-7125-4.ch015

Costello, L., & Fazal, S. (2018). Developing Unique Study Room Reservation Systems: Examples From Teachers College and Stony Brook University. In L. Costello & M. Powers (Eds.), *Developing In-House Digital Tools in Library Spaces* (pp. 168–176). Hershey, PA: IGI Global. doi:10.4018/978-1-5225-2676-6.ch008

Cui, Y. (2017). Research Data Management: Models, Challenges, and Actions. In L. Ruan, Q. Zhu, & Y. Ye (Eds.), *Academic Library Development and Administration in China* (pp. 184–195). Hershey, PA: IGI Global. doi:10.4018/978-1-5225-0550-1.ch011

Dhamdhere, S. N., De Smet, E., & Lihitkar, R. (2017). Web-Based Bibliographic Services Offered by Top World and Indian University Libraries: A Comparative Study. *International Journal of Library and Information Services*, 6(1), 53–72. doi:10.4018/IJLIS.2017010104

Eiriemiokhale, K. A. (2018). Copyright Issues in a Digital Library Environment. In A. Tella & T. Kwanya (Eds.), *Handbook of Research on Managing Intellectual Property in Digital Libraries* (pp. 142–164). Hershey, PA: IGI Global. doi:10.4018/978-1-5225-3093-0.ch008

El Mimouni, H., Anderson, J., Tempelman-Kluit, N. F., & Dolan-Mescal, A. (2018). UX Work in Libraries: How (and Why) to Do It. In L. Costello & M. Powers (Eds.), *Developing In-House Digital Tools in Library Spaces* (pp. 1–36). Hershey, PA: IGI Global. doi:10.4018/978-1-5225-2676-6.ch001

Emiri, O. T. (2017). Digital Literacy Skills Among Librarians in University Libraries In the 21st Century in Edo And Delta States, Nigeria. *International Journal of Library and Information Services*, 6(1), 37–52. doi:10.4018/IJLIS.2017010103

Emmelhainz, C. (2020). Educating the Central Asian Librarian: Considering the International MLIS in Kazakhstan. In R. Alenzuela, H. Kim, & D. Baylen (Eds.), *Internationalization of Library and Information Science Education in the Asia-Pacific Region* (pp. 1–32). IGI Global. https://doi.org/10.4018/978-1-7998-2273-8.ch001

Esguerra, A. C. (2020). Library Education and Librarianship in Japan and the Philippines. In R. Alenzuela, H. Kim, & D. Baylen (Eds.), *Internationalization of Library and Information Science Education in the Asia-Pacific Region* (pp. 131–157). IGI Global. https://doi.org/10.4018/978-1-7998-2273-8.ch006

Esposito, T. (2018). Exploring Opportunities in Health Science Information Instructional Outreach: A Case Study Highlighting One Academic Library's Experience. In S. Bhattacharyya & K. Patnaik (Eds.), *Changing the Scope of Library Instruction in the Digital Age* (pp. 118–135). Hershey, PA: IGI Global. doi:10.4018/978-1-5225-2802-9.ch005

Fagbola, O. O., Smart, A. E., & Oluwaseun, B. O. (2020). Application of Cloud Computing Technologies in Academic Library Management: The National Open University of Nigeria Library in Perspective. In A. Tella (Ed.), *Handbook of Research on Digital Devices for Inclusivity and Engagement in Libraries* (pp. 135–159). IGI Global. https://doi.org/10.4018/978-1-5225-9034-7.ch007

Fan, Y., Zhang, X., & Li, G. (2017). Research Initiatives and Projects in Academic Libraries. In L. Ruan, Q. Zhu, & Y. Ye (Eds.), *Academic Library Development and Administration in China* (pp. 230–252). Hershey, PA: IGI Global. doi:10.4018/978-1-5225-0550-1.ch014

Farmer, L. S. (2017). ICT Literacy Integration: Issues and Sample Efforts. In J. Keengwe & P. Bull (Eds.), *Handbook of Research on Transformative Digital Content and Learning Technologies* (pp. 59–80). Hershey, PA: IGI Global. doi:10.4018/978-1-5225-2000-9.ch004

Farmer, L. S. (2017). Data Analytics for Strategic Management: Getting the Right Data. In V. Wang (Ed.), *Encyclopedia of Strategic Leadership and Management* (pp. 810–822). Hershey, PA: IGI Global. doi:10.4018/978-1-5225-1049-9.ch056

Farmer, L. S. (2017). Managing Portable Technologies for Special Education. In V. Wang (Ed.), *Encyclopedia of Strategic Leadership and Management* (pp. 977–987). Hershey, PA: IGI Global. doi:10.4018/978-1-5225-1049-9.ch068

Fujishima, D., & Kamada, T. (2017). Collective Relocation for Associative Distributed Collections of Objects. *International Journal of Software Innovation*, 5(2), 55–69. doi:10.4018/IJSI.2017040104

Ghani, S. R. (2017). Ontology: Advancing Flawless Library Services. In T. Ashraf & N. Kumar (Eds.), *Interdisciplinary Digital Preservation Tools and Technologies* (pp. 79–102). Hershey, PA: IGI Global. doi:10.4018/978-1-5225-1653-8.ch005

Gu, J. (2017). Library Buildings on New Campuses. In L. Ruan, Q. Zhu, & Y. Ye (Eds.), *Academic Library Development and Administration in China* (pp. 110–124). Hershey, PA: IGI Global. doi:10.4018/978-1-5225-0550-1.ch007

Guan, Z., & Wang, J. (2017). The China Academic Social Sciences and Humanities Library (CASHL). In L. Ruan, Q. Zhu, & Y. Ye (Eds.), *Academic Library Development and Administration in China* (pp. 31–54). Hershey, PA: IGI Global. doi:10.4018/978-1-5225-0550-1.ch003

Gul, S., & Shueb, S. (2018). Confronting/Managing the Crisis of Indian Libraries: E-Consortia Initiatives in India - A Way Forward. In R. Bhardwaj (Ed.), *Digitizing the Modern Library and the Transition From Print to Electronic* (pp. 129–163). Hershey, PA: IGI Global. doi:10.4018/978-1-5225-2119-8.ch006

Gunjal, B. (2017). Managing Knowledge and Scholarly Assets in Academic Libraries: Issues and Challenges. In B. Gunjal (Ed.), *Managing Knowledge and Scholarly Assets in Academic Libraries* (pp. 270–279). Hershey, PA: IGI Global. doi:10.4018/978-1-5225-1741-2.ch013

Guo, J., Zhang, H., & Zong, Y. (2017). Leadership Development and Career Planning. In L. Ruan, Q. Zhu, & Y. Ye (Eds.), *Academic Library Development and Administration in China* (pp. 264–279). Hershey, PA: IGI Global. doi:10.4018/978-1-5225-0550-1.ch016

Hahn, J. (2020). Student Engagement and Smart Spaces: Library Browsing and Internet of Things Technology. In B. Holland (Eds.), *Emerging Trends and Impacts of the Internet of Things in Libraries* (pp. 52-70). IGI Global. https://doi.org/10.4018/978-1-7998-4742-7.ch003

Halder, D. (2020). A Transitional Shift From Traditional Library to Digital Library. In A. Kaushik, A. Kumar, & P. Biswas (Eds.), *Handbook of Research on Emerging Trends and Technologies in Library and Information Science* (pp. 147–155). IGI Global. https://doi.org/10.4018/978-1-5225-9825-1.ch011

Hallis, R. (2018). Leveraging Library Instruction in a Digital Age. In S. Bhattacharyya & K. Patnaik (Eds.), *Changing the Scope of Library Instruction in the Digital Age* (pp. 1–23). Hershey, PA: IGI Global. doi:10.4018/978-1-5225-2802-9.ch001

Halupa, C. (2022). An Introduction to Survey Research. In A. Zimmerman (Ed.), *Methodological Innovations in Research and Academic Writing* (pp. 41–62). IGI Global. https://doi.org/10.4018/978-1-7998-8283-1.ch003

Hartsock, R., & Alemneh, D. G. (2018). Electronic Theses and Dissertations (ETDs). In M. Khosrow-Pour, D.B.A. (Ed.), Encyclopedia of Information Science and Technology, Fourth Edition (pp. 6748-6755). Hershey, PA: IGI Global. https:// doi.org/ doi:10.4018/978-1-5225-2255-3.ch584

Haugh, D. (2018). Mobile Applications for Libraries. In L. Costello & M. Powers (Eds.), *Developing In-House Digital Tools in Library Spaces* (pp. 76–90). Hershey, PA: IGI Global. doi:10.4018/978-1-5225-2676-6.ch004

Hayes, C. (2022). Methodology and Method in Case Study Research: Framing Research Design in Practice. In S. Watson, S. Austin, & J. Bell (Eds.), *Conceptual Analyses of Curriculum Inquiry Methodologies* (pp. 138-154). IGI Global. https:// doi.org/10.4018/978-1-7998-8848-2.ch007

Hayes, C., & Graham, Y. N. (2022). Phenomenology: Conceptually Framing Phenomenological Research Design and Methodology. In S. Watson, S. Austin, & J. Bell (Eds.), *Conceptual Analyses of Curriculum Inquiry Methodologies* (pp. 28-50). IGI Global. https://doi.org/10.4018/978-1-7998-8848-2.ch002

Hill, V. (2017). Digital Citizens as Writers: New Literacies and New Responsibilities. In E. Monske & K. Blair (Eds.), *Handbook of Research on Writing and Composing in the Age of MOOCs* (pp. 56–74). Hershey, PA: IGI Global. doi:10.4018/978-1-5225-1718-4.ch004

Hoh, A. (2019). Expanding the Awareness and Use of Library Collections Through Social Media: A Case Study of the Library of Congress International Collections Social Media Program. In J. Joe & E. Knight (Eds.), *Social Media for Communication and Instruction in Academic Libraries* (pp. 212–236). IGI Global. doi:10.4018/978-1-5225-8097-3.ch013

Holland, B. (2020). Emerging Technology and Today's Libraries. In B. Holland (Eds.), *Emerging Trends and Impacts of the Internet of Things in Libraries* (pp. 1-33). IGI Global. https://doi.org/10.4018/978-1-7998-4742-7.ch001

Homza, A., & Fontno, T. J. (2021). Supporting Teacher Candidates as Social Justice Change-Makers: A Faculty-Librarian Collaboration for Building and Using Diverse Youth Collections. In D. Hartsfield (Ed.), *Handbook of Research on Teaching Diverse Youth Literature to Pre-Service Professionals* (pp. 398–421). IGI Global. https://doi.org/10.4018/978-1-7998-7375-4.ch020

Horne-Popp, L. M., Tessone, E. B., & Welker, J. (2018). If You Build It, They Will Come: Creating a Library Statistics Dashboard for Decision-Making. In L. Costello & M. Powers (Eds.), *Developing In-House Digital Tools in Library Spaces* (pp. 177–203). Hershey, PA: IGI Global. doi:10.4018/978-1-5225-2676-6.ch009

Huang, C., & Xue, H. F. (2017). The China Academic Digital Associative Library (CADAL). In L. Ruan, Q. Zhu, & Y. Ye (Eds.), *Academic Library Development and Administration in China* (pp. 20–30). Hershey, PA: IGI Global. doi:10.4018/978-1-5225-0550-1.ch002

Huang, J., & Vedantham, A. (2019). Cabot Science Library: Creating Transformative Learning Environments in Library Spaces. In A. Darshan Singh, S. Raghunathan, E. Robeck, & B. Sharma (Eds.), *Cases on Smart Learning Environments* (pp. 284–298). IGI Global. doi:10.4018/978-1-5225-6136-1.ch016

Hunsaker, A. J., Majewski, N., & Rocke, L. E. (2018). Pulling Content out the Back Door: Creating an Interactive Digital Collections Experience. In L. Costello & M. Powers (Eds.), *Developing In-House Digital Tools in Library Spaces* (pp. 205–226). Hershey, PA: IGI Global. doi:10.4018/978-1-5225-2676-6.ch010

Hussain, A. (2020). Cutting Edge: Technology's Impact on Library Services. In J. Jesubright & P. Saravanan (Eds.), *Innovations in the Designing and Marketing of Information Services* (pp. 16–27). IGI Global. https://doi.org/10.4018/978-1-7998-1482-5.ch002

Idiegbeyan-ose, J., Owolabi, S. E., Ayooluwa, A., Foluke, O., Toluwani, E., & Sunday, O. (2019). Digital Library and Distance Learning in Developing Countries: Benefits and Challenges. In R. Bhardwaj & P. Banks (Eds.), *Research Data Access and Management in Modern Libraries* (pp. 220–245). IGI Global. https://doi.org/10.4018/978-1-5225-8437-7.ch011

Ifijeh, G., Adebayo, O., Izuagbe, R., & Olawoyin, O. (2018). Institutional Repositories and Libraries in Nigeria: Interrogating the Nexus. *Journal of Cases on Information Technology*, *20*(2), 16–29. doi:10.4018/JCIT.2018040102

Igbinovia, M. O., Solanke, E. O., & Obinyan, O. O. (2020). Building Influence: Strategising for Library Advocacy. In N. Osuigwe (Ed.), *Managing and Adapting Library Information Services for Future Users* (pp. 221–241). IGI Global. https://doi.org/10.4018/978-1-7998-1116-9.ch013

Iglesias, E. (2017). Insourcing and Outsourcing of Library Technology. In E. Iglesias (Ed.), *Library Technology Funding, Planning, and Deployment* (pp. 113–123). Hershey, PA: IGI Global. doi:10.4018/978-1-5225-1735-1.ch007

Ikolo, V. E. (2018). Transformational Leadership for Academic Libraries in Nigeria. In M. Khosrow-Pour, D.B.A. (Ed.), Encyclopedia of Information Science and Technology, Fourth Edition (pp. 5726-5735). Hershey, PA: IGI Global. doi:10.4018/978-1-5225-2255-3.ch497

Ikolo, V. E. (2020). Doctor's Awareness and Perception of Medical Library Resources and Services: A Case Study of Delta State University Teaching Hospital (Delsuth), Nigeria. *International Journal of Library and Information Services*, *9*(2), 58–71. https://doi.org/10.4018/IJLIS.2020070104

Joe, J. A. (2018). Changing Expectations of Academic Libraries. In M. Khosrow-Pour, D.B.A. (Ed.), Encyclopedia of Information Science and Technology, Fourth Edition (pp. 5204-5212). Hershey, PA: IGI Global. doi:10.4018/978-1-5225-2255-3.ch452

Juliana, I., Izuagbe, R., Itsekor, V., Fagbohun, M. O., Asaolu, A., & Nwokeoma, M. N. (2018). The Role of the School Library in Empowering Visually Impaired Children With Lifelong Information Literacy Skills. In P. Epler (Ed.), *Instructional Strategies in General Education and Putting the Individuals With Disabilities Act (IDEA) Into Practice* (pp. 245–271). Hershey, PA: IGI Global. doi:10.4018/978-1-5225-3111-1.ch009

Kalu, C. O., Chidi-Kalu, E. I., & Mafe, T. A. (2021). Research Data Management in an Academic Library. In J. Chigwada & G. Tsvuura (Eds.), *Handbook of Research on Information and Records Management in the Fourth Industrial Revolution* (pp. 38–55). IGI Global. https://doi.org/10.4018/978-1-7998-7740-0.ch003

Kalusopa, T. (2018). Preservation and Access to Digital Materials: Strategic Policy Options for Africa. In P. Ngulube (Ed.), *Handbook of Research on Heritage Management and Preservation* (pp. 150–174). Hershey, PA: IGI Global. doi:10.4018/978-1-5225-3137-1.ch008

Kamau, G. W. (2018). Copyright Challenges in Digital Libraries in Kenya From the Lens of a Librarian. In A. Tella & T. Kwanya (Eds.), *Handbook of Research on Managing Intellectual Property in Digital Libraries* (pp. 312–336). Hershey, PA: IGI Global. doi:10.4018/978-1-5225-3093-0.ch016

Karagöz, E., Güney, L. Ö., & Baran, B. (2022). The Collaborative Digital Content Library Fostering Faculty Members' Collaboratively Building Learning Sources. In G. Durak & S. Çankaya (Eds.), *Handbook of Research on Managing and Designing Online Courses in Synchronous and Asynchronous Environments* (pp. 196–213). IGI Global. doi:10.4018/978-1-7998-8701-0.ch010

Karmakar, R. (2018). Development and Management of Digital Libraries in the Regime of IPR Paradigm. *International Journal of Library and Information Services*, 7(1), 44–57. doi:10.4018/IJLIS.2018010104

Kasemsap, K. (2017). Mastering Knowledge Management in Academic Libraries. In B. Gunjal (Ed.), *Managing Knowledge and Scholarly Assets in Academic Libraries* (pp. 27–55). Hershey, PA: IGI Global. doi:10.4018/978-1-5225-1741-2.ch002

Kehinde, A. (2018). Digital Libraries and the Role of Digital Librarians. In A. Tella & T. Kwanya (Eds.), *Handbook of Research on Managing Intellectual Property in Digital Libraries* (pp. 98–119). Hershey, PA: IGI Global. doi:10.4018/978-1-5225-3093-0.ch006

Kenausis, V., & Herman, D. (2017). Don't Make Us Use the "Get Along Shirt": Communication and Consensus Building in an RFP Process. In E. Iglesias (Ed.), *Library Technology Funding, Planning, and Deployment* (pp. 1–22). Hershey, PA: IGI Global. doi:10.4018/978-1-5225-1735-1.ch001

Kohl, L. E., Lombardi, P., & Moroney, M. (2017). Moving from Local to Global via the Integrated Library System: Cost-Savings, ILS Management, Teams, and End-Users. In E. Iglesias (Ed.), *Library Technology Funding, Planning, and Deployment* (pp. 23–35). Hershey, PA: IGI Global. doi:10.4018/978-1-5225-1735-1.ch002

Kowalsky, M. (2020). School Librarian Experiences of Learning Management Implementation. In A. Tella (Ed.), *Handbook of Research on Digital Devices for Inclusivity and Engagement in Libraries* (pp. 160–184). IGI Global. https://doi.org/10.4018/978-1-5225-9034-7.ch008

Kumar, K. (2018). Library in Your Pocket Delivery of Instruction Service Through Library Mobile Apps: A World in Your Pocket. In S. Bhattacharyya & K. Patnaik (Eds.), *Changing the Scope of Library Instruction in the Digital Age* (pp. 228–249). Hershey, PA: IGI Global. doi:10.4018/978-1-5225-2802-9.ch009

Kwanya, T. (2018). Social Bookmarking in Digital Libraries: Intellectual Property Rights Implications. In A. Tella & T. Kwanya (Eds.), *Handbook of Research on Managing Intellectual Property in Digital Libraries* (pp. 1–18). Hershey, PA: IGI Global. doi:10.4018/978-1-5225-3093-0.ch001

Lapo, P., Makhmudov, G., & Rakhmatullaev, M. (2020). Internationalization of Library and Information Science Education in Tajikistan and Uzbekistan: Implications in Central Asia. In R. Alenzuela, H. Kim, & D. Baylen (Eds.), *Internationalization of Library and Information Science Education in the Asia-Pacific Region* (pp. 225–245). IGI Global. https://doi.org/10.4018/978-1-7998-2273-8.ch010

Lewis, J. K. (2018). Change Leadership Styles and Behaviors in Academic Libraries. In M. Khosrow-Pour, D.B.A. (Ed.), Encyclopedia of Information Science and Technology, Fourth Edition (pp. 5194-5203). Hershey, PA: IGI Global. doi:10.4018/978-1-5225-2255-3.ch451

Lillard, L. L. (2018). Is Interdisciplinary Collaboration in Academia an Elusive Dream?: Can the Institutional Barriers Be Broken Down? A Review of the Literature and the Case of Library Science. In M. Al-Suqri, A. Al-Kindi, S. AlKindi, & N. Saleem (Eds.), *Promoting Interdisciplinarity in Knowledge Generation and Problem Solving* (pp. 139–147). Hershey, PA: IGI Global. doi:10.4018/978-1-5225-3878-3.ch010

Liu, C., Dou, T., Zhou, H., Zhang, B., & Zhang, C. (2021). Library Service Innovation Based on New Information Technology: Taking the Interactive Experience Space "Tsinghua Impression" as an Example. *International Journal of Library and Information Services*, *10*(1), 71–81. https://doi.org/10.4018/IJLIS.2021010106

Long, X., & Yao, B. (2017). The Construction and Development of the Academic Digital Library of Chinese Ancient Collections. In L. Ruan, Q. Zhu, & Y. Ye (Eds.), *Academic Library Development and Administration in China* (pp. 126–135). Hershey, PA: IGI Global. doi:10.4018/978-1-5225-0550-1.ch008

Long, X., & Yao, B. (2020). The Construction and Development of the Academic Digital Library of Chinese Ancient Collections. In I. Management Association (Ed.), *Digital Libraries and Institutional Repositories: Breakthroughs in Research and Practice* (pp. 78-87). IGI Global. https://doi.org/10.4018/978-1-7998-2463-3.ch006

Lowe, M., & Reno, L. M. (2018). Academic Librarianship and Burnout. In *Examining the Emotional Dimensions of Academic Librarianship: Emerging Research and Opportunities* (pp. 72–89). Hershey, PA: IGI Global. doi:10.4018/978-1-5225-3761-8.ch005

Lowe, M., & Reno, L. M. (2018). Emotional Dimensions of Academic Librarianship. In *Examining the Emotional Dimensions of Academic Librarianship: Emerging Research and Opportunities* (pp. 54–71). Hershey, PA: IGI Global. doi:10.4018/978-1-5225-3761-8.ch004

Lowe, M., & Reno, L. M. (2018). Why Isn't This Being Studied? In *Examining the Emotional Dimensions of Academic Librarianship: Emerging Research and Opportunities* (pp. 90–108). Hershey, PA: IGI Global. doi:10.4018/978-1-5225-3761-8.ch006

Lowe, M., & Reno, L. M. (2018). Research Agenda: Research Ideas and Recommendations. In *Examining the Emotional Dimensions of Academic Librarianship: Emerging Research and Opportunities* (pp. 109–125). Hershey, PA: IGI Global. doi:10.4018/978-1-5225-3761-8.ch007

Luyombya, D., Kiyingi, G. W., & Naluwooza, M. (2018). The Nature and Utilisation of Archival Records Deposited in Makerere University Library, Uganda. In P. Ngulube (Ed.), *Handbook of Research on Heritage Management and Preservation* (pp. 96–113). Hershey, PA: IGI Global. doi:10.4018/978-1-5225-3137-1.ch005

Ma, W. Y. (2022). Supporting Indigenous Education From a Distance: Adjusting Strategies to Maintain Access to a Rare Library Collection During a Global Crisis. In P. Pangelinan & T. McVey (Eds.), *Learning and Reconciliation Through Indigenous Education in Oceania* (pp. 197–209). IGI Global. https://doi.org/10.4018/978-1-7998-7736-3.ch012

Mabe, M., & Ashley, E. A. (2017). The Natural Role of the Public Library. In *The Developing Role of Public Libraries in Emergency Management: Emerging Research and Opportunities* (pp. 25–43). Hershey, PA: IGI Global. doi:10.4018/978-1-5225-2196-9.ch003

Mabe, M., & Ashley, E. A. (2017). I'm Trained, Now What? In *The Developing Role of Public Libraries in Emergency Management: Emerging Research and Opportunities* (pp. 87–95). Hershey, PA: IGI Global. doi:10.4018/978-1-5225-2196-9.ch007

Mabe, M., & Ashley, E. A. (2017). Emergency Preparation for the Library and Librarian. In *The Developing Role of Public Libraries in Emergency Management: Emerging Research and Opportunities* (pp. 61–78). Hershey, PA: IGI Global. doi:10.4018/978-1-5225-2196-9.ch005

Mabe, M., & Ashley, E. A. (2017). The CCPL Model. In *The Developing Role of Public Libraries in Emergency Management: Emerging Research and Opportunities* (pp. 15–24). Hershey, PA: IGI Global. doi:10.4018/978-1-5225-2196-9.ch002

Mabe, M., & Ashley, E. A. (2017). The Local Command Structure and How the Library Fits. In *In The Developing Role of Public Libraries in Emergency Management: Emerging Research and Opportunities* (pp. 44–60). Hershey, PA: IGI Global. doi:10.4018/978-1-5225-2196-9.ch004

Mafube, M. A., & Keakopa, S. M. (2019). Customer Services at the Library Archives of the National University of Lesotho. In P. Ngulube (Ed.), *Handbook of Research on Advocacy, Promotion, and Public Programming for Memory Institutions* (pp. 62–76). IGI Global. doi:10.4018/978-1-5225-7429-3.ch004

Majumdar, S. (2022). Community Engagement Through Extension and Outreach Activities: Scope of a College Library. In M. Taher (Ed.), *Handbook of Research on the Role of Libraries, Archives, and Museums in Achieving Civic Engagement and Social Justice in Smart Cities* (pp. 121–138). IGI Global. https://doi.org/10.4018/978-1-7998-8363-0.ch006

Mangone, E. (2022). The Difficult Joining of Theory and Empirical Research: Strengths and Weaknesses of Digital Research Methods. In G. Punziano & A. Delli Paoli (Eds.), *Handbook of Research on Advanced Research Methodologies for a Digital Society* (pp. 11–23). IGI Global. https://doi.org/10.4018/978-1-7998-8473-6.ch002

Manzoor, A. (2018). Social Media: A Librarian's Tool for Instant and Direct Interaction With Library Users. In R. Bhardwaj (Ed.), *Digitizing the Modern Library and the Transition From Print to Electronic* (pp. 112–128). Hershey, PA: IGI Global. doi:10.4018/978-1-5225-2119-8.ch005

Maringanti, H. (2018). A Decision Making Paradigm for Software Development in Libraries. In L. Costello & M. Powers (Eds.), *Developing In-House Digital Tools in Library Spaces* (pp. 59–75). Hershey, PA: IGI Global. doi:10.4018/978-1-5225-2676-6.ch003

Markman, K. M., Ferrarini, M., & Deschenes, A. H. (2018). User Testing and Iterative Design in the Academic Library: A Case Study. In R. Roscoe, S. Craig, & I. Douglas (Eds.), *End-User Considerations in Educational Technology Design* (pp. 160–183). Hershey, PA: IGI Global. doi:10.4018/978-1-5225-2639-1.ch008

Marrazzo, F. (2022). Doing Research With Online Platforms: An Emerging Issue Network. In G. Punziano & A. Delli Paoli (Eds.), *Handbook of Research on Advanced Research Methodologies for a Digital Society* (pp. 65–86). IGI Global. https://doi.org/10.4018/978-1-7998-8473-6.ch006

Mertens, D. (2022). Designing Mixed Methods Studies to Contribute to Social, Economic, and Environmental Justice: Implications for Library and Information Sciences. In P. Ngulube (Ed.), *Handbook of Research on Mixed Methods Research in Information Science* (pp. 173–189). IGI Global. https://doi.org/10.4018/978-1-7998-8844-4.ch009

Moahi, K. H. (2020). The Research Process and Indigenous Epistemologies. In P. Ngulube (Ed.), *Handbook of Research on Connecting Research Methods for Information Science Research* (pp. 245–265). IGI Global. https://doi.org/10.4018/978-1-7998-1471-9.ch013

Mohapatra, N. (2021). Webrarian: A Librarian on the Web. In C. Chisita, R. Enakrire, O. Durodolu, V. Tsabedze, & J. Ngoaketsi (Eds.), *Handbook of Research on Records and Information Management Strategies for Enhanced Knowledge Coordination* (pp. 458–470). IGI Global. https://doi.org/10.4018/978-1-7998-6618-3.ch027

Munatsi, R. (2020). National Research and Knowledge Systems: Role of Libraries. In C. Chisita (Ed.), *Cooperation and Collaboration Initiatives for Libraries and Related Institutions* (pp. 273–293). IGI Global. https://doi.org/10.4018/978-1-7998-0043-9.ch014

Musimbi, W. L., & Mutuku, P. K. (2019). The Future of LIS and Media Training in the Global Era: Challenges and Prospects. In C. Chisita & A. Rusero (Eds.), *Exploring the Relationship Between Media, Libraries, and Archives* (pp. 82–101). IGI Global. doi:10.4018/978-1-5225-5840-8.ch006

Mwanzu, A. (2019). Economics of Resource Sharing via Library Consortia. In C. Chisita & A. Rusero (Eds.), *Exploring the Relationship Between Media, Libraries, and Archives* (pp. 19–34). IGI Global. https://doi.org/10.4018/978-1-5225-5840-8.ch002

Na, L. (2017). Library and Information Science Education and Graduate Programs in Academic Libraries. In L. Ruan, Q. Zhu, & Y. Ye (Eds.), *Academic Library Development and Administration in China* (pp. 218–229). Hershey, PA: IGI Global. doi:10.4018/978-1-5225-0550-1.ch013

Nagarkar, S. P. (2017). Biomedical Librarianship in the Post-Genomic Era. In S. Ram (Ed.), *Library and Information Services for Bioinformatics Education and Research* (pp. 1–17). Hershey, PA: IGI Global. doi:10.4018/978-1-5225-1871-6.ch001

Natarajan, M. (2017). Exploring Knowledge Sharing over Social Media. In R. Chugh (Ed.), *Harnessing Social Media as a Knowledge Management Tool* (pp. 55–73). Hershey, PA: IGI Global. doi:10.4018/978-1-5225-0495-5.ch003

Nazir, T. (2017). Preservation Initiatives in E-Environment to Protect Information Assets. In T. Ashraf & N. Kumar (Eds.), *Interdisciplinary Digital Preservation Tools and Technologies* (pp. 193–208). Hershey, PA: IGI Global. doi:10.4018/978-1-5225-1653-8.ch010

Ngulube, P. (2017). Embedding Indigenous Knowledge in Library and Information Science Education in Anglophone Eastern and Southern Africa. In P. Ngulube (Ed.), *Handbook of Research on Social, Cultural, and Educational Considerations of Indigenous Knowledge in Developing Countries* (pp. 92–115). Hershey, PA: IGI Global. doi:10.4018/978-1-5225-0838-0.ch006

Ngulube, P. (2022). Using Simple and Complex Mixed Methods Research Designs to Understand Research in Information Science. In P. Ngulube (Ed.), *Handbook of Research on Mixed Methods Research in Information Science* (pp. 20–46). IGI Global. doi:10.4018/978-1-7998-8844-4.ch002

Nicolajsen, H. W., Sorensen, F., & Scupola, A. (2018). User Involvement in Service Innovation Processes. In M. Khosrow-Pour (Ed.), *Optimizing Current Practices in E-Services and Mobile Applications* (pp. 42–61). Hershey, PA: IGI Global. doi:10.4018/978-1-5225-5026-6.ch003

Ocholla, D. N. (2022). A Research Dashboard for Aligning Research Components in Research Proposals, Theses, and Dissertations in Library and Information Science. In P. Ngulube (Ed.), *Handbook of Research on Mixed Methods Research in Information Science* (pp. 629–640). IGI Global. https://doi.org/10.4018/978-1-7998-8844-4.ch029

Ochonogor, W. C., & Okite-Amughoro, F. A. (2018). Building an Effective Digital Library in a University Teaching Hospital (UTH) in Nigeria. In A. Tella & T. Kwanya (Eds.), *Handbook of Research on Managing Intellectual Property in Digital Libraries* (pp. 184–204). Hershey, PA: IGI Global. doi:10.4018/978-1-5225-3093-0.ch010

Okada, D. (2020). 10,000 Newly Certified Librarians, 100 Secure Jobs. In R. Alenzuela, H. Kim, & D. Baylen (Eds.), *Internationalization of Library and Information Science Education in the Asia-Pacific Region* (pp. 78–101). IGI Global. https://doi.org/10.4018/978-1-7998-2273-8.ch004

Oladapo, Y. O. (2018). Open Access to Knowledge and Challenges in Digital Libraries. In A. Tella & T. Kwanya (Eds.), *Handbook of Research on Managing Intellectual Property in Digital Libraries* (pp. 260–291). Hershey, PA: IGI Global. doi:10.4018/978-1-5225-3093-0.ch014

Oladokun, O., & Zulu, S. F. (2017). Document Description and Coding as Key Elements in Knowledge, Records, and Information Management. In P. Jain & N. Mnjama (Eds.), *Managing Knowledge Resources and Records in Modern Organizations* (pp. 179–197). Hershey, PA: IGI Global. doi:10.4018/978-1-5225-1965-2.ch011

Olubodun, O. J., & Oye, P. O. (2019). Library: A Tool for Information Dissemination and Creating Awareness in Conflict-Induced Situations. In E. Nyam & F. Idoko (Eds.), *Examining the Social and Economic Impacts of Conflict-Induced Migration* (pp. 55–63). IGI Global. https://doi.org/10.4018/978-1-5225-7615-0.ch003

Omeluzor, S. U., Abayomi, I., & Gbemi-Ogunleye, P. (2018). Contemporary Media for Library Users' Instruction in Academic Libraries in South-West Nigeria: Contemporary Library Instruction in the Digital Age. In S. Bhattacharyya & K. Patnaik (Eds.), *Changing the Scope of Library Instruction in the Digital Age* (pp. 162–185). Hershey, PA: IGI Global. doi:10.4018/978-1-5225-2802-9.ch007

Onwuchekwa, E. O. (2020). Library Signage and Information Graphics: A Communication Tool for Library Users. In A. Tella (Ed.), *Handbook of Research on Digital Devices for Inclusivity and Engagement in Libraries* (pp. 231–237). IGI Global. https://doi.org/10.4018/978-1-5225-9034-7.ch011

Onyancha, O. B. (2020). Informetrics Research Methods Outlined. In P. Ngulube (Ed.), *Handbook of Research on Connecting Research Methods for Information Science Research* (pp. 320–348). IGI Global. https://doi.org/10.4018/978-1-7998-1471-9.ch017

Oshilalu, A. H., & Ogochukwu, E. T. (2017). Modeling a Software for Library and Information Centers. *International Journal of Library and Information Services*, *6*(2), 1–10. doi:10.4018/IJLIS.2017070101

Oswal, S. K. (2017). Institutional, Legal, and Attitudinal Barriers to the Accessibility of University Digital Libraries: Implications for Retention of Disabled Students. In H. Alphin Jr, J. Lavine, & R. Chan (Eds.), *Disability and Equity in Higher Education Accessibility* (pp. 223–241). Hershey, PA: IGI Global. doi:10.4018/978-1-5225-2665-0.ch010

Oukrich, J., & Bouikhalene, B. (2017). A Survey of Users' Satisfaction in the University Library by Using a Pareto Analysis and the Automatic Classification Methods. *International Journal of Library and Information Services*, *6*(1), 17–36. doi:10.4018/IJLIS.2017010102

Oyelude, A. A., & Oluwaniyi, S. A. (2020). Managing Future Library Services for the Medical Sciences: A Pharmacy Library Experience. In N. Osuigwe (Ed.), *Managing and Adapting Library Information Services for Future Users* (pp. 200–220). IGI Global. https://doi.org/10.4018/978-1-7998-1116-9.ch012

Özel, N. (2018). Developing Visual Literacy Skills Through Library Instructions. In V. Osinska & G. Osinski (Eds.), *Information Visualization Techniques in the Social Sciences and Humanities* (pp. 32–48). Hershey, PA: IGI Global. doi:10.4018/978-1-5225-4990-1.ch003

Paganelli, A. L., & Paganelli, A. L. (2021). Blockchain and the Research Libraries: Expanding Interlibrary Loan and Protecting Privacy. In D. Gunter (Ed.), *Transforming Scholarly Publishing With Blockchain Technologies and AI* (pp. 232–250). IGI Global. https://doi.org/10.4018/978-1-7998-5589-7.ch012

Patel, D., & Thakur, D. (2017). Managing Open Access (OA) Scholarly Information Resources in a University. In A. Munigal (Ed.), *Scholarly Communication and the Publish or Perish Pressures of Academia* (pp. 224–255). Hershey, PA: IGI Global. doi:10.4018/978-1-5225-1697-2.ch011

Patnaik, K. R. (2018). Crafting a Framework for Copyright Literacy and Licensed Content: A Case Study at an Advanced Management Education and Research Library. In S. Bhattacharyya & K. Patnaik (Eds.), *Changing the Scope of Library Instruction in the Digital Age* (pp. 136–160). Hershey, PA: IGI Global. doi:10.4018/978-1-5225-2802-9.ch006

Paynter, K. (2017). Elementary Library Media Specialists' Roles in the Implementation of the Common Core State Standards. In M. Grassetti & S. Brookby (Eds.), *Advancing Next-Generation Teacher Education through Digital Tools and Applications* (pp. 262–283). Hershey, PA: IGI Global. doi:10.4018/978-1-5225-0965-3.ch014

Perry, S. C., & Waggoner, J. (2018). Processes for User-Centered Design and Development: The Omeka Curator Dashboard Project. In L. Costello & M. Powers (Eds.), *Developing In-House Digital Tools in Library Spaces* (pp. 37–58). Hershey, PA: IGI Global. doi:10.4018/978-1-5225-2676-6.ch002

Perumalsamy, R., & Kannan, S. P. (2019). User Information Needs in the Public Libraries in India. In S. Thanuskodi (Ed.), *Literacy Skill Development for Library Science Professionals* (pp. 25–53). IGI Global. https://doi.org/10.4018/978-1-5225-7125-4.ch002

Phuritsabam, B., & Devi, A. B. (2017). Information Seeking Behavior of Medical Scientists at Jawaharlal Nehru Institute of Medical Science: A Study. In S. Ram (Ed.), *Library and Information Services for Bioinformatics Education and Research* (pp. 177–187). Hershey, PA: IGI Global. doi:10.4018/978-1-5225-1871-6.ch010

Quadri, R. F., & Sodiq, O. A. (2018). Managing Intellectual Property in Digital Libraries: The Roles of Digital Librarians. In A. Tella & T. Kwanya (Eds.), *Handbook of Research on Managing Intellectual Property in Digital Libraries* (pp. 337–355). Hershey, PA: IGI Global. doi:10.4018/978-1-5225-3093-0.ch017

Quintana, A. J. (2021). Ensuring Research Integrity. In K. Elufiede & C. Barker Stucky (Eds.), *Strategies and Tactics for Multidisciplinary Writing* (pp. 192–201). IGI Global. https://doi.org/10.4018/978-1-7998-4477-8.ch015

Qutab, S., Adil, S. A., Gardner, L. A., & Ullah, F. S. (2022). The Role of Libraries, Archives, and Museums for Metaliteracy in Smart Cities: Implications, Challenges, and Opportunities. In M. Taher (Ed.), *Handbook of Research on the Role of Libraries, Archives, and Museums in Achieving Civic Engagement and Social Justice in Smart Cities* (pp. 355–375). IGI Global. https://doi.org/10.4018/978-1-7998-8363-0.ch019

Raj, S. K., & De, K. (2020). Electronic Resource Management and Digitisation: Library System of the University of Calcutta. In A. Kaushik, A. Kumar, & P. Biswas (Eds.), *Handbook of Research on Emerging Trends and Technologies in Library and Information Science* (pp. 231–265). IGI Global. https://doi.org/10.4018/978-1-5225-9825-1.ch017

Ram, S. (2017). Library Services for Bioinformatics: Establishing Synergy Data Information and Knowledge. In S. Ram (Ed.), *Library and Information Services for Bioinformatics Education and Research* (pp. 18–33). Hershey, PA: IGI Global. doi:10.4018/978-1-5225-1871-6.ch002

Rao, M. (2017). Use of Institutional Repository for Information Dissemination and Knowledge Management. In B. Gunjal (Ed.), *Managing Knowledge and Scholarly Assets in Academic Libraries* (pp. 156–173). Hershey, PA: IGI Global. doi:10.4018/978-1-5225-1741-2.ch007

Rao, Y., & Zhang, Y. (2017). The Construction and Development of Academic Library Digital Special Subject Databases. In L. Ruan, Q. Zhu, & Y. Ye (Eds.), *Academic Library Development and Administration in China* (pp. 163–183). Hershey, PA: IGI Global. doi:10.4018/978-1-5225-0550-1.ch010

Rao, Y., & Zhang, Y. (2020). The Construction and Development of Academic Library Digital Special Subject Databases. In I. Management Association (Ed.), *Digital Libraries and Institutional Repositories: Breakthroughs in Research and Practice* (pp. 24-44). IGI Global. https://doi.org/10.4018/978-1-7998-2463-3.ch002

Razip, S. N., Kadir, S. F., Saim, S. N., Dolhan, F. N., Jarmil, N., Salleh, N. H., & Rajin, G. (2017). Predicting Users' Intention towards Using Library Self-Issue and Return Systems. In N. Suki (Ed.), *Handbook of Research on Leveraging Consumer Psychology for Effective Customer Engagement* (pp. 102–115). Hershey, PA: IGI Global. doi:10.4018/978-1-5225-0746-8.ch007

Rothwell, S. L. (2018). Librarians and Instructional Design Challenges: Concepts, Examples, and a Flexible Design Framework. In S. Bhattacharyya & K. Patnaik (Eds.), *Changing the Scope of Library Instruction in the Digital Age* (pp. 24–59). Hershey, PA: IGI Global. doi:10.4018/978-1-5225-2802-9.ch002

Roy, L., & Frydman, A. (2018). Community Outreach. In M. Khosrow-Pour, D.B.A. (Ed.), Encyclopedia of Information Science and Technology, Fourth Edition (pp. 6685-6694). Hershey, PA: IGI Global. doi:10.4018/978-1-5225-2255-3.ch579

Rutto, D., & Yudah, O. (2018). E-Books in University Libraries in Kenya: Trends, Usage, and Intellectual Property Issues. In A. Tella & T. Kwanya (Eds.), *Handbook of Research on Managing Intellectual Property in Digital Libraries* (pp. 120–141). Hershey, PA: IGI Global. doi:10.4018/978-1-5225-3093-0.ch007

Sabharwal, A. (2017). The Transformative Role of Institutional Repositories in Academic Knowledge Management. In B. Gunjal (Ed.), *Managing Knowledge and Scholarly Assets in Academic Libraries* (pp. 127–155). Hershey, PA: IGI Global. doi:10.4018/978-1-5225-1741-2.ch006

Sadiku, S. A., Kpakiko, M. M., & Tsafe, A. G. (2018). Institutional Digital Repository and the Challenges of Global Visibility in Nigeria. In A. Tella & T. Kwanya (Eds.), *Handbook of Research on Managing Intellectual Property in Digital Libraries* (pp. 356–376). Hershey, PA: IGI Global. doi:10.4018/978-1-5225-3093-0.ch018

Sahu, M. K. (2018). Web-Scale Discovery Service in Academic Library Environment: A Birds Eye View. *International Journal of Library and Information Services*, 7(1), 1–14. doi:10.4018/IJLIS.2018010101

Salim, F., Saigar, B., Armoham, P. K., Gobalakrishnan, S., Jap, M. Y., & Lim, N. A. (2017). Students' Information-Seeking Intention in Academic Digital Libraries. In N. Suki (Ed.), *Handbook of Research on Leveraging Consumer Psychology for Effective Customer Engagement* (pp. 259–273). Hershey, PA: IGI Global. doi:10.4018/978-1-5225-0746-8.ch017

Saroja, G. (2017). Changing Face of Scholarly Communication and Its Impact on Library and Information Centres. In A. Munigal (Ed.), *Scholarly Communication and the Publish or Perish Pressures of Academia* (pp. 100–117). Hershey, PA: IGI Global. doi:10.4018/978-1-5225-1697-2.ch006

Sauti, L. (2020). Social Media and Library Collaboration: Analysis of Government Libraries (Kaguvi Building). In C. Chisita (Ed.), *Cooperation and Collaboration Initiatives for Libraries and Related Institutions* (pp. 312–334). IGI Global. https://doi.org/10.4018/978-1-7998-0043-9.ch016

Schuster, D. W. (2017). Selection Process for Free Open Source Software. In E. Iglesias (Ed.), *Library Technology Funding, Planning, and Deployment* (pp. 55–71). Hershey, PA: IGI Global. doi:10.4018/978-1-5225-1735-1.ch004

Seagraves, K., & Weyand, L. (2021). From Bake-Alongs to Tech Talks: How One Public Library System Pivoted to Virtual Programming. In B. Holland (Ed.), *Handbook of Research on Library Response to the COVID-19 Pandemic* (pp. 447-480). IGI Global. https://doi.org/10.4018/978-1-7998-6449-3.ch023

Segaetsho, T. (2018). Environmental Consideration in the Preservation of Paper Materials in Heritage Institutions in the East and Southern African Region. In P. Ngulube (Ed.), *Handbook of Research on Heritage Management and Preservation* (pp. 183–212). Hershey, PA: IGI Global. doi:10.4018/978-1-5225-3137-1.ch010

Shakhsi, L. (2017). Cataloging Images in Library, Archive, and Museum. In T. Ashraf & N. Kumar (Eds.), *Interdisciplinary Digital Preservation Tools and Technologies* (pp. 119–141). Hershey, PA: IGI Global. doi:10.4018/978-1-5225-1653-8.ch007

Sharma, C. (2017). Digital Initiatives of the Indian Council of World Affairs' Library. In T. Ashraf & N. Kumar (Eds.), *Interdisciplinary Digital Preservation Tools and Technologies* (pp. 231–241). Hershey, PA: IGI Global. doi:10.4018/978-1-5225-1653-8.ch012

Shook, R. (2022). Achieving Balance Through Fundamentals of Digital Librarianship. In P. Pangelinan & T. McVey (Eds.), *Learning and Reconciliation Through Indigenous Education in Oceania* (pp. 185–196). IGI Global. https://doi.org/10.4018/978-1-7998-7736-3.ch011

Shukla, P., & Das, C. (2020). Plagiarism: The Role of Librarian and Teachers in Combating It. In J. Jesubright & P. Saravanan (Eds.), *Innovations in the Designing and Marketing of Information Services* (pp. 148–158). IGI Global. https://doi.org/10.4018/978-1-7998-1482-5.ch011

Siddaiah, D. K. (2018). Commonwealth Professional Fellowship: A Gateway for the Strategic Development of Libraries in India. In R. Bhardwaj (Ed.), *Digitizing the Modern Library and the Transition From Print to Electronic* (pp. 270–286). Hershey, PA: IGI Global. doi:10.4018/978-1-5225-2119-8.ch012

Silvana de Rosa, A. (2018). Mission, Tools, and Ongoing Developments in the So.Re. Com. "A.S. de Rosa" @-library. In M. Khosrow-Pour, D.B.A. (Ed.), Encyclopedia of Information Science and Technology, Fourth Edition (pp. 5237-5251). Hershey, PA: IGI Global. https://doi.org/ doi:10.4018/978-1-5225-2255-3.ch455

Smolenski, N., Kostic, M., & Sofronijevic, A. M. (2018). Intrapreneurship and Enterprise 2.0 as Grounds for Developing In-House Digital Tools for Handling METS/ALTO Files at the University Library Belgrade. In L. Costello & M. Powers (Eds.), *Developing In-House Digital Tools in Library Spaces* (pp. 92–116). Hershey, PA: IGI Global. doi:10.4018/978-1-5225-2676-6.ch005

Sochay, L., & Junus, R. (2017). From Summon to SearchPlus: The RFP Process for a Discovery Tool at the MSU Libraries. In E. Iglesias (Ed.), *Library Technology Funding, Planning, and Deployment* (pp. 72–98). Hershey, PA: IGI Global. doi:10.4018/978-1-5225-1735-1.ch005

Soliudeen, M. J. (2021). The Relevance of Feedback Mechanisms to Library Databases in Academic Libraries. In A. Maake, B. Maake, & F. Awuor (Eds.), *Digital Solutions and the Case for Africa's Sustainable Development* (pp. 116–130). IGI Global. doi:10.4018/978-1-7998-2967-6.ch008

Sonawane, C. S. (2018). Library Catalogue in the Internet Age. In R. Bhardwaj (Ed.), *Digitizing the Modern Library and the Transition From Print to Electronic* (pp. 204–223). Hershey, PA: IGI Global. doi:10.4018/978-1-5225-2119-8.ch009

Sorokhaibam, S. D., & Mathabela, N. N. (2017). Information Needs and Assessment of Bioinformatics Students at the University of Swaziland: Librarian View. In S. Ram (Ed.), *Library and Information Services for Bioinformatics Education and Research* (pp. 188–198). IGI Global. https://doi.org/10.4018/978-1-5225-1871-6.ch011

Staley, C., Kenyon, R. S., & Marcovitz, D. M. (2018). Embedded Services: Going Beyond the Field of Dreams Model for Online Programs. In D. Polly, M. Putman, T. Petty, & A. Good (Eds.), *Innovative Practices in Teacher Preparation and Graduate-Level Teacher Education Programs* (pp. 368–381). Hershey, PA: IGI Global. doi:10.4018/978-1-5225-3068-8.ch020

Stevenson, C. N. (2020). Data Speaks: Use of Poems and Photography in Qualitative Research. In M. Baran & J. Jones (Eds.), *Applied Social Science Approaches to Mixed Methods Research* (pp. 119–144). IGI Global. doi:10.4018/978-1-7998-1025-4.ch006

Stewart, M. C., Atilano, M., & Arnold, C. L. (2017). Improving Customer Relations with Social Listening: A Case Study of an American Academic Library. *International Journal of Customer Relationship Marketing and Management*, 8(1), 49–63. doi:10.4018/IJCRMM.2017010104

Sukula, S. K., & Bhardwaj, R. K. (2018). An Extensive Discussion on Transition of Libraries: The Panoramic View of Library Resources, Services, and Evolved Librarianship. In R. Bhardwaj (Ed.), *Digitizing the Modern Library and the Transition From Print to Electronic* (pp. 255–269). Hershey, PA: IGI Global. doi:10.4018/978-1-5225-2119-8.ch011

Surendran, B., & Kumar, K. (2020). Implementing Information Literacy Skills and Soft Skills for Better Use of Library Resources and Services. In S. Thanuskodi (Ed.), *Handbook of Research on Digital Content Management and Development in Modern Libraries* (pp. 214–224). IGI Global. doi:10.4018/978-1-7998-2201-1.ch012

Suresh, M., & Ravi, S. (2020). Online Database Use by Science Research Scholars of Alagappa University, Karaikudi: A Study. In S. Thanuskodi (Ed.), *Handbook of Research on Digital Content Management and Development in Modern Libraries* (pp. 86-102). IGI Global. https://doi.org/10.4018/978-1-7998-2201-1.ch006

Tella, A., & Babatunde, B. J. (2017). Determinants of Continuance Intention of Facebook Usage Among Library and Information Science Female Undergraduates in Selected Nigerian Universities. *International Journal of E-Adoption*, 9(2), 59–76. doi:10.4018/IJEA.2017070104

Tella, A., Okojie, V., & Olaniyi, O. T. (2018). Social Bookmarking Tools and Digital Libraries. In A. Tella & T. Kwanya (Eds.), *Handbook of Research on Managing Intellectual Property in Digital Libraries* (pp. 396–409). Hershey, PA: IGI Global. doi:10.4018/978-1-5225-3093-0.ch020

Thobane, M. S., & Jansen van Rensburg, S. K. (2022). Transforming Methods for Research With Indigenous Communities: An African Social Sciences Perspective. In P. Ngulube (Ed.), *Handbook of Research on Mixed Methods Research in Information Science* (pp. 190–203). IGI Global. doi:10.4018/978-1-7998-8844-4.ch010

Thull, J. J. (2018). Librarians and the Evolving Research Needs of Distance Students. In I. Oncioiu (Ed.), *Ethics and Decision-Making for Sustainable Business Practices* (pp. 203–216). Hershey, PA: IGI Global. doi:10.4018/978-1-5225-3773-1.ch012

Titilope, A. O. (2017). Ethical Issues in Library and Information Science Profession in Nigeria: An Appraisal. *International Journal of Library and Information Services*, 6(2), 11–22. doi:10.4018/IJLIS.2017070102

Tutu, J. M. (2018). Intellectual Property Challenges in Digital Library Environments. In A. Tella & T. Kwanya (Eds.), *Handbook of Research on Managing Intellectual Property in Digital Libraries* (pp. 225–240). Hershey, PA: IGI Global. doi:10.4018/978-1-5225-3093-0.ch012

Udo-Anyanwu, A. J., & Alor, A. R. (2020). Library Associations, Leadership, and Programmes: IFLA, AfLIA, and NLA. In N. Osuigwe (Ed.), *Managing and Adapting Library Information Services for Future Users* (pp. 89–102). IGI Global. https://doi.org/10.4018/978-1-7998-1116-9.ch006

Wallace, D., & Hemment, M. (2018). Enabling Scholarship in the Digital Age: A Case for Libraries Creating Value at HBS. In S. Bhattacharyya & K. Patnaik (Eds.), *Changing the Scope of Library Instruction in the Digital Age* (pp. 86–117). Hershey, PA: IGI Global. doi:10.4018/978-1-5225-2802-9.ch004

Wang, W., & Wei, Z. (2021). Tongwei County Library: Practices of Social Cooperation in Grassroots Libraries in Western China. *International Journal of Library and Information Services*, *10*(1), 48–60. https://doi.org/10.4018/IJLIS.2021010104

Wani, Z. A., Zainab, T., & Hussain, S. (2018). Web 2.0 From Evolution to Revolutionary Impact in Library and Information Centers. In M. Khosrow-Pour, D.B.A. (Ed.), Encyclopedia of Information Science and Technology, Fourth Edition (pp. 5262-5271). Hershey, PA: IGI Global. https://doi.org/ doi:10.4018/978-1-5225-2255-3.ch457

Watkins, K. E., Nicolaides, A., & Marsick, V. J. (2021). Action Research Approaches. In V. Wang (Ed.), *Promoting Qualitative Research Methods for Critical Reflection and Change* (pp. 119–139). IGI Global. https://doi.org/10.4018/978-1-7998-7600-7.ch007

Weiss, A. P. (2018). Massive Digital Libraries (MDLs). In M. Khosrow-Pour, D.B.A. (Ed.), Encyclopedia of Information Science and Technology, Fourth Edition (pp. 5226-5236). Hershey, PA: IGI Global. https://doi.org/ doi:10.4018/978-1-5225-2255-3.ch454

Wu, S. K., Bess, M., & Price, B. R. (2018). Digitizing Library Outreach: Leveraging Bluetooth Beacons and Mobile Applications to Expand Library Outreach. In R. Bhardwaj (Ed.), *Digitizing the Modern Library and the Transition From Print to Electronic* (pp. 193–203). Hershey, PA: IGI Global. doi:10.4018/978-1-5225-2119-8.ch008

Wulff, E. (2018). Evaluation of Digital Collections and Political Visibility of the Library. In R. Bhardwaj (Ed.), *Digitizing the Modern Library and the Transition From Print to Electronic* (pp. 64–89). Hershey, PA: IGI Global. doi:10.4018/978-1-5225-2119-8.ch003

Wulff, E. (2019). Research Data Access and Management in National Libraries. In R. Bhardwaj & P. Banks (Eds.), *Research Data Access and Management in Modern Libraries* (pp. 1–28). IGI Global. https://doi.org/10.4018/978-1-5225-8437-7.ch001

Xiao, L., & Liu, Y. (2017). Development of Innovative User Services. In L. Ruan, Q. Zhu, & Y. Ye (Eds.), *Academic Library Development and Administration in China* (pp. 56–73). Hershey, PA: IGI Global. doi:10.4018/978-1-5225-0550-1.ch004

Xin, X., & Wu, X. (2017). The Practice of Outreach Services in Chinese Special Libraries. In L. Ruan, Q. Zhu, & Y. Ye (Eds.), *Academic Library Development and Administration in China* (pp. 74–89). Hershey, PA: IGI Global. doi:10.4018/978-1-5225-0550-1.ch005

Yao, X., Zhu, Q., & Liu, J. (2017). The China Academic Library and Information System (CALIS). In L. Ruan, Q. Zhu, & Y. Ye (Eds.), *Academic Library Development and Administration in China* (pp. 1–19). Hershey, PA: IGI Global. doi:10.4018/978-1-5225-0550-1.ch001

Yin, Q., Yingying, W., Yan, Z., & Xiaojia, M. (2017). Resource Sharing and Mutually Beneficial Cooperation: A Look at the New United Model in Public and College Libraries. In L. Ruan, Q. Zhu, & Y. Ye (Eds.), *Academic Library Development and Administration in China* (pp. 334–352). Hershey, PA: IGI Global. doi:10.4018/978-1-5225-0550-1.ch019

Yuhua, F. (2018). Computer Information Library Clusters. In M. Khosrow-Pour, D.B.A. (Ed.), Encyclopedia of Information Science and Technology, Fourth Edition (pp. 4399-4403). Hershey, PA: IGI Global. doi:10.4018/978-1-5225-2255-3.ch382

Yusuf, F., & Owolabi, S. E. (2018). Open Access to Knowledge and Challenges in Digital Libraries: Nigeria's Peculiarity. In A. Tella & T. Kwanya (Eds.), *Handbook of Research on Managing Intellectual Property in Digital Libraries* (pp. 241–259). Hershey, PA: IGI Global. doi:10.4018/978-1-5225-3093-0.ch013

Zhang, Q., Zhang, C., & Zhang, Z. (2021). Open Data Services in the Library: Case Study of the Shanghai Library. *International Journal of Library and Information Services*, *10*(1), 1–17. https://doi.org/10.4018/IJLIS.2021010101

Zhang, W., Zou, W., & Qiu, X. (2019). A Unique Development Road of Urban Public Libraries of China: Practice and Exploration of Pudong Library. *International Journal of Library and Information Services*, 8(2), 51–71. https://doi.org/10.4018/IJLIS.2019070104

Zhu, S., & Shi, W. (2017). A Bibliometric Analysis of Research and Services in Chinese Academic Libraries. In L. Ruan, Q. Zhu, & Y. Ye (Eds.), *Academic Library Development and Administration in China* (pp. 253–262). Hershey, PA: IGI Global. doi:10.4018/978-1-5225-0550-1.ch015

About the Contributors

Dickson K.W. Chiu received the M.Sc. (1994) and Ph.D. (2000) degrees in Computer Science from the Hong Kong University of Science and Technology. His research interests are Library & Information Management, Service Computing, and E-learning with a cross-disciplinary approach. The results have been widely published in over 400 international publications (most have been indexed by SCI/-E, SSCI, and EI, including many taught master and undergraduate project results and around 20 edited books. He is an Editor (-in-chief) of *Library Hi Tech*, a prestigious journal indexed by SSCI. He is the Editor-in-chief Emeritus of the *International Journal on Systems and Service-Oriented Engineering* (founding) and the *International Journal of Organizational and Collective Intelligence* and serves on the editorial boards of several international journals. According to Google Scholar, he has over 7,300 citations, h-index=44, i-10 index=166, and ranked worldwide 1st in "LIS," "m-learning," and "e-services."

Kevin K. W. Ho is a professor of management information systems at the University of Tsukuba, Tokyo, Japan. His research interests include electronic services, information systems strategy, social media, and fake news and misinformation. Ho received his Ph.D. in information systems from the Hong Kong University of Science and Technology. He is a senior member of the IEEE, a member of the Association for Computing Machinery, the Association for Information Systems, and the Society of American Military Engineers, and a Certified Management & Business Educator. He is currently the Co-Editor(-in-Chief) of Library Hi Tech and the Editor-in-Chief of the *Journal of Organizational Computing and Electronic Commerce*.

* * *

Daniel Akwasi Afrane is a Systems Librarian at University of Media, Arts and Communication, Ghana Institute of Journalism Campus. Has Master of Information Science degree from University of South Africa. He has certifications in IT including Cisco CCNA, Google IT Support Specialist certificate, Linux server administrator and has attended workshops and conferences in India and South Africa.

Asefeh Asemi is a faculty member at the Institute of Data Analytics and Information Systems at the Corvinus University of Budapest. Outside of her educational role at CUB, she is also doing research at the Doctoral School of Economics, Business, and Informatics, Department of Information Systems. Her research areas of interest include Information Science, Information Systems, AI, IoT, IoB, and HCI. Her recent focus is on Investment recommender systems. In particular, data analytics, machine learning, and fuzzy inference system concepts are applied to recommender systems to enable investor data analysis. Asefeh began her second doctoral studies at the Doctoral School of Economics, Business, and Informatics at the Corvinus University of Budapest in 2018, where she was awarded the SH Scholarship award. She worked as a visiting scholar at the Victoria University of Wellington, New Zealand, and the University of Malaya, Malaysia before joining CUB. She received her first doctoral degree in 2007 from Pune University, India by defending her thesis subject: "Impact of Information Technology on the Medical Information Centers", where she was awarded the Ministry of Science, Research, and Technology scholarship award from Iran. She joined the University of Isfahan, Iran as a full-time professor from 2007 until 2018 where she was awarded the best researcher and the best HoD several times. She currently upholds a high h index rank in the Department of Information Systems at the Corvinus University of Budapest where she was awarded the Research Excellence Award (CKK) for the third time. She has been appointed as an associate professor at CUB.

Arshia Ayoub received her MS in Library and Information Science from the University of Kashmir, India. She is currently pursuing PhD in Library and Information Science. Her area of research interest is social Q&A sites, open access, repositories and information management.

Mimi Mei Wa Chan received the B.Sc. in Information Management and M.Sc. in Library and Information Management degrees from the University of Hong Kong. She is currently working as a Library Assistant in an academic library in Hong Kong. Her research interests are in Librarianship and Information Science.

Samuel Kin Fung Chan received the Master of Science in Library and Information Management from the University of Hong Kong in 2020. His research interest is in reading promotion and library management.

Cho Yiu Cheung received the MSc degree in Library and Information Management from the University of Hong Kong in 2020. Her research interests are in technology adoption and library management.

Antonia Bernadette Donkor is an Assistant Librarian at the University of Ghana, Balme Library and the Head of the Reference Services Department. She attained her PhD in Information Science from the University of South Africa, Pretoria and has certificates in Academic Peer Reviews, Publons Academy; Inferential Statistics–R Specialisation, Duke University; Linear Regression and Modeling, Duke University; Digital Marketing, and Gender Studies, from the University of Ghana. She assumes great roles when it comes to her family and also has a keen interest in the well-being of her staff and colleagues. She has served on boards and Committees such as the Academic Board-University of Ghana, Member; Sub-committee to Review and Update Policy for Students and Staff with Special Needs: facilities and services, Member; Ghana Library Association, Secretary; Library Re-opening Committee, (UGLS), Chair. She serves as a Reviewer for some internationally peer-reviewed journals and has published a number of research articles in Peer Reviewed Journals. Her hobbies are reading novels, working, making new friends, and imparting knowledge to others.

Helen Chiu Ling Kee received the Master of Science in Library and Information Management from the University of Hong Kong in 2019. Her research interest is in social capital and library management.

Apple Hiu Ching Lam obtained her degree of Bachelor of Business Administration (Honours) in International Business from the City University of Hong Kong (2016) and degree of Master of Science in Library and Information Management with distinction from the University of Hong Kong (2020). She is a doctoral candidate in Education at the University of Hong Kong. Her current research interests are social media in libraries, user education, and the 5E Instructional Model.

Binbin Liu is currently studying at Yanshan University and is pursuing the master's degree. He enjoys researching informetrics and scientific evaluation.

Xu Wang received his Ph.D. in information science from School of Information Management of Wuhan University in 2020. He was a postdoctoral researcher at the Institute of Scientific and Technical Information of China. He is now working at the School of Economics and Management of Yanshan University. His research interests include Informetrics and Science Evaluation, Altmetrics and Impact Evaluation, Information resources management, Data Science, Scientific research management, Scientometrics and Social Network Analysis. He has published more than 60 academic papers and 2 academic books.

Javaid Ahmad Wani received his MS in Library and Information Science from the University of Kashmir, India in 2017. In 2018 he was hired as the Librarian by the Higher Education Department of J & K, India for one year. He writes and presents widely on the themes like bibliometrics, altmetrics, human resource management, open access, and grey literature. He has published many research articles in reputed journals like Library Hi Tech, Information Discovery and Delivery, Global Knowledge, Memory and Communication. He is currently perusing PhD in the Department of Library and Information Science at the University of Kashmir on "Human resource management in agriculture university libraries of North India".

Sze Wing Wong received the Master of Science in Library and Information Management from the University of Hong Kong in 2022. Her research interest is in digital art and culture management.

Index

Milton Keynes UK
Ingram Content Group UK Ltd.
UKHW031918010923
427938UK00006B/117

9 781668 486719